James Francis Katherinus Hewitt

The Ruling Races of Prehistoric Times in India, South-western Asia

and Southern Europe

James Francis Katherinus Hewitt

The Ruling Races of Prehistoric Times in India, South-western Asia and Southern Europe

ISBN/EAN: 9783744757782

Printed in Europe, USA, Canada, Australia, Japan

Cover: Foto ©berggeist007 / pixelio.de

More available books at **www.hansebooks.com**

THE RULING RACES

OF PREHISTORIC TIMES

IN INDIA, SOUTH-WESTERN ASIA

AND SOUTHERN EUROPE

BY

J. F. HEWITT

LATE COMMISSIONER OF CHOTA NAGPORE

VOLUME II

Westminster

ARCHIBALD CONSTABLE AND COMPANY

PUBLISHERS TO THE INDIA OFFICE

14 PARLIAMENT STREET, S.W.

MDCCCXCV

CONTENTS OF VOL. II

PREFACE

In the first volume of this treatise I have tried to tell the story of the divine education of civilised man, and to show how our forefathers, who founded the primaeval village, which grew into the State formed of united provinces, gradually fought their way out of the darkness of ignorance, and the twilight of imperfect civilisation, and how, in the contest, they made law victorious over individual passion and self-will. I have shown how this victory was gained, and the rule of law made permanent by a national system of education begun by the village elders and schoolmasters, who first taught the village children the lessons learned from their own personal experience in agriculture and practical life, and who made their lessons lasting national possessions by embodying them in proverbs, and short, practical, impressive stories, easy to be remembered by learners whose interest was roused by their dramatic power. The actors in these stories were never persons who had lived on earth, and the names and descriptive attributes assigned to them were, like the catch-words in proverbs, which are as old as the native tales, key-notes to their meaning, and sign-marks to fix the narratives in the memory of those to whom they were taught. A striking instance of the methods of these early story-tellers is to be found in the nature-myth, telling how the Sleeping Beauty, the earth, was lulled to sleep by winter, and left ready to be

vii

awakened to fresh life by the kiss of spring. The progress
of these early pioneers is marked by the evidence given by
the diffusion of myths, customs, and ritual, which I have
shown to be common to all countries, from India to Western
Europe, and these show that the primitive village farmers,
who made their way from the southern forests of India to
South-western Asia and Europe, and the hunters, shepherds,
and artisans, who went southwards from Northern Europe
and Asia, were impelled forwards, first by the land-hunger of
the crop-growing forest races, and afterwards by the trading
instincts of the combined agricultural and pastoral tribes,
which increased in strength as commerce grew from the
interchange of the products of their fields and flocks by the
farmers and shepherds, to the more profitable barter of the
weavers and potters of the North, and of the metallic smiths
of the early Bronze Age, the sons of the German Wieland,
the master-smith, who became Danu, the primaeval father and
judge of the Greeks, Semites, Akkadians, Turanians, and
Hindus. In these early ages, when the first fear of strangers
which beset the primitive hunting races passed away, the
desire for land and gain led the tribes, who ultimately united
to form the ruling races of the infant world, to mix together
freely, to cement their union by national alliances, and to
increase the national strength by making blood-brotherhood
with all stranger immigrants received into territories occupied
by previous settlers. During the subsequent process of forming new dialects from the languages of intermingling races
from the North, South, and East, there was also a coalescence
of national customs, ritual, and symbols. As the complexity
of national affairs increased under these processes, historical
knowledge was found to be a necessary equipment for the
rulers and leaders of the growing confederacies, and to meet

this want the teaching of the past practical experiences of the nation was added to the work consigned to the technical schoolmasters of the infant communities, who had become the judges and law-givers of the federated provinces into which India, South-western Asia, Greece, and Italy were divided in the dawn of civilisation. It was then that the original nature-myths became histories, telling of the stages of the nation's progress. The stories thus altered, like the nature-myth of Nala and Damayantī, which, as I have shown, became the plot of the great historical poem of the Mahābhārata,[1] were framed according to carefully matured rules laid down by the national historians, called Prashastri by the Hindus, Asipu by the Akkadians, and Exegetae by the Greeks, who had gained teaching experience by their own training, and the lessons bequeathed to them by their forefathers. Therefore, their primitive epitomes of national history were never biographies of individuals, which would have overburdened the memories of teachers and learners, but dramatic narratives condensed so as to be easily remembered by the use, as algebraic signs, of symbolic names and sacred numbers, the meanings of which were universally understood by those who had learned their interpretation in the national schools, where these lessons were taught as a preliminary introduction to knowledge similar to our alphabets and numerical signs. These stories depicted, in broad outlines, the history of ages, and were moulded into a form which their makers knew would coincide with the national taste, and would thus ensure that they would remain indelibly fixed in the minds of those who learned them. It is a consequence of these rules that in India the history of the country down to the victory of the Pāṇḍavas

[1] *The Ruling Races of Prehistoric Times*, Essay ii., pp. 64-76.

is told, as the events of the year of destiny, in accordance
with the instincts of the agricultural village communities, who
looked on the year as the universal measure of time, while in
Greece, where Northern individualism predominated, history,
as in the myth of Peleus, told in Essay VI., became the record
of scenes in the supposed life of the hero, who, in the Peleus
myth, ended his career as the father of the sun-god Achilles.[1]
Among the Semite Jews history was told, as in the early
stories of Genesis, as a genealogical record, in which the
successive epochs appear as fathers of the descendants of
Adam, meaning the red man, the founder of the red race,
and of Shem, meaning the name, who were two variant forms
of the first father of life in the birth-story which traced the
origin of man,[2] as distinguished from the savage, to the age
when he first compared and differentiated natural objects and
phenomena by the use of names.   As these stories were, like
the instructive nature-myths which preceded them, intended
to be strictly accurate, one of the chief objects of those who
framed these national historical lessons was to make them,
though short, scrupulously exact.   They, therefore, compiled
them in the same spirit which made the primitive Roman
law-givers, who inherited the traditions of the Dravidian
founders of villages, require that all sales of Res Mancipi,
that is, of land, slaves, horses, and oxen, the peculiar property
of the farming and pastoral races could only be legally
binding when the due performance of the prescribed gestures
and words declared to be necessary to make the sale valid
had been proved.   Also, according to their legal rulings, the
sale was vitiated, 'if any formula was mispronounced or

---

[1] *The Ruling Races of Prehistoric Times*, Essay vi., pp. 522-532.
[2] Similarly the Hindus call Kashyapa the father of the Kushite race,
the father of men.

symbolical act omitted.'[1]   It was the descendants of the
Latin race, who founded the village communities of Italy on
the pattern of those in Southern India, who showed in the
Roman rule that ' no religious service can be acceptable to
the gods unless performed without a flaw,' that this spirit,
which reproduced the indomitable obstinacy of the Indian
Dravidians,[2] permeated the whole organisation of the State,
for this rule was so rigorously adhered to that sacrifices had
to be repeated thirty times in succession, on account of
mistakes again and again committed.[3] This same anxiety
for scrupulous exactness appears in the elaborate rules of the
Indian Brāhmaṇas, and has left indelible traces of its presence
in the society of all countries which have been governed by
the agricultural and pastoral races. The authors of these
dramatic histories believed themselves, and were believed by
those who appointed them, to be divinely inspired, and they,
therefore, took care that their work should descend to future
generations in an unaltered form by forbidding, under the
penalties imposed upon all who transgressed a national
taboo, all tampering with these sacred records.

In India we have distinct evidence that these histories
were recited at the national festivals just as our Bible history
is read in churches, and the Jewish Thora in their synagogues;
for they formed the contents of the Itahāsa Purāna, called
the fifth Veda, which were afterwards disseminated among
the people in the national poems of the Mahābhārata and
Ramāyana, and in the histories told in the Brāhmaṇas, just
as the similar histories of the mythic age in Greece and

---

[1] Maine, *Ancient Law*, chap. viii., p. 276 ; also pp. 273, 274, 277, chap. ix.
p. 320.

[2] *The Ruling Races of Prehistoric Times*, Essay ii., pp. 45, 61.

[3] Mommsen, *History of Rome*, translated by W. P. Dickson, People's
Edition, vol. ii. p. 400.

Persia became the sources whence the epic poets of these countries and the dramatists of Greece drew their plots. The Itahāsa Purāna is said in the Upanishads to contain the historical hymns (*itahas*) of the Atharvāṅgiras, that is, of the two classes of priests who succeeded the primæval unsexed Bhrigus, or fire-priests. The Atharvans, or priests of the light and sun-god Atri, who instituted the bloodless ritual of the Rigveda, in which milk was offered to Indra, and barley, the divine seed, mixed in cups of pure running water, was consecrated and drunk in the sacramental Soma sacrifice. Their predecessors were the Aṅgiras, who offered burnt sacrifices of animal victims.[1] These ancient sacred histories, preserved from the days when the Northern totemistic tribes introduced into India the annual sacrifices of their totem animals, were recited[2] by the priests during the Ashvamedha sacrifice of the sons of the sun-horse, held in the month of October. The time set apart for the recital was called the Pariplava nights, or those of the circling (*pari*) boat (*plava*). This boat was the moon-boat, guiding in their monthly and annual circuits round the pole, the stars led in the infant astronomy of the Akkadians and Egyptians by Dumu-zi, the son (*dumu*) of life (*zi*), and Smati-Osiris, the barley-god, the names by which they called the hunting-star Orion, and which carried afterwards, in his annual circuit of the heavens, the sun, called in Northern mythology the knight Lohengrin, the bearer of the blazing

---

[1] *The Ruling Races of Prehistoric Times*, Part I. Preface, pp. xv-xvii; Essay iii., pp. 169, 170, 274, 301.

[2] This practice of recitation brought by the Druids, the sons of the tree (*dru*), from India, where all lessons required to be repeated verbally in the exact words in which they were taught, was continued by them long after the introduction of writing, for Cæsar (*De Bello Gallico*, book vi.) tells us that the Druids forbade their pupils to commit to writing the sacred stories they learned.

flame (*lohe*),[1] who, as Sigurd, made his annual journey, not in
the moon-boat, but as the rider on the cloud-horse Gräni.
It was the history of the nation during the previous annual
voyages of the sun-warrior which was recited during the
nights especially dedicated to him.[2]

This custom of reciting the national history as part of the
ritual of the annual sacrifice of the totem of the race must
be of immense antiquity, and of Northern origin, as the
Ashvamedha sacrifice, which is, as I have shown, connected
with the Equiria at Rome, held at the same time of the year,[3]
belonged, in its primitive form, when the horse was killed
and eaten in honour of the Northern sun-god, instead of being
set free to wander at its will as in post-Vedic Indian ritual,
to the age of animal sacrifices preceding the purer rites of the
Vedic Soma sacrifice of the united sons of the tree, the river,
and the barley. It was for the Soma sacrifice to the sun
that ripened the barley, the seed of life, and to the rain-god,
who matured it, that the greater number of the hymns of
the Rigveda were written, and the chanting of these hymns
of prayer and praise was only gradually made the principal
part of the national public worship when the silent ritual of
Prajāpati,[4] the invisible god of life dwelling in the seed, and
the ruler of the year, with its animal totemistic sacrifices, was
made subordinate to the bloodless sacrifices of the newer

[1] Lohengrin, the bearer of the Holy Grail, the water and blood of life,
preserved in its cloud casket, was drawn in the moon-boat by the moon-swan.
The Ruling Races of Prehistoric Times, Essay iii., p. 302 ; iv., p. 351.
[2] Max Müller, *Upanishads Chāndogya Upanishad*, iii. 4. 1 and 2 ;
S.B.E. vol. i. pp. 39 note 1, 40.
[3] *The Ruling Races of Prehistoric Times*, Essay iv. pp. 335, 336, 395.
[4] Eggeling, *Sat. Brāh.*, i. 4, 5, 12 ; S.B.E. vol. xii. pp. 130 note 2, 131,
where we are told how, in the contest between Mind and Speech, Prajāpati
decides in favour of the superiority of Mind, and, therefore, Speech refused
to assist at his ritual.

belief in the sun-god as the messenger, image, and son of the invisible creating god of light, the ever-revolving pole or fire-drill of the universe, whose existence was proved by the perpetual revolution of the stars. We also find a surviving relic of the reverence with which these stories were regarded in the rule forbidding the American-Indians—who still, like the primitive people of Europe and Asia, preserve the greater part of their knowledge in the form of oral tradition—to tell them except after fasting and prayer.[1]

But when we, who have been taught to look on history as a detailed record of actual facts, and not as a summary of results, told in symbolical language, first hear, and examine these stories which were looked on with such reverence by our ancestors, they appear to us as they did to Plato, to be so full of unutterable absurdities, that we cannot realise how they were ever invented or valued by the shrewd intellects of the pioneers of civilisation, or how they have since remained fixed in the popular memory as prized memorials of the past. This riddle can only be solved when we realise the fact that the historical methods of the framers of these stories were totally different from those to which we are accustomed, that their rules for recording history have been obliterated, and the meaning of the names, symbols, sacred numbers, and memory-saving abbreviations they used, ignored and forgotten by the Aryan bards who founded our modern histories, and who thought that the only records of the past worth preserving were those which told of the actual deeds and conquests of the national warriors and leaders. In order to understand the real meaning of the ancient stories, which have been derided by philosophers, but preserved by popular conservative instincts, and to convince ourselves that they are

---

[1] *The Ruling Races of Prehistoric Times*, Part II. Essay ix., p. 244.

genuine epitomes of past history, giving a summary of the leading facts in the national career, we must recover the knowledge that has been lost by ages of neglect, learn to interpret the ancient symbols and language in their original meaning, think the thoughts, and see with the eyes of the primitive chroniclers, and give up the erroneous system of looking on the heroes of their historical tales as individuals, and of interpreting them as biographies of living men, who have influenced the progress of the nation.

But though these sacred verbal records give us outlines of the national history, they only preserve the salient incidents chosen by the national historiographers to mark the stages of national growth, and give us little insight into the inner life of the consecutive ruling races, whose story they tell. It is to ritual and national customs that we must look if we want a living picture of the childhood of the human race. It is these which tell us how carefully the primitive agriculturists marked the recurrence of the annual seasons by greeting the year's changes with festivals and dances; of the union of the people of the North and South by the marriage rite of making blood-brotherhood between the bride and bridegroom, which is common to almost all Indian tribes, and of which a trace appears to have survived in the Roman rule, requiring the bridegroom to part the bride's hair with a spear-point, and also of the significant marriages of the bride and bridegroom to trees, which are so prevalent in India.[1] Ritual and custom also tell us how the house-mother of the Finnish race, the prototype of the Roman Vestal virgins, was the family priestess, who offered daily libations to the household-fire.[2] It is through ritual that we can trace the stages

---

[1] *The Ruling Races of Prehistoric Times*, Essay iii., pp. 174, 175, 152, 153, 209, 278.   [2] *Ibid.* Essay iii., p. 200; iv., p. 361.

of the sacrifice of the sheep-totem of the shepherds, who, as sun-worshippers, substituted the sun-sheep, Rachel, the ewe-mother of the Jews, for the moon-goat, and can see how it was first the black ewes offered to Hekate, the goddess of the cross-roads in Asia Minor,[1] which became the wether sacrificed to Ptahiel, the opening (*patah*) god, the Egyptian Ptah, the Jewish Japhet, at the autumnal equinox, by the worshippers of the polar star, whose ritual is preserved by the Sabæan Mandanites;[2] how this unsexed parent-god of the early fire-worshippers became, on the rise to power of the twin races, the ram of Varuṇa, which, as we see in Abram's sacrifice of the ram in the place of Isaac, was the substitute for the sacrifice of the eldest son formerly offered by one of the branches of the twin confederacy; and we can see how this ram became the lamb[3] eaten by the Jews at the Passover, instead of the eldest son, the child eaten by the Sabæan Haranites, and we can also trace this lamb as a sacrifice to St. George, the plough-god of spring of Asia Minor and Western Europe, for the Bulgarians still sacrifice lambs on his day, and tell a story of Abraham sacrificing his child to St. George instead of a lamb, which, as he was a poor man in the story, he did not possess, and how its life was miraculously preserved.[4] From this sacrifice to St. George, and the Jewish Passover, we are led by ritual to the Greek sacrifice of the lamb at Easter, when a lamb is eaten in every household in Athens. It is also through the ritual of the Palilia and Equiria festivals of the corn-growing races of Italy, who worshipped the moon-cow and the sun-horse, through the sacrifice of horses by the Northern worshippers

---

[1] *The Ruling Races of Prehistoric Times*, Essay iii., pp. 197, 215, 216.

[2] *Ibid.* Part II. Essay viii., pp. 163, 164.

[3] *Ibid.* Essay vii., pp. 55, 56.

[4] Garnett and Stuart-Glennie, *Women of Turkey*, chap. xii., pp. 332, 333.

of Odin, those of the Massagetæ on the Caspian Sea, the Ashvamedha, or horse-sacrifice, in India, and the use of the Ashva-vāla, or horse-tail grass, as the magic rain-wand in the Indian Soma festival, that we can trace the gradual movements of the North-western races of Europe to the South-east, and the establishment of their dominion in South-western Asia and India,[1] after that of the sons of the antelope and the cow, who were the successors of the worshippers of Varuṇa and the sun-rain. But by far the most valuable evidence given by ritual, as to the inner life of our primitive ancestors, is that which reveals to us the stages of the gradual evolution of a pure and spiritual faith out of the early sacrifices of the Northern races, in which the united tribe feasted on the national totem animal at their chief annual festivals held at the winter and summer solstices. We can trace in Indian ritual the change from the totem feasts, accompanied by copious draughts of intoxicating drink, to the offerings of milk, sour milk, barley, and rice, to the rain-gods, Indra and Pūshān.[2] This was the work of the corn-growing races, sons of the Northern animal totems, and of the Southern village grove, who, when united, called themselves the sons of the mother-bird, who brought the showers of spring. They changed their name from that of the sons of the bird to the sons of the rivers, which supplied water for their crops when they left the hill slopes, chosen as the site of the earliest terraced cultivation of the Northern agriculturists, and of the primitive Indian forest cultivators.[3]

---

[1] *The Ruling Races of Prehistoric Times*, Essay ii., pp. 132, 133; iii., pp. 166, 232, 321-323; iv. p. 336; v. pp. 483-485.

[2] *Ibid.* Essay iii., pp. 162, 201, 202, 204, 205, 242, 243; v., pp. 435, 436.

[3] All nomad forest agriculturists in India, who clear the forest for their cultivation by burning the trees, and who move from one forest clearing to

and came down to the river-valleys as the race who used the waters of the river for irrigation. It was when the Basques of Asia Minor, called Iberians or Ibai-erri, the people (*erri*) of the rivers (*ibai*), who became the Irāvata, or sons of Irā or Idā in India,[1] and of the rivers called Irāvati, were named the sons of the rivers,[2] instead of the children of animal totems and the mother-bird, that they changed the custom of making blood-brotherhood with all strangers admitted into the tribal confederacy to that which required all who joined the corn-growing communities to consecrate their admission by the baptismal bath of regeneration. This, in India, made the new-born brother a son of the black sun-antelope, whose skin the neophyte wore in the bath as the Jarayu or afterbirth, and it was when he came out of the bath as a new-born disciple, the twice-born[3] son of the regenerated race, that he received the staff of Udumbara wood (*Ficus glomerata*), which marked him as a member of the gardening races, the sons of the Syrian fig-tree.[4]

The waters in which the new birth was consummated were those of the sacred reservoirs, which preserved in the thirsty lands of semi-tropical Asia the regenerating gifts of the life-giving rain which filled them. It was in these waters that

another every two or three years, make their temporary settlements on the upper hill-slopes, and always avoid the river valleys. For terraced cultivation in Europe, see Essay viii., pp. 129, 130.

[1] *The Ruling Races of Prehistoric Times*, Part I. Preface, p. xviii; Essay i., p. 11; iii., p. 213; v., p. 493.

[2] Thus the Jews of the seed of Abram were descended from his grandfather Nahor, or Nahr, the river Euphrates, and his son Terah, the antelope.

[3] All Hindus belonging to the three higher castes of the Brahmins, Kshatriyas, and Vaishyas call themselves 'twice born.'

[4] *The Ruling Races of Prehistoric Times*, Part I. Preface, pp. xliv-xlvi; Essay iv., p. 367.

the never-dying fish [1] dwelt as the symbol of the soul of life
to the Kushite sons of the bird-mother Gandhārī, she who
wets (dhāra) the land (gan).[2] This was originally the
Indian carp or Rohu, the sun-fish, which hides itself in the
mud when the waters dry up towards the end of the long
season of drought, between the end of the rains in autumn
and their reappearance at the summer solstice with the South-
west monsoon, and emerges when the rain refills the water
reservoirs. This immortal fish became to the sons of the
rivers, in more temperate climes, the eel, which, like the
mother-bird, regulated its annual disappearances and re-
appearances by the coming of winter and the return of
spring. This eel was called by the Finns Il-ja, the sacred
fish born (ja) of Il; that is, of the mother of the year
of three (iru) seasons, the mother Idā, Irā, or Ilā, while by
the Hindus it was called Indu, the root whence the name of
Indra, the rain-god, was formed. It was the sacred eel which
gave place to the mother-dolphin, the Indian Makara, the
Greek Delphus, when the encircling ocean became the mother
of life, the home of the life-giving waters, in place of the
rivers and the holy pools,[3] which, as I show in Essay VIII.,

---

[1] For the belief in the undying fish, which travelled from India to England,
see Wordsworth's Song at the Feast of Brougham Castle—

'And both the undying fish that swim
Through Bowscale-tarn did wait on him.'

[2] *The Ruling Races of Prehistoric Times*, Essay i. p. 22 ; iii. p. 249.

[3] *Ibid.* Part I. Preface, pp. xl-xlv ; Essay i., pp. 22, 23; ii., pp. 125,
126; iii., pp. 284, 286 ; iv., pp. 374, 375. With regard to the derivation of
Indra from Indu, which all Sanskrit scholars admit to be its root, and the
assignment of the eel, as the meaning of Indu, it must be remembered that
Agni, the fire-god, always associated with Indra in the Rigveda, is the Wend
god Ogon, and this makes it probable that it is to Northern sources that we
must look for Indu, the root of Indra. It is, as I have shown in the
Preface to Part I., one of the names of the eel-totem of the Kharwars and
Cheroos, who are sons of the hawk, the bird-mother of the mining races, who

were first the springs brought to light by the wonder-working hoofs of the sun-horse.[1]

Thus it was the sons of the river-worshippers of the fish-god, as the symbol of the soul of life, who instituted the ceremony of baptism, which began with the bathing in the sacred spring dew at the festival of the Palilia, celebrated by the ploughing races in the month of April, and culminated in the elaborate baptismal ceremonial of the Hindu Soma festival, the still more ancient baptism of the Sabæan Mandanites of the Euphrates valley, that of the Greek penitents of the Eleusinian festival to Demēter, the barley-mother, and of the American-Indians of Mexico.[2]

It was these same sons of the barley, the seed of life, symbolised in the eight-rayed star, and of the more primitive Sesamum, or sacred oil-seed of the Mandanite Sabæans and Indian Telis,[3] who added to the baptismal bath of incorporation into the brotherhood and of moral regeneration, a substitute for the sacrificial feast on the totem ancestor of the tribe. This was the sacramental meal on the barley, the seed of life, mixed with the running waters of the parent-rivers, eaten with the variations I have described in the accounts of each ritual by the Hindus, Zends, Greeks, Euphratean Mandanites, and the Mexican Indians.[4] Only

lived with the Wends in their first home in Northern Europe. Indra is the rain-god who succeeded the Finno-Akkadian Sukra, who was the Indian form of the Akkadian mother-goddess, Shuk-us, who became better known as Istar, and of the Finnic thunder and rain-god Ukko, the storm-bird parent of Linda, the Esthonian mother-bird. *The Ruling Races of Prehistoric Times*, Part I. Essay iii., pp. 147, 148; Part II. Essay viii., p. 156.

[1] *Ibid.* Part II. Essay viii., p. 176.

[2] *Ibid.* Part I. Preface, pp. xliv-xlviii; Essay iii., p. 232; Part II. Essay viii., p. 159; ix., p. 244.

[3] *Ibid.* Part I. Preface, p. xxviii; Essay ii., p. 86; Part II. Essay viii., pp. 161, 162.

[4] *Ibid.* Part I. Preface, p. xlviii; Essay iii., pp. 206, 242; v., p. 435; Part II. Essay viii., pp. 161-164; ix., pp. 286-288.

those who had been cleansed from their sins by the purifying
baptismal bath of regeneration were allowed to partake of
this meal, and they believed that in eating the sacred food
they were incorporating into their body and soul the living
son of the god of light, who ripened the grain, and breathed
into it the spirit of his son as the germ of future life.[1]

It was these corn-growing races, united by a common
ritual and a common symbolic worship of the two sun-crosses,
combined in the eight-rayed star, which formed the great
trading confederacy of the merchant sons of the seed of the
sun Rā or Ram, who became the Phœnicians and the
Semites, who claimed to be born of the seed of Abram, that
is, of the father (ab) Ram, the sun-god, and the seed grain
he ripened. They, as I have shown in Essays v. and ix.,[2]
spread themselves as the leaders of trade over the whole
civilised world, and instituted the rigid rule of law, and of a
complicated and burdensome ritual which, with its accompanying social tyranny, led the way to the great revolt of
the subject races, headed by the Aryan Celts, sons of the wine-
god, which I have described in Essay vi.[3] It was this revolt
which substituted for the Unitarian faith of the Semite

---

[1] That this statement that those who ate the sacramental barley believed
that they ate the living Son of God is not exaggerated, is proved by the
existence of this belief among the Banū Ḥanīfa of Arabia, the sons of the
date palm-tree. Their god, called Hais, was made of dates, butter, and
dried curd ; and an Arab poet describing the view taken of this god, by
those outside the tribe, says, 'The Banū Ḥanīfa have eaten their Lord for
hunger.' Sachau's Alberuni's *Chronology of Ancient Nations*, chap. viii. p.
193. This sacramental meal is also spoken of in the *Arabian Nights* by the
chief of the tribe called Jamrkhan. He, when conquered by Gharib, the
sun-god, says, 'I worship a god made of dates, butter, and honey, and I eat
him' (Burton's *Arabian Nights*, 'Story of Gharib and his brother Ajib'
vol. v. pp. 213, 216). See Essay ix. of this volume, p. 338.

[2] *The Ruling Races of Prehistoric Times*, Part I. Essay v., pp. 471-481 ;
Part II. Essay ix., pp. 307-319.

[3] *Ibid*. Essay vi., pp. 539 ff.

confederacy, the Greek polytheistic worship of individual
gods, represented in human forms, and which broke up the
Kusho-Semite system of communal government, and of confederated states and provinces superintended by monarchic
kings. It was then, as I have shown in the Preface to Part
I., that the ancient mythic histories and genealogies, telling
history under the names of supposed ancestors,[1] were distorted
into individualistic biographies similar to those recited by
the Celtic bards.[2]

But though the evidence of ancient mythic history, ritual,
symbols, language, antiquarian remains, and historical botany
and zoology, enables us to reproduce graphic pictures of the
early history of mankind, they give us but very uncertain
guidance as to the chronological order in which the various
stages of growth succeeded one another. This is best shown
by the successive methods of computing time which mark
the gradual growth of civilisation by the progress of astronomical and weather knowledge, which was always a favourite
study of the agricultural races, who found knowledge of the
changes of the seasons essentially necessary to success in
farming. It was these farmers, who called themselves sons
of the seed-grain, who, as I have shown in Essays IV., VIII.,
and IX., wrote the history of mankind on the heavens in the
names and attributes they assigned to the stars and constellations. (1.) The first method of measuring time by the
stars which they used was that of the year of the Pleiades,

---

[1] Such as the genealogies of Genesis iv. 17, 18, in which the city of Erech,
the Akkadian Unuk, becomes Enoch, the son of Cain, and the father of Irad
or Eridu, the holy city of Ia, and of Genesis xi. 12 ff., in which Arpachsad, the
land of Armenia, is the son of Shem, and the father of Shelah, the spear, the
symbol of the fire-god, which led the early Semite emigrants. *The Ruling
Races of Prehistoric Times*, Part I. Essay iii., pp. 141, 150, 179, 189 ; Part
II. Essay ix., p. 243.

[2] *Ibid.* Part I. Preface, pp. lii-lv.

called the Mothers and the Spinners, divided into two
periods of six months each, beginning in November and
April. This was the year which the founders of village
communities brought with them from the Southern Hemisphere to Europe,[1] and that represented in the cosmogony
of the Sia Mexican Indians as the year of the world's
creation by Sus-sistinnako, the spider, which I have described
in Essay IX. (2.) It was followed by the year of the three
seasons of the growing seed, beginning with the birth of the
young sun-god at the winter solstice, symbolised by the
equilateral triangle, the first symbol of the triune creating
god, the one god who rules the year under three forms of
the three recurring seasons.[2] This was the year of Orion,
the hunter, the father star of the Indian Brahmins, and of
the corn-growing races of Asia Minor, the star which led
the stars round the pole in their daily revolutions, divided
into longer periods by the lunar phases of the moon-hare,
symbolised by the constellation Lepus, at the foot of Orion,
which he, as the wild hunter, or wind-god, hunts round the
sky in her monthly courses. It was also the year of the
Akkadian and Egyptian sun-gods Dumu-zi (Tammuz) and
Smati-Osiris, both of whom gave their names to the star
Orion.[3] In this year the three seasons are represented by
the three stars forming the belt of Orion, and these were
one of the three constellations of the Pleiades, the three
stars of Orion and the seven stars of the Great Bear, which
were placed in the sky by Ŭt'set, the mother of corn in
the Sia cosmogony, when she emerged from the nether earth
to the corn-growing plateau on the top of the mother

---

[1] *The Ruling Races of Prehistoric Times*, Essay ii., pp. 123 ff.
[2] *Ibid.* Part I. Preface, pp. xii, xiii.
[3] *Ibid.* Part I. Essay ii., p. 127 ; iv., pp. 401, 402 ; Part II. Essay vii.,
pp. 18, 20-23.

mountain.[1] The beginning of the age of the years of Orion
and that of the Pleiades was also the time of the supremacy in the south of the ape-god, ruler of the winds to the
matriarchal Dravidians, the Hindu Hanuman, who became
in Egypt Set, the god of the South, whose first name was
Hapi, the Egyptian form of the Hindu Kapi, the ape,
the earliest Nile-god, and the Wild Hunter of the North.
He was, as I have shown in Essay viii., the god raised to
heaven as the constellation Cepheus. The Stars a Cepheus
Alderamin and γ Cepheus were the pole-stars from 21,000 to
19,000 B.C., and it was while they ruled the heavens that the
ape-god was looked on as the ruler of the winds, who preceded the mother-bird, the pole-stars in Cygnus and Lyra.
It was he who turned the pole, and his form was depicted in
the stars round the pole, which were first in Hindu astronomy
called the stars of the ape, and afterwards those of Sisumāra,
the alligator.[2]   (3.) This year was succeeded by that of four
seasons, the year of the corn and fruit growing races of
Syria, Asia Minor, Macedonia, and the Greek Peloponnesus,
who all began their year with the autumnal equinox.[3]   This
was the year of weeks of seven days in place of the previous
five days' week reckoned by the Northern Scandinavians and
the Zend Mazdeans, and still used by the Shans of Northern
Burmah, who hold their weekly markets at intervals of five
days.[4]   This seven days' week was symbolised by the seven

---

[1] The Ruling Races of Prehistoric Times, Part II. Essay ix., p. 260.
[2] See diagram of the polar circle, O'Neill, Night of the Gods, vol. i. p. 500.
The description of the constellation Sisumāra in the Vishnu Dharma is that
of the figure of a man with hands, and in the Rāmāyana Tarā, the star, the
pole-star, is the wife of Su-griva, the ape king.   Essay viii., pp. 214-225.
[3] The Ruling Races of Prehistoric Times, Essay iv., p. 330.
[4] Darmesteter, Zendavesta Mah Yasht, 3 ; S.B.E. vol. xxiii. p. 90, note 5 ;
Sir D. Forsyth, Autobiography and Reminiscences, p. 188 ; The Ruling Races
of Prehistoric Times, Part II. Essay viii., pp. 138, note 4, 139.

stars of the Great Bear, the heavenly axle which turned round the pole, and which was driven by the Greek son of Phlegyas, the fire god, called Ixion or Axi-fon, the driver of the axle (*aksha*).[1] This was the year of the sun-antelope, the heavenly god of light, who, when he was the earthly totem of the corn-growing races, showed his worshippers the best corn-lands in the gently undulating river valleys, and who, when he was raised to heaven as the antelope or deer-sun,[2] circled in his annual course, beginning in the west at the autumnal equinox, the four points of the compass marked by the Latin cross of St. George, the rain and plough-god of Cappadocia and of the Indian Gonds. This cross was crowned in the north by the pole-star, the god Njord of the Edda, and the star worshipped by the Indian Kushika or Kushites, the Sabæan Mandanites of the Euphrates valley, the Sabæans of Southern Arabia, and the Egyptians.[3] (4.) When these worshippers of the sun-antelope, who first worshipped the Phrygian and Hittite goat, Tur, the Greek αἴξ, or wind-goat, who under his other name of Uz, derived from the Finnic Uk-ko, became the Jewish Esau,[4] reached India and the countries of the Persian Gulf within the influence of the South-west monsoon, they added the rainy season to the four seasons of their predecessors, and began their year with the summer solstice.[5] It was with the year

---

[1] *The Ruling Races of Prehistoric Times*, Essay ii., pp. 83, 84.

[2] The guardian star of this year was the Indian star Marichi, meaning the spark of fire, who was the father of the Kushite race, and was raised to heaven as one of the stars in the tail of the Great Bear, when he was slain by Râma, the ploughing-god, as the deer who enticed him away from Sitâ, the furrow. *The Ruling Races of Prehistoric Times*, Part I. Essay iii., p. 261 ; iv., pp. 343, 357.

[3] *Ibid.* Part I. Essay i., p. 8 ff. ; iii., pp. 230-232 ; iv., pp. 378, note 3, 379, 412 ; Part II. Essay vii., pp. 24-28, 45.

[4] *Ibid.* Essay iv., pp. 364, 365 ; iii., pp. 145, 149 ; vi., p. 544.

[5] *Ibid.* Part I. Essay iv., pp. 364-367, 369-372 ; Part II. Essay vii., pp. 28-30.

opening with the summer solstice that the age of the twins, originally Day and Night, the children of Saranyu, the corn-mother, began. This age of the twin-parents of the Hittites, called the Khati, or joined race, in Assyria and India,[1] was that marked by the worship of the solstitial sun, which brought the rains of the Persian Gulf and Northern India at the summer solstice. But this season was first reverenced, not by the nations of the South, but by the Northern wor-shippers of the sun-horse and its rider, whose mythic history is given in Essay viii.[2] He was the god born at the winter solstice, who celebrated at midsummer, when fires were lighted in his honour throughout North-western Europe, his annual victory over the powers of frost and darkness. The course of the solstitial sun was symbolised by the transverse St. Andrew's cross, sacred to the sun and rain-god, and it was the union

of this symbol $\times$, described in the Rigveda as the 'four-pointed weapon' with which Indra, the bull, brings the rains,[3] with the Latin Cross of St. George $+$, which made the eight-rayed star the symbol of god and seed to all the barley-growing races.[4] (5.) This year of the solstitial sun was followed by the year of six seasons, the year of the Zend and Western Hindu sons of the date palm tree, described in Essays ii. and iv., and also in Essays vii. and ix. in this volume,[5] where it is shown to be the wheel-year of Hindu astronomy, beginning, like the Hindu year of

[1] *The Ruling Races of Prehistoric Times*, Essay iii., pp. 210, 211 ; iv., p. 339.

[2] *Ibid.* Part ii. Essay viii., pp. 110 ff.

[3] Rigveda vi. 22. 1, 2.

[4] *The Ruling Races of Prehistoric Times*, Part i. Preface, p. xxviii.

[5] *Ibid.* Part i. Essay ii., pp. 78, 115 ; iv., p. 405 ; Part ii. Essay vii., pp. 7, 30-32 ; ix., p. 343.

three seasons, at the winter solstice with the month sacred
to the bull-god Push or Pushān, the constellation Taurus,
who was, as I show in Essay v., originally the barley-growing
rain-god of the Lithuanian races,[1] who also gave the gods
Agni and Indra to Hindu mythology. This was the year of
the six Āditya of the Rigveda, the year of Pushān, under his
Vedic name of Aryaman, the ploughing-god, who became
Airyaman in the Zendavesta.[2] The early astronomers, who
calculated this year of the perfect circle of 360 degrees or
days, looked on it as that which measured the annual circuit
of the sun revolving round the pole, as the oil-press revolves
round its beam. In the Babylonian astronomy of the sons
of the date palm-tree, who made the year begin with the
vernal equinox, it became the year of the Star Dilgan, our
Capella a Aurigae, the patron star of Babylon, the heavenly
charioteer which drove round the pole, the sun drawn by the
stars of the bull (*Taurus*), which, according to the earlier
myth, trod out in their circular path, the ripened seed of
life, as the oxen circling the central pole of the threshing-
floor tread out the grain. This threshing-floor became in
the later myth, which I have described in Essay ix., the
eight-sided enclosure of the sons of the eight-rayed star, the
earth over-arched by the heavens, which are the home of the
invisible father of life, the revolving fire-drill of heaven only
seen in the effluence of his brightness, shown in the pole-
star and its attendant stars.[3] (6.) This year was followed
by the lunar-solar year of the fish-sun-god, described in
Essay iv., and calculated by the Akkadian astronomical
observers, who tracked the course of the moon through the

---

[1] *The Ruling Races of Prehistoric Times*, Essay v., pp. 437-439.

[2] *Ibid.* Essay v., pp. 415-422.

[3] *Ibid.* Part I. Essay v., pp. 419, note 2 ; Part II. Essay vii., pp. 61-63 ;
ix., pp. 278-282.

thirty stars,[1] marking its course from the middle of November to the middle of February, and that of the sun through the ten stars called the Ten Kings of Babylon, the sign-marks of the road it traversed from the middle of February till November,[2] when it again entered into and emerged from the baptismal bath of regeneration in Aquarius. It is the record of these successive measurements of the year, as told in the Rigveda, Brāhmaṇas, the Mahābhārata, and Rāmayana, which I have examined in Essay VII. of this volume. The evidence of the Rigveda on this subject is especially valuable, as it is the oldest official liturgical book in the world except the Egyptian Book of the Dead, having certainly been known in its present form sometime between 1400 and 1100 B.C. But its contents date back to a very much earlier period, as is shown by the arrangement of the hymns into sections, each of which contains the hymns authorised before the whole collection was united into one national hymn-book by the separate colleges of national priests, who were guardians of the ritual in each of the primitive kingdoms of Northern India.[3] But this proof of the great antiquity of the earliest hymns of these separate national collections is very far from showing the age of the Vedic ritual, for it rests on the national traditions preserved in the Itihāsa Purāna, of which I have already spoken, and derived from pre-Sanskrit

[1] Thirty stars were chosen as the number of the days of the month in the wheel-year of twelve months and 360 days. They became, when the stars were divided into thirty-six constellations, as in Egyptian astronomy, the thirty-six steps of the Hindu Vishnu through the starry host of heaven ; the footmarks of his monthly course in each of the ten months of gestation of thirty-six days, the tenths of the year of 360 days preceding the birth of the new year from its predecessor. Maspero, *Dawn of Civilisation*, p. 205 note. This corrects the interpretation given in Essay viii., pp. 139, 140.

[2] *The Ruling Races of Prehistoric Times*, Part I. Essay iv., pp. 374-387 ; Part II. Essay vii., pp. 59-61.

[3] *Ibid.* Part II. Essay vii., p. 4.

Dravidian rulers, while the language in which the hymns are
written is shown by the Dravidian cerebral letters to be an
Aryan dialect, moulded by pupils of Aryan instructors, whose
native speech, before they learnt Sanskrit, was a Dravidian
agglutinative language. Also, after this Dravidianised
Sanskrit became the national dialect of the upper classes,
the metres forming the stanzas, into which the Vedic hymns
are divided, had to be invented, and these, as I have shown,
are proved by the Brāhmaṇas and the hymn in the Rigveda
describing the national measurement of time to be reproductions of the previous time-reckonings used by the Turano-
Dravidian priests, who had, from the dawn of ritual, fixed
the dates of the popular religious festivals.[1] I have shown
in the latter part of Essay vii. how the Indian conception of
history as the records of events in the cycle year of destiny,
which appears in the plot of the Mahābhārata, and makes the
history there told culminate at the close of the year in the
victory of the Pāṇḍavas,[2] is repeated in Buddhist history and
theology. which is based first on the cycles of the twenty-
eight Buddhas, the twenty-eight days of the lunar month,
of which the fourteenth, corresponding to the day of the full-
moon, is Su-medha, the sacrifice (medha) of Su, the root of
Soma, whom I have shown to be the original sun-god,
worshipped as the moral regenerator of mankind :[3] secondly,
on the lunar-solar year of the thirteen Theris, or thirteen
lunar months of twenty-eight days each, and on the wheel-
year symbolised in the Buddhist wheel of the law. The
period covered by the history of the epochs represented by
these two years is that elapsing between the adoption of

---

[1] *The Ruling Races of Prehistoric Times*, Part I. Essay vi., pp. 553-554 ;
Part II. Essay vii., p. 9.
[2] *Ibid.* Essay ii., pp. 74-76.   [3] *Ibid.* Essay iv., pp. 398, 399.

the sun as the ruler of the wheel-year, or the beginning of the fifth epoch into which I have divided the prehistoric ages, and the birth of the Buddha Sidarttha Gautuma, who was not, like his predecessors, a mythic hero, but a living man. It was the age of the growth of Jain theology and its diffusion over India by the conquering Licchavis, or sons of the lion (*lig*), also called the Ikshvākus, or sons of the sugar-cane (*iksha*), whose career I have traced in Essay III.[1] In Essay VIII. I have depicted the rise of Norse mythology, and the mutual interaction on each other of the individualising tendencies of Northern temperaments and the communistic organisation of Southern society. I have shown how the beliefs of the totemistic hunting races of the North were superseded by the theology of the inventive and self-reliant workers in metal, whose father-god was the master-smith, the Norse Völundr, the German Wieland, the English Weyland, who made the shoes of the horse of the sun. I have traced his mythic history from the time when he wielded the creating hammer, and made the year-rings, which he hung on the wall of his dwelling, through his defeat by Nidung, king of the nether (*nid*) earth, who made him the maimed, or one-legged, turner of the pole, which made the stars revolve, to his subsequent return to active rule, when, after killing the sons of Nidung, he made himself wings and became the mother-bird-parent of the god, who is called in the Rigveda the Aja ekapad or one-footed goat, who watched the revolutions of the solar disc and in Russian mythology the one-legged chicken supporting the revolving house of Baba Yaga, the witch-mother of the sons of the fire-drill, whose home is the heavenly vault or tent, the rain-god who made the meadows green, the bird who brought the showers of spring. It was this bird

---

[1] *The Ruling Races of Prehistoric Times*, Essay iii., pp. 323 ff.

who became the flying sun-horse, who made, by his footsteps, fountains of living and healing water to rise from the earth, the holy wells and springs worshipped in Europe and Asia. This cloud-horse of the sun was the heavenly steed ridden by the sun-knight Sigurd, who was first the sun-warrior with the red hair and beard, called Hadding, the hairy (*hadd-r*) one, the god of the red dawn and the gloaming, and afterwards the gnomon stone, the pillar (*urd-r*) of victory (*sig*), the prophet's stone, telling by its shadows the path of the sun through the day and year. It is this sun-knight and prophet who is worshipped by the Sabæan Mandanites of the Euphrates (whose supreme god is the pole-star) as Avather Ramo, the sun-god, Ra or Rai, and Ptahiel, the opener (*patāh*) the morning dawn, heralding the rising sun of the East; and it is to him they sacrifice at the autumnal equinox a wether, the unsexed form of the ram, sacred to the sun and to the god Varuṇa, the flying ram of the story of Jason. I have shown, from an examination of the ritual of the Sabæan baptismal ceremony, and the sacramental meal preceding the sacrifice of the wether, that it is a survival of a most ancient form of worship, the precursor of the Hindu Soma sacrifice, which again was the parent-form of the baptism and sacramental meal of the Greek Eleusinian mysteries.

I have traced the worship of the sun-horse in its earlier and later form to England, and have shown how, in the temple of the sun-horse at Stonehenge, we find that the Indian and Asiatic astronomy of the horned horse of the sun, ruling the wheel-year of twelve months, and three hundred and sixty days, was transferred from Asia to Western England. I next proceed to show how the Northern knight riding on the sun-horse became in Greece and Asia Minor the physician,

Jason, sailing round the heavens in the Southern star-ship, the constellation Argo, and have traced, by an examination of the Argo story, that of Perseus, the fish-god, and of Herakles, the history they tell of the rise of experimental and scientific knowledge. Towards the end of the Essay, I have proved that the myths of Sir Galahad and the Holy Grail, and of the Arthurian cycle, are a reproduction, in a variant and greatly embellished form, of the myth of Sigurd in the Nibelungen Lied, with the addition of the belief of the Eastern sons of the barley and the eight-rayed star in the sanctity of the water of life, the heaven-sent rain, the life-giving blood of god, borne to earth in its cloud casket to sustain and reproduce life on earth.

In Essay IX., of which a considerable portion has appeared in essays written by me in the *Westminster Review* of February and March, 1895. I have given the history of the worship of Ia or Yah, the all-wise fish-sun-god, the man-fish, who, in American tradition, led the Indians to America. I have set forth the identity between American totemism and that of Northern Europe and Asia, and have shown how the sanctity of St. George's cross, the sacred sign of the fire and sun worshippers of Asia Minor and Syria is retained among the American Indian tribes, who lay their tribal fires in the form of the sun-cross of St. George. The rules for laying these fires by the peace and warlike sections of the tribes tell of the use of two forms of year-reckonings, both beginning, like those of the Syrian Semites, with the equinoxes, one, like the official Hebrew and ancient Syrian year, with the autumnal, and the other, like that of Joshua, with the vernal equinox. This reproduction of Syrian reckonings of time by the American Indians points, like the invocation of the buffalo and deer in the laying of the fires, to an ancient

connection between the American Indians and the Indian
Dravidian worship of the buffalo and the worship of the deer-
god, Orion, in India and Asia Minor; and an additional
proof of community of origin is given by the reverence paid
in America and India to the sacred fire-pole, made of the
united wood of two trees, which are looked on in both
countries as parent-trees. The American-Indian custom of
using this pole as the sacred sign which precedes the tribes
on their marches, throws, as I have pointed out, light on the
Hebrew story of the nation's march southward from Mount
Ararat, under the guidance of Shelah, the pole or lance.

Identity between the American-Indian beliefs and those of
Asia and Europe is shown in the common worship of the
stone-god, the creating fire-stone, and the storm-bird, in the
great similarity between the cosmogonic myth of the Sia
Indians with the history of mankind as told in the Asiatic
and European mythic histories described in these Essays, and
in the very close approach to identity of ritual in the sacramental feast of the rain-god and the Hindu Soma sacrifice.
I have shown also that the Mexican Indians, like the European and Asiatic sons of the rivers, baptize their children,
and that they and the nomad American Indians have adopted
Asiatic and self-torturing customs exactly similar to those
still surviving among the Hindus; also that the last emana-
tion or avatar of the deity, told of in the Sia cosmogony, is
the god Poshai-yänne, the sun-fish-god, whose story recalls
that of the sun-fish-god of Akkadian and Hindu astronomy,
the god called Ia, Assur, and Pradyumna, the supreme
(*pra*) bright one (*dyumna*), and that of the first Buddha,
called Sumēdha, or the sacrifice (*medha*) of the Su—that is,
the god born of the Soma sacrifice. I have traced the worship of this god through the Arabian legends of Shakr, the

Indian rain-god, Sukra, Solomon, or Salmanu, the sun-fish-god of summer, and Aminah, his female substitute, the winter sun, and also through the story of the Persian triad of Frashaostra, Jamāspa, and Vīstāspa, who formed the sacred Bahrām fire, the perpetually burning fire of the Parsis. It was they who were the divine assistants of Zarathustra, the Zend reformer, who substituted the inspiration of Bhang or Hashish for the intoxicating drink drunk by the earlier inspired priests.

The identity of the Arabian mythic history with other Asiatic and European world-stories of the evolution of religious beliefs which led to the bloodless sacrifices of Soma worship, and of the Zend Haoma ritual, and the rigidly moral creed of Zarathustra, and of the attendant changes in year-reckonings, is set forth in the analysis of the Arabian stories of Bulukiya and Janshah, and of Gharib, the poor (*gharib*) sun-god and his brother Ajib. It is shown how the converging proofs, thence derived, together with others taken from Chinese and Japanese mythology, prove that the age of the worship of the fish-sun-god, and of his father, the god of light, dwelling unseen in his eight-sided temple of the heavens, the home of the eight winds, crowned with the central stone of the heavenly dome, called Solomon's seal, the symbol of the Masonic Royal Arch, and of the nine gods of heaven—was one of great maritime activity and trade, extending from China on the east to the Atlantic Ocean on the west. It was this age which preceded the warlike period introduced by the invasion and conquests of the Aryans, who have been the creators of our modern society in Europe, and have in Asia left the old world to die slowly through the disintegration and stagnation of the living force which once made South-western Asia the ruler of the world. But this slow death, or death-like sleep, is a torpor which will cease with the arrival of the ruler, or ruling race, who can, like the Knight of the Holy Ghost in Heine's *Berg-Idyll*, speak the word of power which is to awaken the people of the East from the sleep of ages; who can assume and retain command over the patient and industrious races who laid the foundations of our modern civilisation; who can stimulate their intellects, rouse their aspirations, and make them once more active agents in the regeneration of the world. These people, who lack the individual ambition of the men of the North, are not distracted with the wish to distinguish themselves, and not disposed to turn the world upside down in order to make themselves famous.   But they are essentially obedient, and will steadily follow, though after long hesitation and doubt, the guidance of leaders who gain their confidence, and show them by results that obedience to their rulers will better their condition.   But leaders and rulers who assert their authority resolutely and wisely they will always want, and without them they will remain asleep.

ERRATA

PREFACE.

Page xii, note 2 line 2 *for* ' lessons required' *read* ' lessons were required.'

ESSAY VII.

Page 17, line 27—*strike out semicolon* (;) *after* race.
   ,, 43, line 5—*for* (dru) *read* (druh).
   ,, 50, line 7—*for* Simargh *read* Simurgh.
   ,, 56, line 7—*for* Pascal *read* Paschal.
   ,, 70, line 23 —*for* Sujātā *read* Sujāta.
   ,, 72, line 34 —*for* Kundalukesā *read* Kundalakesā.
   ,, 72, line 35—*for* (kundula) *read* (kundala).
   ,, 74, line 9 *for* Karmis *read* Kurmis.
   ,, 79, line 23—*for* Jatuvedas *read* Jatavedas.

ESSAY VIII.

Page 108, line 7 *for* Tonngnistr *read* Tanngnistr.
   ,, 167, *here the top line of the page has been omitted, insert* ' the ground was watered with water mixed with barley con-'.
   ,, 167, line 21—*for* Udumbārā *read* Udumbara.
   ,, 168, note 1—*for* Pilner *read* Pierres.
   ,, 169, note 2 line 1—*for* Hekale *read* Hekate.
   ,, 172, line 25—*for* Veda *read* Vedi.
   ,, 190, line 12—*for* rain *read* ram.
   ,, 190, line 13 —*omit* of the.

ESSAY IX.

Page 234, line 10—*for* one *read* our.
   ,, 246, note 1 line 5—*for* Shaptano *read* Shaptanu.
   ,, 257, line 14—*for* Tamel *read* Tamil.
   ,, 274, line 6—*for* seats *read* sixty.
   ,, 299, line 10—*for* Solmon, Ia, fish-god *read* Solomon Ia, the fish-god.
   ,, 299, line 13—*omit* as.
   ,, 324, line 24—*for* India *read* Indra.
   ,, 337, lines 1 and 2—*for* promise Hásib *read* Hásib promise.
   ,, 338, line 2—*after* Sheikh *insert comma*.
   ,, 338, line 16 —*for* Marad *read* Murad.

ESSAY VII

In considering this subject we must, first of all, always remember that the Rigveda is a national collection of Indian ritualistic hymns, written as accompaniments of the worship of the national gods, and chiefly of Indra, the rain-god, and of Agni, the god of the household and sacrificial fire; that the worship of these two gods, and of the other Vedic deities, all culminated in that of the god Soma, the god of the Su, the sap, or essence of life, residing in the mother-tree and the mother-grain, and that Agni and Indra were adored as the gods who, by distributing, at their appointed seasons, the life-generating heat and rain, made plants to grow, and thus secured the continuance of human existence, which, as was believed, originally owed its origin to the indwelling and in-born essence of life, the primordial and unseen god, whose home was in the mother-earth, and the seed thence generated by the mother-tree of the early matriarchal agriculturists, and the sacred barley of the corn-growing races. All the gods of the Vedic Pantheon were gods who represented some form of the annual manifestations of natural phenomena: the twins, Day and Night, called the Ashvins; the Maruts, or the storm winds; Rudra, the storm-god; Pūshan, like Indra, the rain-god; Tvashtar, the maker of Two (*tva*), and the Ribhus, makers of more seasons; Varuṇa, the god of rain (*var*), and the dark heaven of night; Mitra, the moon-god; Savitar, the sun-god. The authors of these hymns drew their inspira-

1

tion and their methods of interpreting the mysteries of life, not directly from traditions framed in the far North, whence the ruling tribes of the land had for the most part emigrated, but from popular Indian methods of measuring time, from the popular ritual and the popular mythology; and these had, in the course of uncounted ages, been moulded from the indigenous belief of the earliest cultivators of the soil, and founders of village communities, and from those contributed by the successive immigrating races into a uniform system of national religion and natural philosophy. This included all the original and contributed axioms, which had shown their vitality by their incorporation into the theology and history framed by the long succession of priestly teachers, and historiographers, who traced their origin in unbroken succession from the remote ages when the rulers of the earliest village communities first began the systematic education of the young, by imparting to them the laws regulating the succession of the seasons, the times of sowing and reaping, the methods of securing good crops, and the political rules which secured the continuance of national well-being.

Though the hymns of the Rigveda were written in a language which marks its meanings by inflections of the root-words, formed according to rules originally elaborated in the far-distant lands of the north-west of Europe, and in one totally distinct from the aboriginal agglutinative tongues of India,—yet this foreign speech, before it became the language of the national religious hymnals, had been altered by the retention of the indigenous cerebral letters, bequeathed to the Sanskrit-speaking priests by their Dravidian predecessors, in whose agglutinative word-workshop these letters had been formed. With these letters they inherited also an elaborate ritual and a national history, preserved in the myths transmitted verbally by countless generations of national historiographers, who handed down to their children the lamp of light and knowledge they had received from their parents, together with the additional learning they had themselves

3

added, to ensure the maintenance of an undimmed and steadily growing flame. These myths told, on the one side, of the descent of the Indian races from the Kushika mother-bird, Gan-dhārī, the wetter (dhāra) of the Gan, or arable land, who laid the egg whence the sons of the bird (khū), or tortoise (kush), the Kaurāvya, Kurus (Khū-rāvya, or Khū-rus) or Kushika were born, and, on the other, from the corn-growing sons of the mother-cow and the sun-antelope. The sons of the bird were the trading races, the Paṇis of the Rigveda, and the sons of the cow and antelope the corn-growing agriculturists, who combined with the traders to form the race of the city-building Purus, the eastern rulers of pre-Vedic India, who, before the myth of the sun-antelope was formed, were the sons of the house-pole and household-fire, the offspring of the mother sun-god Rā, the Rā-hu of the Dosadhs, or fire-priests, of Māgadha. But all these foreign and pre-Sanskrit mythologies of the Finns, the Turanian fire-worshipping and corn-growing races of Asia Minor, who formed the confederacy of the Kushika, grouped round the mother-mountain of the East, were ultimately based upon the primæval belief in the origin of life from the village-tree, the growing-plant, its sap and seed, drawing their vital energy from the mother-earth, symbolised in the national earth-altar, which was the indigenous faith of the farming tribes of South India, who founded the earliest village communities, leaving in the centre of each the Sarna, or mother-grove of primæval forest trees. This belief culminated in the national Vedic worship of Soma, the life-giving sap, born from the Su, or essence of life, brought from heaven to earth by the rain-bird Khū, which bequeathed its name to its offspring, called the Khū-rāvya, or sons of the bird (khū), and Kushika, or sons of the tortoise Kush. This name, after the generating processes of thought forming the Vedic theology had been completed, became in the mouths of the Southern races, who finally formulated its conclusions, and changed the Aryan and Finnic gutturals into sibilants, Su.

It was for the worship of the god Su or Soma that the Vedic ritual was added to that of the previous national festivals of the rain-god and fire-god, and it was as accompaniments of this ritual that the Vedic hymns were composed. These, as we learn from the form in which the national hymnal, the Rig-veda, has been handed down to us, were the ritualistic chants of the distinct priestly colleges, distributed among the separate kingdoms into which Northern India was divided; and before these, originally local chants, were collected into the hymnals used within the area in which each priestly college of authors was predominant, and before they were finally incorporated into the great national collection, they must have been carefully sifted and selected, and only those hymns which obtained the approval of a most conservative and critical priesthood were allowed to survive the ordeal. In the eyes of these priestly authors and critics, who hated innovation, as tending to throw into confusion the efforts of ages spent in securing order and agreement, astronomy was only useful as a guide to determine the days to be fixed for the national festivals; and as these had all been deter-mined in accordance with the recurrent annual procession of the seasons before the Soma ritual was elaborated, Vedic astronomy only concerned itself with the determination of the length of the revolving year of recurring seasons, with their appropriate sacrifices, inaugurated by the worship of Su, the god of life, the rain-bird of the summer solstice, which brought the rains of Northern India. This bird, which was originally the rain stork, the Adjutant (*Ciconia argala*), which yearly announces the coming of the rains in North-eastern India, had become, in the theology of the authors of the Soma ritual, Soma, the moon-god. But this theology of the worship of the rain-god of Northern India was not that which sprang from the very different arrangement of seasonal changes found in Southern India, the original home of the Dravidian founders of villages, who, as practical farmers, devoted their earliest efforts to the correct computation of annual time, and this theology and its accompanying ritual started from a much earlier series of attempts made to measure the year recorded in the national traditions, received by the Vedic theologians as an inheritance bequeathed by their forefathers.

The worship of the divine sap, or essence of life, rested upon a calculation of the order and times of the succession of the annual processes of the sowing, growth, and ripening of the divine seed, which supplied the agricultural races with their yearly crops, and thus the year within which these processes were annually completed became the mother-cow of the Northern corn-growers, and the mother-tree of the Southern rice-growers. The several changes of the seasons by which the gradual growth of plant-life was brought about, had been from time immemorial celebrated by the village communities descended from the earliest rice-growing races, by dances and sacrifices, commemorating the successive stages in the evolution of time; and the times at which the dances for each season were to be held were determined partly by the flowering of the national parent-trees, such as the Sāl tree, which gave the signal for the Sar-hul festival in March-April, but chiefly in Northern India by the regular arrival of the annual rains, which almost invariably, in the Gangetic valley, begin with the summer solstice, when the rice was sown. This, together with observations made by the Gnomon, or rain-pole of the god Vasu, determined the times of the solstices, and the intervals were determined by the traditional rules for calculating time by the lunar phases. The equinoxes, owing to the nearness of the country to the equator, did not attract much observation in early times in Northern India. But this primæval age, though it is referred to in the Vedic poems, which describe the seasonal dances, and tell us of the Soma, or rice-beer, made in every house, to be drunk at these festivals in accordance with the custom of the Ho Kols and Mundas of Western Bengal,[1]

[1] Rigveda, i. 28, 3, 4, 5; viii. 5 (69), 7-9.

belongs to a much earlier phase of national existence than that on which the Vedic cosmogony is founded. For this last tells us of the time when the agricultural age had become one of manufactures, and when farmers looked for other profits from their labours than those derived from grain and edible roots. In the Vedic age, the annual round of seasonal changes was compared to the revolution of the oil-press, with its turning beam and central pole, which was, in the early theology of the sons of the North, the house-pole. This simile appears in the ritual of the Vāja-peya sacrifice, in which the pure running water, the produce of the rains of heaven, the heavenly Soma, made above the axle of the sacrificial press, becomes the life-generating Sūra, or intoxicating spirit, taken from under the axle by the Neshtri, the priest of Tvashtar, the early god of the races who attributed the origin of life to pairs, to the fire-drill and the fire-socket, the father and the mother. This doctrine of the seed of life, generated in the heavenly oil-press, became in the Rigveda that which likened the revolution of time to the circling of the meridian-pole by the twin Ashvins, the twins Day and Night, who became the stars Gemini, and which are said to drive in a chariot drawn, according to one account, by stallion asses,[1] or in another, by the constellations of the Bull and the Alligator,[2] round the pole, with one wheel of the chariot in the sea and the other in heaven.[3] From this we learn that the Vedic poets believed that the rain, the mother of the Soma-plant, called Vārshabhu, the rain (*vārsha*) born (*bhu*) plant, was generated by the ceaseless revolution of the pole, the heavenly fire-drill, turned round by the twins Day and Night, according to the newer, and by the Maruts, or wind-goddesses, of the south-west rain-bringing wind, according to the earliest cosmogony, as preserved in the Vāyu Purāna.[4] The final astronomical conclusions arrived at by the Vedic

[1] Rigveda, i. 34, 9, 116, 2 ; iii. 57, 5.
[2] *Ibid.* i. 116, 18.   [3] *Ibid.* i. 30, 18, 19.
[4] Sachau's Alberuni's *India*, vol. i. chap. xxii. p. 241.

writers, after they had successively adopted, as ritualistic
standards, the reckonings of annual time, measured by the
revolutions of the pole and the phases of the moon, included
in the year or ring of ten lunar months of gestation, and the
eleven months, sacred to the gods of generation, were the
lunar year of thirteen months, and the year of the perfect
circle of 360 degrees or days, divided into twelve months of
thirty days each.   These are described in the cosmological
hymn,[1] ending the series of twenty-five hymns,[2] said to be
the work of the writer called Dīrgha-tamas, or, in other words,
of the long (dīrgha) ages of darkness (tamas).   This series
includes the hymn belonging to the ritual of the Ashvamedha, or horse-sacrifice,[3] and the hymn recording its chrono-
metrical conclusions is thus shown to be an exposition of the
ritualistic astronomy of the sons of the horse, who instituted
the final form of the Soma sacrifice, in which the *prastara*,
or rain-compelling magic wand, was made of Ashva-vāla,
or horsetail-grass (*Saccharum spontaneum*),[4] instead of the
Kusha grass (*Poa cynosuroides*), used in the ritual of the
early Kushika sacrificers.   In this hymn the first ten stanzas,
symbolising the ten lunar months of gestation, tell of the
birth of the year-calf, the child of the totem parents of the
race, who have seven sons (v. 1).   Their year-car is drawn
by the father-horse, with seven names, the seven days of the
week; and its one wheel, the horizontal wheel of the oil-press,
has three navels, the three seasons of the original year of the
barley-growing races (v. 2). the year of the fathers, whose
spirits were supposed to sit on the seats of Kusha grass, the
parent-grass of the Kushika race, at the autumnal Pitri-
yajña, or sacrifice to the fathers, to whom parched barley
was offered.   This primæval one-wheeled chariot, revolving
round the pole, becomes in v. 3 the seven-wheeled car,
drawn by seven horses, to whom seven sisters, who bear the
names of the seven cows, sing—that is to say, the fourteen

[1] Rigveda, i. 164.   [2] *Ibid.* i. 140-164.   [3] *Ibid.* i. 162.
[4] Eggeling, *Sat. Brāh.* iii. 4, 1, 17, 18; S.B.E. vol. xxvi. p. 89, note 3.

days of the lunar phases, formed by the union of the seven
sons of the horse, the Gandharva, with the seven mothers
of the cow race, descended from Rohini, the cow-mother, who
were, according to the Mahābhārata, descended from the king
Surabhi.[1]   His name, meaning the firm (*su*) encloser (*rabhi*),
is a synonym for the enclosing snake, the ring of arable land,
surrounding the mother-grove of the primaeval village; and
the children of the horse-father and the cow-mother, who are
said to draw or turn the revolving wheel of the year, are thus
the twin races, whose children are born in the grove consecrated to the parent-gods of the confederacy.   The birth and
generation of the year-calf, from whom these seven and fourteen measurers were born, is described (v. 8, 9, 10) as happen-
ing (v. 7) in the place of the 'beautiful mother-bird.'   She
was made pregnant by the thought of the creator, generated
in the life-giving moisture of the atmosphere, and the
maternal and paternal elements in the act of generation were
each three in number (*trirīn-pitrīn, trisras-mātrās*), making
up the six seasons or *ritu* of two months each, reckoned by
the oldest race of fathers, the Pitarah-somavantah, to whom
rice, and not barley, was offered at the Pitri-yajña upon six
platters.[2]   The year of twelve months or spokes thus conceived and brought into being is said (v. 11) to have 720
sons—that is, 360 days and 360 nights.   Its father is he who
dwells in the depths of heaven, who is divided into five parts,
the five seasons of the sacrificial year of the worshippers of
the new and full moons, but who is called by the people of
the land the far-seeing one (*Vichaksana*).   He stands upon
seven wheels with six spokes (v. 12), the seven days of the
week, and the six seasons of the Southern year.   All living
beings rest on the five-spoked wheel, whose axles do not
become heated (v. 13).   Ten horses draw the never-ageing
wheel (through space), whence the eye of the sun, on which
all life depends, looks down (v. 14).   The seventh of those

---

[1] Mahābhārata Ādi (*Sambhava*) Parva, lxvi. p. 193.
[2] Eggeling, *Śat. Brāh.* ii. 6, 1, 4 ; S.B.E. vol. xxvi. p. 421.

born together they call 'that born alone': this is the self-created thirteenth or central month; the six twinned months are said to be those begotten of the gods. They are arranged in their order (six on each side of the central month) by the leader who dwells above (v. 15).

The hymn then goes on to tell how all are ignorant whither the year-cow, with her calf, goes, and whence the divine spirit of life came; but all the days made by Indra, the rain-god, and Soma, the plant of life, are drawn by the horses of the year-car (v. 17-19). It then describes how the guardian-god of the world dwells in the tree of immortality, on which sit the two birds (Day and Night), one of which eats the sweet fruit of the tree, while the other assimilates it by contemplation (v. 20-22). The hymn next (v. 23-25) speaks of the influence of the metres called Gāyatrī, Trishtubh, and Jagatī, consisting of lines of eight, eleven, and twelve syllables each, in the computation of time, and thus proves that these metres, together with that of ten syllables, the Virāj, consecrated to the matriarchal age, which is here ignored, were adopted with direct reference to the various methods of computing time, commemorated in (1) the eight-sided altar of the Kushite race, dedicated to Agni, and formed by the superposition of the sun and rain-cross ✕ on the fire-cross ┼, the whole forming the eight-rayed star ✳, called by the ancient Akkadians, and also by the earlier Indian races, the sign of Anu, meaning god or seed; (2) the eleven months sacred to the gods of generation; and (3) the twelve months of the solar year of 360 days. Thus, these three metres are called (v. 25) the three pieces of wood which light the sacrificial flame: the three *paradhis*, forming, on the sacrificial altar in the form of a woman, the three sides of the triangular image of the three seasons of the mother-year.

It is for the two Ashvins, or heavenly twins, Day and Night, that the year-cow lets her milk flow (v. 27), and the hymn then proceeds through a series of very obscure stanzas to the forty-eighth, which sums up its teachings in saying that, though the wheel is one (the year), yet it has twelve divisions (the twelve months), and they bring together 360 pegs, the days of the year.

The remaining four stanzas of the hymn, added to complete the perfect number of four times thirteen, the fifty-two weeks of the lunar year of thirteen months and 364 days, tell how Sarasvatī, the mother of the waters, nourishes all things from her breast, and how the spirit of life rises from earth to heaven through the offerings of the faithful. And thus, in the course of days the water of life soars aloft, and hence the rain-clouds which nourish the earth and the fire which nourishes heaven, the rain and fire cross, are generated; and the hymn closes with stanza fifty-two invoking the heavenly eagle-child of the waters and of the plants, which brings the rain.   In this hymn the life generated by the revolving years is said to spring ultimately from fire and water, that is, from the rains churned out by the revolving pole, the father of fire: the mother-rains called Idāh, which are the third or middle season in the year of the new and full-moon worshippers,[1] who called the spring the Samidhs, or kindling season ; the summer the self-begotten Tanū-napāt or Narashaṃsa, the praised (*shaṃsa*) of men (*nara*), the burning fire, the father of all things lighted on the altar of the mother-earth, the national hearth-stone ; the autumn the Barhis, or sacred seats of Kuṣha grass, consecrated to the fathers ; and the winter the Agni Svishtakrit, the most effective (*svishta*) of the fires, the god Rudra of the sacrificial stake,[2] to which the animal victims offered at this season to the twin gods of the twin races of the North and South, the twins Day and Night, were bound.

---

[1] Eggeling, *Sat. Brāh.* i. 5, 3, 4 ; S.B.E. vol. xii. p. 147.
[2] *Ibid.* i. 5, 3, 9, 10, 12, 13, 23 ; i. 7, 3, 1-7 ; S.B.E. vol. xii. pp. 146-148, 149, 199-201.

This year-reckoning depicted the year as turned round the meridian pole (*Tur*), represented by the solstices called Tur-āyana or times (*āyana*) of the Tur. This was originally the house-pole of the races who worshipped the household fire, and it became the gnomon by which the times of the turning of the sun from south to north and north to south could be determined. This national time-pillar or obelisk, by which the path of the sun was measured and the time of the solstices determined, had in Vedic times become an instrument of astronomical prediction called 'Turīya Brahma,' the Tur of knowledge, and it was by means of this that Atri, the generic name for the priest of the sun-god, the priest of the latest priestly guild, called the Athārvans, predicted, as we are told in the Rigveda, an eclipse of the sun, and its temporary effacement by Svar-bhanu, the Assura or Assyrio-Semite god of light, the moon-god.[1]

But the year hymn which tells us that the earliest solstitial and equinoctial year, measured by days, contained 360 days, does not mention the Nakshatras, which were undoubtedly known to the Vedic writers, as in the hymn telling of the marriage of the moon-god Soma to the sun-maiden, who was brought by the Ashvins, as groomsmen, on their three-wheeled car, the three divisions of the day into darkness, dawn, and daylight, or the year of three seasons, the wedding-oxen are said to have been slain in the signs of Maghā, and the wedding to have taken place in the signs of Arjuni or Phalgun.[2] This wedding, consummated in the month ending with the vernal equinox, tells us of a year in which, as in modern Hindu popular chronometry, the Huli or Saturnalia of the red races[3] was celebrated at the full moon of Phalgun, and when the birth of the young sun-god of the next year, who was to be married on the full moon of the next Phalgun, took place at the full moon Margasirsha, the

---

[1] Rigveda, v. 40, 5, 6, 9.   [2] *Ibid.* x. 85, 13, 9, 14.
[3] It is marked as the Saturnalia of the red races by the red powder thrown by the women upon the clothes of the young men.

month of the winter solstice, which is, as I shall show presently, always called in Hindu ritual the Agrahāyaṇi, or beginning of the year.

But the year of 360 days, as well as that measured by the lunar phases of the twelve months reckoned in the year, gave a wrong measurement of time, though the errors could be corrected by observations made with the gnomon at the solstices and equinoxes. In order to correct those occurring during the intervals between these seasonal starting-points, and to make lunar and solar time coincide in a cycle of five years, the heavenly star-circle originally formed to mark the daily positions of the moon during each lunar month was divided into twenty-seven equal parts of 13° 20' each. This circle had no connection with the later ecliptic, which does not appear in Hindu astronomy till the days of Varāha mihira, who calls the ecliptic stars after names derived from their Greek titles, thus showing that the method of time measurement by these stars was a foreign custom, brought to India by the Greek settlers who had made it their home after the conquests of Alexander the Great.

The stars denoting each sign of the original twenty-eight Nakshatras are thus given by Brahma Gupta, and his list, in which the stars coincide with those in the Chinese and Arabian lunar mansions,[1] begins with the stars consecrated to the Ashvins, or heavenly twins, which, in the original astronomy of the circumpolar stars, were the stars Gemini, and not with those appropriated to the Pleiades, which are third in Brahma Gupta's list, and, as I shall show later on, head the list in the earlier Hindu and in the Soghdian and Chorasmian astronomy of the fire-worshippers.

1. Ashvini or Ashvayujau.[2]	β Arietis.
2. Bharani or Apa Bharani.	α Muscœ.

[1] Whitney on 'Jacobi and Tilak, or the Age of the Veda,' *Proceedings of American Oriental Society* for March 1894, p. 92.

[2] Of these stars Nos. 28, 1, 4, 7, 8, 10, 12, 14, 16, 17, 18 are identical with stars Nos. 1, 2, 5, 12, 13, 15, 17, 20, 21, 24 of the Babylonian ecliptic

3. Krittakā or Krittakas.
4. Rohini Aldebaran.
5. Mrigasirsha, Andhakā, Aryikā, Invikā or Ilvala.
6. Ardrā or Bāhu.
7. Punarvasu.
8. Pushya, Tishya or Sidhya.
9. Āshleshā, Āsreshā or Āshleshās.
10. Maghā or Maghās.
11. Pūrva, Phalguni or Arjuni.
12. Uttara Phalguni.
13. Hastā.
14. Chitrā.
15. Svāti or Nishtya.
16. Visakhā or Visakhe.
17. Anurādha.
18. Jyeshtha or Rohini.
19. Mūla or Vichritau.
20. Pūrva, Ashādhā or Apya.
21. Uttara, Ashādhā or Vaishoa.
22. Abhijit, meaning now (abhi) conquered (jit); this sign omitted to make the annual 27 Nakshatra.
23. Shravana, Shrona or Ashvattha.
24. Shravishtha or Dhanishthā.
25. Sata bhisaj.
26. Purva Bhādrapadā, Proshthapadā or Pratishāna.
27. Uttara Bhādrapada, etc.
28. Revati.

23 Tauri.
a Tauri.
λ Orionis.

a Orionis (?).
β Geminorum.
δ Cancri.
ε Hydræ.
Regulus.
δ Leonis.
β Leonis Alsarfa.
γ or δ Corvi.
Spica Virginis.
Arcturus.
ι Libræ.
δ Scorpionis.
Antares.
λ Scorpionis.
δ Sagittarii.
σ Sagittarii.

Vega, Al nasr alwaqi.
a Aquilæ, Al nasr altāir.
β Delphini.
λ Aquarii.
a Pegasi.

γ Pegasi or a Andromedæ.
ζ Piscium.[1]

But before discussing fully the evidence as to the use made of the Nakshatra in Vedic times, I must first speak of the

of 28 stars situated in the constellations: Pisces, one star; Aries, two; Taurus, four; Gemini, five; Cancer, one; Leo, four; Virgo, three; Libra, two; Scorpio, two; Ophiuchi, one; Capricornus, three.—Epping and Strass-meyer, *Astronomisches aus Babylon*, pp. 117-133; Norman Lockyer, *The Dawn of Astronomy*, chap. xxxvii. p. 407. The whole series evidently marks a transition stage from the original list of lunar mansions to the later solar zodiac.

[1] J. Burgess, C.I.E., on 'Hindu Astronomy,' *Journal Royal Asiatic Society*, Oct. 1893, p. 756.

Vedic evidence as to the existence of earlier computations of
time than those which culminated in the elaborate calculations of the literary age of the Athārvans indicated by the
systematic observation of the heavens, and the record of
results which must have preceded the adoption of the star-
marked lunar stations and the solar-lunar twenty-seven
Nakshatra.

The earliest chronological system contemplated by the
Vedic writers was that of the god Tvashtar. He is the most
complete two (*tva*), the father of Saranyu, the mother of the
heavenly twins Ushāsā-nakta, day and night.[1] The year of
Tvashtar was divided into two parts of six months each,
called the Dev-ayāna or times (*ayāna*) of the bright gods
(*dev*) and the Pitriyāna or times of the fathers. They are
spoken of in the Rigveda in a stanza which tells how Tvashtar had created the earth, heavens, and waters from fore-
knowledge of the light which streams from the path traversed
by the fathers (*Pitriyāna*),[2] thus showing that Tvashtar's
year began with the six months dedicated to the Pitris. In
other hymns the path of the gods Devayāna is said to lie
in the contrary direction to the path of death,[3] and Agni is
said to know the Devayāna by the seasons.[4] These two
divisions of the year are said in the Brāhmaṇas to mark the
path of the sun between the solstices: the Devayāna, the six
months during which he goes north, beginning with the
winter solstice, and the Pitriyāna, those after the summer
solstice when he goes south.[5] But this distribution of time
makes the year begin with the winter solstice and the Devayāna six months, instead of with the Pitriyāna, according to
the Vedic description of Tvashtar's year. But in one of
the descriptions of the year measured by the Nakshatras
in the Taittirīya Brāhmaṇa, it is said that 'the Nakshatras
are the houses of the gods.' The Nakshatras of the Devas

---

[1] Rigveda, x. 17, 2.   [2] *Ibid.* x. 2, 7.
[3] *Ibid.* x. 18, 1.   [4] *Ibid.* x. 9, 8, 11.
[5] Eggeling, *Sat. Brāh.* ii. 1, 33; S.B.E. vol. xii. p. 289.

begin with the Krittakas and end with Vishākhā, whereas
the Nakshatras of Yama, that is, of the twins (*yama*), the
dead ancestral fathers of the twin races, begin with the Anu-
rādhās, a name meaning the arc or half circle (*rādha*) of
non-Aryan men (*anu*),[1] and end with the Apa Bharanis.[2]
Elaborate arguments, of which the most remarkable is that
of Bentley,[3] have been framed to prove that this passage, and
the authoritative lists of the Nakshatras, beginning with the
Krittakas, speak of a time when the sun was in the Pleiades
at the time of the vernal equinox, thus making them begin
the Nakshatra year. In proof of this conclusion Bentley has
proved that the occultation of the planets Mercury, Venus,
Mars, and Jupiter, which he thinks to be denoted by the
story in the Linga and Vāyu Purāṇas that they were born
from the Nakshatra mothers, took place in 1424 and 1425
B.C., or the year after, 1426 B.C., when the sun was in the
Pleiades at the vernal equinox. But whatever chronological
results may be derived from myths telling of early astronomical observations made when the science was so far advanced as to record unusual positions of the planets, yet it
must be remembered that it was through observations made
as to the movements of the wandering stars among the fixed
stars, chosen as marking the mile-stones of the heavenly
circle, that the laws of planetary motion were first discovered, and that they were thus said to be born, that is,
to be first understood as children of the Nakshatras. But
the fatal objection to the acceptance of this deduction,
which makes the place of the Pleiades among the Nakshatras
to depend on the position of the sun at the vernal equinox,
arises from the fact that this conclusion does not explain how
it is that the people of Western India have from time immemorial begun their year in the month Khārtik, named

[1] Grassmann, *Wörterbuch zum Rigveda*, s.v. ' Anu.'

[2] *Taitt. Brāh.* i. 5, 2, 7.

[3] *Historical Review of Hindu Astronomy* (Calcutta, 1823), quoted by Max
Müller, Preface to vol. iv. of the Rigveda, pp. 30-33.

after the Pleiades, or that the Gonds and all Western Hindus celebrate the festival of lights, the Hindu Debālī, on the new moon of Khartik, and the fact that this festival is the same as that of Osiris, which began the year in ancient Egypt, and the present festival of lanterns in Japan, both of which were and are kept in November. Nor does it explain how it is that the Soghdians or ancient Sabæan fire-worshippers began their list of lunar stations, describing the monthly circuit of the moon, with the Pleiades, called the Parwe, a name meaning the conceiving (*peru*) mothers.[1] These people, who measured their year by lunar months, must, in choosing their lunar stations, have been led to fix upon that which they placed first in the list and called the conceiving mothers, by the previous use of these stars as marking the beginning of the year. This year, the beginning of which was marked by the Pleiades, was that which has been from time immemorial used by the people of the Southern hemisphere, and divided into two parts by the appearance of the Pleiades above the horizon at sunset in the beginning of November, and their disappearance at the same time in April. The rising of the Pleiades in November is the time when the fathers are worshipped throughout the whole Southern hemisphere, and it is this custom which has been brought by the Southern founders of the village community to Europe, and has given birth to the festivals of All Hallow Eve, All Saints' Day, and All Souls' Day, kept on the 31st October and the 1st and 2nd November. It is at this season that the dead were anciently worshipped by the Sabæans in Persia, who placed the Pleiades at the head of their lunar stations; by the Peruvians and Druids, or priests of the mother-tree (*dru*);[2] and as the extinguishing of the old year's fires and the lighting of the new was marked in the Druid year by the appearance of the Pleiades above the horizon at sunset in November, this must have been the beginning of the year of the early fire-

[1] Sachau's Alberuni's *Chronology of Ancient Nations*, chap. xi. p. 227.
[2] *The Ruling Races of Prehistoric Times*, Essay ii. pp. 123-133.

worshipping Sabæans or Soghdians of Central Asia, who called the Pleiades the Parwe, or conceiving-mothers, and placed them at the head of the lunar stations. The Pleiades have always been looked on as the mother-stars of all the Southern Indian tribes, and the Brāhmaṇas give the mythic history of the union in India of the Northern barley-growers and Southern rice-growing races by describing it as the marriage of the Krittakas, the Pleiades, the Southern mothers, with the stars of the Great Bear, called the Rikshas, the bears, and Rishya, the antelopes. The Pleiades also, under the name of the Turanyā, become the mother-stars of the Southern Arabian or Sabæan tribes.[1] Their name, the Krittakas, meaning the spinners, shows that they were looked on as the mother-weavers of the web of measured time, and this name coincides in meaning with that of the Parvis, Parven, and Parwe, the conceiving (*peru*) mothers, by which they were known to the Zends, Persians,[2] and Chorasmian Sabæans.

It is through the seasons of the Pleiades' year beginning in November and April that we are able to explain the origin of the name of the Hindu month Vishākhā, meaning the two (*vi*) branches (*shākha*), for it is in this month that we are told, in the passage of the Taittirīya Brāhmaṇa quoted above, that the Nakshatras of the Devas, beginning with the Krittakas, and those of Yama begin. As in the primæval age, the ancestors were the creating-gods or totems of the race; the six months dedicated to them were the months sacred to the gods, while the next six months of sowing and rice-harvest were those sacred to the twins (*yama*), their offspring, the fructifying months of summer and autumn when the seed of the next year was produced.

This original year of the Pleiades and the rice-growers

---

[1] Tiele, *Outlines of the History of Ancient Religions*, ' Primitive Arabian Religion,' § 40, p. 63.

[2] West, *Bundahish*, ii. 3 ; S.B.E. vol. v. p. 11, note 3.

was followed by the year of three seasons of the Northern
millet- and corn-growing races, beginning at the winter solstice; and we learn the mythical history of the origin of
this year in the stories which tell how Prajā, the lord (*pati*)
of former (*pra*) generations (*ja*), transformed into a Rishya
or antelope, pursued Rohinī, his daughter, the star Aldebaran,
the companion of the Pleiades, in the shape of a doe, and
how when he was engaged in the pursuit he was pierced by
the arrow of Rudra, the god of the sacrificial stake of the
Northern sacrifices of totemistic animal victims. He then
mounted up into the sky and became the constellation
Mṛigasirsha (*Orion*), while the arrow with which he was
pierced became the belt or girdle of Orion, called in Hindu
astronomy the 'three-knotted arrow,'[1] the three seasons of
the year described in the Brāhmaṇas as the arrow of Vishnu,
who was first Kṛishṇa, the black antelope, consisting of the
point (*anīka*), Agni, the wing feathers (*Shalya*) or Soma,
and the connecting piece (*kulmala*),[2] the year beginning
with the spring, the season of the winged mother-bird,
which hatches through the summer the seeds of autumn falling and flying into the winter earth, whence the Soma plant
springs. This same story is told in two other forms in
Hindu epic legend; one of these is that in the Rāmāyana,
which tells how Rāma, when decoyed from Sitā by Marīchi,
the spark of light, in the form of a deer, killed him, and how
the slain Marīchi went up to heaven as a star, and became
one of the stars of the Great Bear, while Sitā, who had up to
that time been Sitā the furrow, the bride of Rāma, the sexless ploughing-ox, became Sitā the crescent moon, carried off
by Rāvana, the god of storms, as his virgin wife; destined after
her delivery from him to remarry Rāma, who became the sun
and moon-god.   In the other story Paṇḍu, the sun-antelope,
reputed father of the Pāṇḍavas, while hunting in the forest,

---

[1] Eggeling, *Sat. Brāh.* ii. 1, 2, 8, 9; S.B.E. vol. xii. p. 284, note 1;
Ait. Brāh. iii. 33.

[2] Eggeling, *Sat. Brāh.* iii. 4, 4, 15: S.B.E. vol. xxvi. p. 108.

like the hunter Orion kills a Rishi, an antelope, called a
Brahmin, while having intercourse with a deer, and at once
became a sexless father-god, that is to say, he went up to
heaven, while his sons the Pāṇḍavas were begotten by the
gods of heaven. This story also appears in another form in
the Rigveda, where it is said that the gods created from the
union of Prajā-pati with his daughter the Brahma Vāstoṣh-
pati, the lord (*pati*) of the house (*vastu*), the guardian of
good works, and this son became, in a later part of the same
hymn, the sacrificial fire in the centre of the altar called
Nābhā-nediṣhthā, that nearest (*nedishthā*) to the navel
(*nābhā*),[1] the Agni Jatavedas, which knows (*vedas*) the secret
of birth (*jata*) of the Rigveda. This evidence that the son
of Prajā-pati was the god of the sacrificial flame is confirmed by the statement in the Mahābhārata that Rudra
pierced the heart of Yajña, the sacrifice, with an arrow.[2]
The death of the antelope also appears in the Rigveda in the
account of how Indra cut off the head of Namuchi, the demon
who keeps back the rain, who is elsewhere called an antelope,[3]
with the foam of the water,[4] and this expression is explained
in the Taṇḍya Brāhmaṇa, xii. 6, 8, to denote a compact made by Indra with Namuchi that he could not kill
him by day or by night or with any weapon wet or dry, so
he killed him at dawn with the foam of the waters—which
foam is said to have proceeded from the mouth of Brahma-
Vāstoṣh-pati, Prajā-pati's son by Rohinī.[5] This foam was
the breath of the south-west monsoon, which at once, on its
arrival, makes the god of the burning summer haze, brought
by the dry west winds of the hot season, give up the rain he
kept back : and in this form of the myth the reign of Prajā-
pati, the star Orion, who rules the six months beginning with
the winter solstice, is brought to an end by the rains of the
summer solstice.

[1] Rigveda, x. 61, 7, 8, 18, 19.
[2] Mahābhārata Sauptika Parva, xviii. 13, 14.
[3] Rigveda, v. 34, 2.   [4] *Ibid.* vii. 14, 13.   [5] *Ibid.* x. 61, 8.

The foam of the waters is also explained by Bāl Gangādhar Tilak, in his learned work called *Orion*, from which I have taken many of the quotations in this Essay, to mean the Milky Way; but while I think that he has, in the part of his treatise in which he works out the meaning of the legend of Prajā-pati as the constellation, proved his contention conclusively, I cannot agree with his further conclusions, in which he tries to prove that the year of Prajā-pati and the Kṛittakas began when the sun was in these constellations at the vernal equinox. For the reasons I have already stated, and for those which I shall bring forward in the sequel, I believe that the years ruled by the Pleiades and Orion were time-measurements of an age very much older than that in which the position of the sun in the heavens at any time in the year was one of the factors used in computing time. That the story which tells how the antelope-star, Orion, was made the ruler of the year of the Northern corn-growing races belongs to a very early period in Hindu national history, is proved conclusively by Bāl Gangādhar Tilak in his account of the ceremonial of the investiture of young Brahmins with the sacred thread or girdle. The officiating Brahmin, when placing the girdle on the youth, says: ' I invest you with the Upavīta or waist-cloth of sacrifice' (*yajña*); and that this Upavīta or waist-cloth is the belt of Orion is proved by the rule that the Mekhalā or grass girdle, made of three strands of Muñja grass (*Saccharum munja*) (showing that the wearer of it belonged to the Ikshvāku or sugar (*iksha*) cane race, sons of the Ashva-vāla, horse-tail grass (*Saccharum spontaneum*)), is tied with three knots over the navel to represent the three-knotted arrow or belt of Orion. Also the neophyte has at the same time to put on the ancestral antelope-skin, and to bear the staff—the magic wand, the sacrificial stake, the gnomon or indicator of time, and the sceptre of office, which have in the mythological astronomy of all nations been assigned to Orion.[1] The star-

---

[1] Bāl Gangādhar Tilak, *Orion*, chap. vi. 145-150.

bespangled girdle of Orion, called the Aivyaonghana, the
Zend name of the *Kūsti*, or sacred girdle of the Parsis,
appears in the Zendavesta as the child of Mazda and Haoma
(*Soma*), said, according to the translation proposed by Dr.
Haug, to lead the Pleiades (*Pourvanim*),[1] and this confirms
the Hindu evidence, telling that the deification of the girdle
of Orion followed that of the Pleiades. This legend, telling
of the meaning of the arrow, or three stars of Orion, appears
also in the Greek story, which tells how Orion, the hunter,
wearing a girdle, sword, a lion's (not an antelope's) skin, and
club, was placed among the stars after he had been killed
by Artemis, in the morning, in Ortygia, meaning the land
of the Quails,[2] the Vartika of the Rigveda, sacred to the
Ashvins, the twin-fathers of the corn-growing race, who
made Orion their year-star. Another account says that
Artemis, the bear-mother, killed him at the suggestion of
Apollo, the god of day ; while the relation between the deer-
god and Artemis, who began by being the bear-mother, the
constellation of the Great Bear, the plough of heaven, and
became Elaphia, the moon-deer goddess of Elis, is similar to
that disclosed in the story in the Rāmāyana, which tells how
Sitā, the crescent-moon, ruled the year, when the deer-god
was taken up to heaven as a star. That is to say, that the
legends of India and Greece both tell us that the year of
Orion was a lunar year measured by the lunar phases, the
moon-hare, or Indian fox of the year, symbolised by the constellation Lepus, at the foot of Orion, hunted by him
through each month of the year, and always, like the Indian
fox or hare, returning to her first form on earth when she
escapes from her pursuer. We also find in the legend of
Orion the Hindu and Greek equivalent of the Akkadian
and Egyptian year of Orion, called Dumu-zi, the son (*dumu*)
of life (*zi*), by the Akkadians, and Smati-Osiris by the
Egyptians, who travel through the year, beginning in Novem-

---

[1] Mill, *Yasna Haoma Yasht*, ix. 26 ; S.B.E. vol. xxxi. p. 238.
[2] Homer, *Odyssey*, v. 121-124.

ber in Egypt, in the moon-boat, the crescent moon. This sea-rover Orion, who ruled the year succeeding that of the Pleiades, appears also in Greek mythology as the wandering sun-god, Odusseus, the Orwandil of the North, whose toe was the star Rigel in Orion, and whose wife was Penelope, the weaver of the web (πήνη) of time, the Greek form of the Hindu Ambā, the chief star of the Pleiades, the Spinners.

But when this year of twelve months, measured by the lunar phases, came to be compared with that marked by the Pleiades and solstices, it was found to be nearly twelve days too short, and hence arose the system of intercalating twelve days between the close of the lunar year of twelve months and the date of the solstice, as ascertained by the gnomon. This interval of twelve days is said, in the Rigveda, to be the days during which the Ribhus, the makers of the seasons, slept in the house of Agohya, meaning 'he who cannot be concealed,' the Polar star.[1] These were the twelve days during which, according to the Atharva-Veda iv. 11, 11, Prajā-pati or Orion, the god of sacrifices, after finishing the lunar sacrifices of the old year, prepares himself for the sacrifices of the new, and it is these twelve days which survive in the Dvādasāha, or twelve days' sacrifice of the Brāhmaṇas, which intervenes before the beginning of the annual Sattras or sacrificial session.[2] These intercalated twelve days survive in the German national festivals of the Wild Hunter, the star Orion, which are held, according to Professor Kuhn, during the twelve days' interval between the close of the old year and the beginning of the new one.[3] In this festival, as in the Indian legend, the chief actors are a man dressed as a stag, and a woman as a hind, who sing unchaste songs, and the stag is shot by the Wild Hunter. We thus find that

---

[1] Rigveda, iv. 33, 7 ; 161, 11.

[2] Eggeling, *Sat. Brāh.* iv. 5, 4, 1, ff. the Dvādasāha ; S.B.E. vol. xxvi. pp. 402 ff.

[3] Letter from Prof. Kuhn to Dr. Rajendralal Mitra, *Indo-Aryans*, vol. ii. pp. 300-302. Bāl Gangādhar Tilak, *Orion*, chap. vi. pp. 138-140.

very early mythic historical legends, current both in India
and Germany, tell us that in the days of the lunar year of
Orion, when sacrifices were first offered, a method had been
discovered of adjusting the differences between solar and
lunar time, and of measuring a year almost with the same
accuracy as that given by astronomical observation of the
path of the sun through the ecliptic. It was when these
twelve days were over that the great animal sacrifices to
the year-gods took place, at which the tribal totem was
sacrificed and eaten. The earliest form of these sacrifices,
when the totem god was the goat and not the antelope, survives in those offered round the house-pole by the Indian
Dravidian Mālēs and Mal Paharias, who place round it
balls of clay representing their ancestors, and pour upon
them the blood of fowls, the mother sun-bird, and of goats,
which are afterwards eaten. It was at these festivals that
the deer, the old year, the father-totem of the corn-growing
sons of the antelope, was killed and eaten.

But though it was the goat or antelope which awoke the
new year in the current mythology of the year of three
seasons, the office was, at the advent to power of the new
ruling race of the fire-worshipping Māghadas, sons of the
dog, transferred to the dog; and hence, in one of the hymns
of the Rigveda, telling of the sleeping Ribhus, they ask who
has awoke them, and are told by the goat it was the dog.[1]
This was the dog-star, Sirius, called by Homer the dog of
Orion,[2] and Mṛiga-vyādha, the deer-hunter, by the Hindus;
the dog called, in the Rigveda, Arjuna, the fair one, and
Saramāya, the yellow dog, the father-totem of the yellow
race, who shows his teeth, and is invoked as Vāstoṣh-pati, the
lord or guardian (*pati*) of the house (*vastu*),[3] the dog Argus,
who guards the house of Odusseus during his absence. This
name, Vāstoṣh-pati, is also, as I have shown above, given to
the son of Prajā-pati (*Orion*) and Rohinī, the star Aldebaran,

---

[1] Rigveda, i. 161, 13.			[2] Homer, *Iliad*, xxii. 29.
[3] Rigveda, vii. 55, 1, 2.

the father and mother of the deer race, and this son is also
called the sacrificial fire, lighted on the altar of the fire-
worshippers. This son and successor of Prajā-pati Orion,
called Sarameya, appears in another form as Saramā, the
sacred bitch of the Rigveda, the messenger of Indra, the
rain-god ; and the Añgiras, the priests, sons of charcoal
(añga), who offered burnt-offerings to the fire-god, in which
the blood was poured out at the foot of the altar to vitalise
the mother-earth, instead of being drunk by those who
wished to unite themselves in blood brotherhood with the
ancestral totem and inherit his spirit. Saramā was sent by
Indra and the Añgiras to get the cows of light from the
Paṇis or traders,[1] the sons of Orion. She failed to bring
them back, having been bribed by the Paṇis with a drink of
milk ; and, according to the Brihaddevata, when she returned
to Indra and told him she had not seen the cows, he kicked
her, and she vomited the milk given by the Paṇis.[2] This
became the Milky Way, and it was this that Saramā was
placed to guard, for she became one of the two four-eyed
dogs of Yama, the twins who, in the Rigveda, are called the
dogs of Saramā,[3] who guard the path, that is, the path from
earth to heaven, the bridge of the gods, the Milky Way.
These are the two four-eyed dogs of the Zendavesta,[4] the
twin dog-stars of the Greeks, Sirius or Kuōn and Procyon,
the Sanskrit Shvan and Prashvan, the dog (shvan), and the
fore (pra) dog, Canis major and Canis minor, one on each
side of the Milky Way. Every Brahmin, as Bāl Gangādhar
Tilak tells us, has every day to give two small offerings of
cooked rice to the dogs of Yama Shyama and Shabala ; they
are placed outside the circle appropriated to the offerings to
the Vaishvadeva gods, that to Shabala being placed at the

---

[1] Rigveda, x. 108, 3, 8, 10.

[2] Max Müller, *Lectures on the Science of Languages*, Second Series, first
edition, pp. 466, 467.

[3] Rigveda, x. 14, 10, 11.

[4] Darmesteter, *Zendavesta*, Introduction, v. 3-4 ; S.B.E. vol. iv. pp. lxxxvi-
xxxviii.

north-west, and that to Shyama at the south-west corner of
the consecrated area. In the Zendavesta they are the yellow
dogs who guard the Chinvat bridge.[1] They are the totem
animals of the yellow twin races descended from the Zend
father-gods Yima, the twin (*yama*) son of Vīvanghat, and his
twin brother Takhma Urupa, the cleansing fire, the offspring
of the fire-drill and socket, who called themselves the sons
of the dog. It was they who introduced the year of four
seasons marked by the equinoxes and solstices, and consecrated to Indra by the transverse cross, marked on the Hindu
altar by the line from south-west to north-east, denoting the
path travelled by the south-west monsoon; and that from
the north-west to the south-east showing the road travelled
by the worshippers of the household fire after they entered
India, on which the fire-socket called Urvasi was placed in
the Soma ritual. Hence they gave four eyes to their guardian dogs, and it was they who placed a celestial ship in the
heavens, the constellation Argo, on which the wandering
sun could embark on his annual voyage round the four points
of the compass, instead of confining his voyage to the moon-
boat. It was these yellow twin races, the Hittites or Khati
of Semitic and Akkadian history—the Khati, who are still a
ruling tribe in the Panjāb, and who ruled the Indian trading country of Khātīawār—who introduced the Zend year
beginning with the rains ushered in by Tishtrya Sirius, and
called the rains of Tishtrya.[2] This was the dog who awoke
the three Ribhus, the original three seasons, when they had
decided to make four cups or seasons instead of three,[3] and
had changed the beginning of the year from the winter to
the summer solstice. The stars marking the points of the sun's
yearly voyage were, according to Zend astronomy, Tishtrya
(Sirius), ruling the east; Satavaēsa (Argo), the south: Vanaṇt

[1] Darmesteter, *Zendavesta Vendīdād Fargard*, viii. 16, 17 ; S.B.E. vol. iv.
p. 97 ; Introduction, v. 4, pp. lxxvii, lxxviii.
[2] Darmesteter, *Zendavesta Tir Yasht*, vi. 12 ; S.B.E. vol. xxiii. p. 97.
[3] Rigveda, iv. 33, 5.

(Corvus), the west; and the Hapto-iringas, or seven bulls
(the Great Bear), the north; and the course of the sun,

as marked in the heavenly Svastika , the sacred

sign of the voyage, was from east to west by south and
west to east by north. This was the year marked at the
summer solstice by the rising of Sirius, called Tishtrya by
the Zends, and Sukra, or the rain-star, and the Sanskrit
Tishya by the Hindus; for in the Soma ritual of the year of
Prajā-pati, when Indra, the rain-god, is invoked at the beginning of the ceremonies with the cry, Brihat, calling on him to
create (*bri*) life, the first season or cup is that of Sukra, the
rain-god, while the third is the Agrāyana, or winter season—
the Agrahāyaṇi or beginning of the year, which, I shall show
later on, began on the winter solstice dedicated to Orion.[1]
It was this astronomy of the twin races whose parent-gods
were the wolf-mother, called Lēto by the Greeks and Lithuanians, who became Apollo, the wolf-god of day, and
Artemis, the Great Bear, goddess of night, which made the
Egyptian as well as the Hindu and Zend year to begin with
the rising of Sirius, called Isis-Satēt, at the summer solstice;
and it is this reckoning of the year which we find retained in
the Olympian chronology, which made the year of the Olym-
piads to begin with the first new moon after the summer
solstice. This was the year of the 'black antelope,' the year
of the race who worshipped Artemis, not as the bear-mother
of the older mythology, but as the Olympian goddess called
Elaphia, the deer-goddess, and who changed the name of
the constellation of the Great Bear into that of the seven
Rishya, or antelopes, of Indian mythology. In the mythological astronomy of this year the sun-god, which is the ship
Argo, steered by the Hindu father-god Agastya, the star
Canopus, who, in the Rigveda, brought from out the light-

---

[1] Eggeling, *Sat. Brāh.* iv. 3, 3, 1, 2; S.B.E. vol. xxvi. pp. 331-332.

ning Vasishṭha, the most creating (vasu) fire, the perpetual
fire burning on the altar of the fire-worshippers, the son of
Mitrā, the moon-god, and Varuṇa, the god of rain (var) and
the dark night.[1] This steersman star was, by his wife Lopā
Mudra, the moon-fox (lopā), the father of the three Dasyu
or Dravidian Tamil races,[2] the Cheroos, sons of the bird
(chirya); the Cholas, sons of the mountain (kol); and the
Pandyas, the yellow sons of the sun-antelope Paṇḍu, the
father of the fair (paṇḍu) race. He steered the sun-ship to
the house of the bird of winter, Corvus, and the sun-god
thence climbed up the mother-mountain of the Kuṣhika race
as the constellation Hercules, who is depicted in the old
traditional pictorial astronomy as climbing painfully up the
hill to reach the constellation of the Tortoise, now called
Lyra, and thus attain the polar star Vega, which was the
polar star from 10,000 to 8000 B.C. On the other side of the
polar constellation of the Tortoise, the sun-god became transformed into the rain-bird, the constellation of the Bird Ornis of
the Greeks, and Cygnus, the Swan of Latin astronomical mythology, who brought Soma, the seed of life, to earth; while the
mother-mountain bird, who laid the egg whence the hundred
parent sons of the Kuṣhite race were born, the hundred (satā)
creators (vaēsa), who formed the crew of the ship Argo in
the Zendavesta, was the Vulture, the Gṛidhra, or sacred bird
of the Rigveda, who, in Egyptian mythology, ruled the year
and gave her name to this constellation.[3] This pictorial
astronomy telling us the history of the sun-god, the polar
star, and the mother-bird, must, as we know from its agree-

---

[1] Rigveda, vii. 33, 10, 11.

[2] Mahābhārata Vana (Tirtha-Yatra) Parva, xcvi.-xcviii. pp. 307, 314.

[3] Professor Romieu, Sur an Decan, etc., p. 39, has identified the Egyptian
star of the Vulture with the constellation Lyra, the star of the goddess Ma-at,
the mother of law and order; and in Egyptian mythology the vulture ruled
the year. In the Rigveda the vulture Gṛidhra is represented as a rival ruler
of time with the Ashvins, or twins, who are invoked to come and drink the
Soma cup early in the morning before the greedy vulture (Rigveda, v. 77, 1),
to whom the Marka or Soma cup of the dead (Mahrka) was offered.

ment with the *Phainomena* of Aratus, have come down to us
from the traditional picture-writing of the Akkadians, from
whom Aratus got his facts through Eudoxus.

It was these early astronomical reckoners of the equinoctial and solstitial year who added a fifth season to the year
consecrated to the god of the meridian pole, the Egyptian

Horus, the centre pole of the five-rayed Egyptian star

standing in the middle of the rain-cross   , the Tur,

who was the meridian-pole or gnomon of the Akkadians, and the father-god of the Indian Tur-vasu, sons
of the rain-creating (*vasu* or *varsu*) Tur.   This autumn
season, as being that in which barley was sown in the
countries of the Indian rains, they made their father-season,
instead of that beginning in November; and it was to this
season that they transferred the dates of the festival to the
fathers, to the month Bhadon, or Bhadra-pada, the month of
the goat-alligator or fish, in which the autumnal equinox
takes place; and it is in the dark half of this month that the
Pitrī-yajña or sacrifices to the fathers are celebrated throughout India.   It was in the corresponding month that the
Nekusia were celebrated in Athens, and it was these people
who, as the Semites, began their year after the autumnal
equinox with the month Tisri, the year current in Syria,
Asia Minor, Macedonia, and the Peloponnesus.   It was
through the influence of their national traditions that the
stars consecrated to the Ashvins, or twins, were placed at the
head of the list of lunar stations instead of the Pleiades, who
were looked on as mother-stars during the year of Orion, and
which had been placed first in the original list of lunar mansions prepared by the fire-worshippers, and preserved by the
fire-worshipping Soghdians and Chorasmians of Central Asia.[1]

---

[1] Sachau's Alberuni's *Chronology of Ancient Nations*, chap. xi. p. 227.

They made Varuna, the god of rain (*var*) and the dark heaven, the Greek Ouranos, their father-god, and consecrated their year of five seasons to the five Vareṇya Devas of Zend theology called (1) Indra, the rainy season, who, as we have seen, began the year of Prajā-pati as Indra Sukra; (2) Sauru, the autumn season, called by the Hindus Shar-adas, sacred to the cloud-mother Sar and the rain-constellation Hydra; (3) the Nāuṅghaïtya, also called the Nā-satya, or they who do not deceive, a Vedic and Zend name for the heavenly twins, the winter; (4) Tauro, the pole (*Tur*), the spring; and (5) Zairi, the Indian Hari, the yellow storm-god, father of the yellow twin races born on the Yamona, or river of the twins (*yama*), the summer.

It was these five seasons which made up the Gond year; only their year began with the summer in April, when the Nagur or plough-god, the rain Nāga, is worshipped, a survival from the earlier agricultural year of the Pleiades, in which the second period of six months began in April. It is then that everywhere throughout Europe and Western Asia the rain-god is worshipped, under the varying forms of Zeus Ombrios, the showery Zeus, St. George, the Greek plough-god called Geourgos, or the worker (*Ourgos*) of the earth (*Ge*), and El-ia or El-ias, the god (*El*) of the house (*i*) of the waters (*a*). It was the worshippers of the year-gods of the year of five seasons who reckoned the gods of time of the Rigveda as thirty-three in number, the five seasons of the year and the twenty-eight days of the lunar months, and who called these gods in the Zendavesta 'the thirty-three lords of the ritual order.' It is these gods who appear in the Egyptian illustrations to the Book of the Dead, preserved in the Papyrus of Ani,[1] as the thirty-three judges of the dead seated in the hall of the Ma'at, the mother-goddess of the invariable order of natural phenomena. It was from this source that two reckonings of time descended; one of these

---

[1] Vignette xxxi. in the illustrations of the Book of the Dead, called the Papyrus of Ani, in the British Museum.

was the lunar year of thirteen lunar months, called in the
Mahābhārata the thirteen wives of Kashyapa, the father of
the tortoise race, ending with Kadrū, the tree (*dru*) of Ka,
a name of the creating god Prajā-pati, and she is said in the
Mahābhārata to be the mother of the Nāgas, or worshippers
of the rain plough-god, and in the Rigveda she is the mother
of the Soma or Su, the essence (*su*) of life, which made Indra
strong in battle.[1]   She becomes in the Brāhmaṇas the queen
of the serpents, and the goddess-mother who received the
Soma brought from heaven by the mother-bird ; the Soma
which was enclosed in the two caskets of consecration
(*dīksha*) and penance (*tapas*),[2] the two original seasons of
the year, which she gave to Indra and Agni.   This year of
thirteen lunar months is the year which is still reckoned by
the Santals.

But besides this year, there was that of which I have
spoken at the beginning of this essay—the year of twelve
months of thirty days each, making 360 days in the year.
This, as we are told in the Brāhmaṇas, was divided into pairs
of months, beginning with the two spring months, called Madhu
and Mādhava, from the honey, Soma (*madhu*), which was
drunk by the Ashvins, or heavenly twins.   This was the Hindu
year of six *ritus* or seasons of six months each, called in the
Brāhmaṇas, Vasanta, spring ; Grishma, summer ; Varshā, the
rainy season, which are said to be the seasons of the Devas :
and Sharad, autumn ; Hemanta, winter ; and Shishira, the
season of the winter rains, which are the seasons of the
Pitṛis.[3]   This year is denoted by the six-rayed star ✳ of the
twin races, still borne, with the lunar-crescent, on the banners
of the Turks, the modern representatives of the Asiatic
Parthians or twin races.   It was the one-wheel year, with six
spokes of the cosmological Vedic hymn, turned, like the wheel
of the oil-press, by the revolving pole and its beam, the Ash-

---

[1] Rigveda, viii. 36, 26.
[2] Eggeling, *Sat. Brāh.* iii. 6, 2, 6-11 ; S.B.E. vol. xxvi. pp. 150-151.
[3] *Ibid.*, *Sat. Brāh.* ii. 1, 3, 1 ; S.B.E. vol. xii. p. 289.

vins, the twin stars, and day and night. It was to this
year that the Hindus likened their universal monarch, the
Chakra-varta, king, who sits, like the Kushite monarch, as the
father of his subject tribes, in the central province of his
dominions, and directs his satellites, the rulers of the seasons,
who became the ruling stars of the frontier provinces—the
Nakshatra stars—to turn the wheel (*chakra*) of time in its
yearly round.[1] This became the Zend year of six seasons,
and it is this Zend year which is commemorated in the
Kūsti, or girdle, with which all young Parsis boys and girls
are invested, and which is a direct descendant of the girdle of
Orion (*Prajā-pati*), which is worn by all the Hindu twin-
born castes. This Zend girdle is made of six strands, each
of which has twelve very fine white woollen threads, or
seventy-two threads in all. These strands are braided at the
ends into three separate string ends of two strands each, containing twenty-four threads. These numbers, six, twelve,
twenty-four, and seventy-two, are all component parts of the
perfect circle of 360 degrees, and the seventy-two threads
forming the fifth part of the circle are reproduced in the
seventy-two assistants of Set, who took possession of the
body of Osiris, the Egyptian year-god of the barley-growers,
after it was brought back from Byblus by Isis, the moon-cow,
and cut it up into fourteen pieces, the fourteen days of the
lunar phases. It is this year of six seasons, measured by the
circle described in the heavens by the sun and moon, which
tells of an age which showed, by the development of a
literary class of studious observers of natural phenomena,
that it had advanced beyond the imperfect civilisation of the
early cultivators now represented in India by the Gonds and
Santals; and we learn from the results of their work, stereo-
typed in their ritual, that the priestly astronomers who

[1] But these stars were first probably the stars of the year of the bull, driven
by Capella in Aurigæ, the heavenly charioteer, and patron star of Babylon,
and of the year of Marduk, the young bull-sun of Babylon, and of the Hindu
Phalguni, the year in which, as I show in the sequel, the sun attained his
majority at the vernal equinox.

measured time had learned by their studies how to trace
among the stars a perfect circle, in which the daily and
monthly changes of the position of the moving heavenly
bodies, the sun, moon, and planets, could be noted. The
first of these circles was doubtless that of the Lunar Mansions, showing the daily position of the moon during each
lunar month, and it was this circle that the Vedic astronomers, the sons of Atri, who, as we saw above, predicted an
eclipse of the sun by a portable *tur* or gnomon, made into
that of twenty-seven parts, by which they were enabled to
detect the position of the sun during the intervals between
the solstices and equinoxes, and to make solar and lunar
time coincide in a five years' cycle. The rules laid down for
calculating time by the 'tithis' or lunar days, of which there
are thirty in the month, and which differ in length from
solar days, show that it was upon the right adjustment of
time differences by arithmetical calculation, and not on observation of the sun's place in the heavens, that they chiefly
relied to correct the differences of solar and lunar time.[1]   It
was only as a means of correcting any errors in the prescribed
calculations that the circle was used at first, and it was not
till the later days of the Vedānga-jyotisha that these Nakshatras were used as a means of fixing dates by stating the
sun's position at the time the observation was taken.   When
the Vedānga-jyotisha was written, the sun was in the winter
solstice in the beginning of Shravishtha, and the vernal
equinox took place in 10° Bharani, the summer solstice in
the middle of Āshleshā, and the autumnal equinox in 3° 20'
of Vishāka.   The date indicated by these positions is either
1269 B.C. or 1181 B.C., according to the mean reckoned as the
rate of the precession of the equinoxes,[2] and it is impossible
to believe that the practice of fixing beforehand the dates of
festivals and sacrifices originated at so late a time as this.
These festivals and sacrifices were all founded and their dates

[1] Sachau's Alberuni's *India*, vol. i. chaps. xvi., lxxvii., pp. 194-197.
[2] Bāl Gangādhar Tilak, *Orion*, chap. iii. p. 38.

fixed ages before the Veda was thought of, or a Vedic hymn composed by the Rishis; and they all, as the Brāhmaṇas tell us, date back in their original forms to the days of the silent worship of Prajā-pati, the god who preferred mind to speech,[1] the supreme god of time of the age of the year of Orion.

The names of the Hindu months, which are much earlier than those of the Nakshatras, all point to an age in which the star-gods of the early mythology were worshipped as guardian gods, in connection with the magical astrology still preserved in the universal custom of making horoscopes at births. They belong to a time when the Ashvins or heavenly twins, the stars Gemini, who gave their name to the month Ashva-yujau or Assin, were part of the constellation of the Alligator, Shimshu-māra, which helped the bulls to draw their car in the Rigveda, for they are named among the stars in this constellation in the Vishnu Dharma, where they are called the Ashvini, the physicians of the gods, the two hands of the constellation which make the pole revolve.[2] They, as their name Ashvins shows, were the parent-stars of the sons of the horse (*Ashva*),[3] who first founded the elaborate ritual of the Soma sacrifice, and the month of Assin consecrated to them is that between Bhādra-pada, the month of the autumnal equinox, sacred to the goat and alligator, the goat-fish, Capricornus, of astronomy, and Khartik, the month of the Pleiades. The month Assin, after the autumnal equinox,

---

[1] Eggeling's *Sat. Brāh.* i. 4, 5, 11, 12 ; S.B.E. vol. xii. p. 131.

[2] Sachau's Alberuni's *India*, vol. i. chap. xxii. p. 242.

[3] The Eastern horse, the Ashva of the Hindus, and the Ashpa of the Zends, became the European ass, the Lat. *asinus*; Icelandic, *asni*; Welsh, *asyn*; German, *esel*; a name which is universally acknowledged to be derived from Eastern tongues. The Hindu father of horses is Indra's horse Ucchai-shravas, the long-eared ass, and the chariot of the Ashvins is, in the Rigveda, drawn by stallion asses. Hence the twin heavenly horsemen were first the father-stars of the race who sowed barley with the plough, the Phrygian, Manassite, and Ooraon sons of and riders on the ass, the animal consecrated to the prophets, the ass of Balaam.—*The Ruling Races of Prehistoric Times*, Essay ii. p. 91 ; iii. pp. 255, 256.

was that especially sacred to the horse-father, and it was on the
15th of this month, October, that the Roman horse-sacrifice,
the Equiria, took place. It was these people who hallowed
the eleven lunar months of the gods of generation, that being
the period of the gestation of the horse. The Hindu names
of months taken from the stars are not taken from any circle
in the heavens, but from those of the guardian stars of the
successive ruling races. The sons of the goat or alligator,
the Muggur father of the Māghadas, the sons of the horse,
the sons of the Pleiades, the sons of the antelope, the deer-
god, Orion, father-star of the Brahmins, the sons of the
mother-antelope or red cow, Rohinī (*Aldebaran*), which
afterwards became the black bull, Push, the constellation
Taurus, and the year reckoned by these months was that
begun at the winter solstice, the sacred season following
the meeting of Mrigasirsha, Orion or Prajā-pati, and his
daughter, Rohinī, the red antelope or cow. In the
naming of the months we find also a reference to the
earlier year ruled by the motions of the Pleiades in November and April, the latter being the month called
Vishākhā, or the two (*vi*) branches when the Pleiades year
divides.

The gods worshipped by the sons of the horse were those
summoned to their sacrifices in the eleven stanzas of the
Āpri hymns, recited at the animal sacrifices offered at the
solstices. In these hymns, the first four stanzas summon to
the sacrifice (1) the god of the sacrificial flame, the spring-
god ; (2) the wind-god of summer, called Tanū-napāt, the
son (*napāt*) of the alone one (*Tanū*), the self-begotten or
Narā-shamsha, the perpetual fire on the altar ; (3) the Iḍ
or Idah, the mother-goddesses of the rains ; (4) the Barhis
or sacrificial seats of Kusha grass, allotted to the fathers
ruling the autumn seasons at the Pitri-yajña sacrifices.
These stanzas are, in short, addressed to the gods of the
seasons, who unite to bring the rains which cause the
Kusha grass, the parent-grass of the Kushika, to grow. In

the fifth stanza, the gates of the sacrificial hall, the sacred door-posts, the Semitic Bab-el, the gates (*bab*) of god (*el*), the temple pillars Boaz and Jachin, and the propylons of the Egyptian and Greek temples are called to the sacrifice; in the sixth, the twins, Dawn and Night; in the seventh, the two Hotars, pourers (*hu*) of the libations of Manu the thinker,[1] the twin-race stars of the Kushika, the Ashvins; in the eighth, the three goddess-mothers of the year of three seasons; in the ninth, Tvashtar, the father-god of the year of two (*tva*) seasons; in the tenth, Vanas-pati, the lord (*pati*) of the wood (*vana*), the primæval author of life, the parent-tree of the Dravidian races; and in the eleventh, the god of the fifth or winter season, the god Rudra Svishtakrit of the sacrificial stake.[2]

[1] Rigveda, v. 5, 7.

[2] This is an abstract of the meaning of each stanza of the Ápri hymn with eleven stanzas, Rigveda i. 188; ii. 3; iii. 4; v. 5; vii. 2; ix. 5; x. 70; x. 110. The god invoked in the eleventh stanza of these hymns is the god called Sváhá, meaning he who gives the blessing. He is called in the Bráhmaṇas (Eggeling, *Śat. Bráh.* i, 5, 4, 5; S.B.E. vol. xii. p. 153) the god of the winter season; and Sváhá is explained to mean Rudra, the god of the sacrificial stake (Eggeling, *Śat. Bráh.* i. 7, 3, 1-8; vol. xii. pp. 199-202). From this it is clear that the great national animal sacrifices, in which the tribal totem was killed and eaten, took place at the winter solstice. At the sacrifices to the five seasons of the year of Prajá-pati at the Mádhyandina or midday pressing of Soma, beginning with the sacrifice to Sukra, the star Sirius ushering in the summer solstice (Eggeling, *Śat. Bráh.* iv. 3, 3, 1, ff.; S.B.E. vol. xxvi. pp. 331-340), no animal victims were offered. Nor were animal sacrifices offered to the twin-gods, Mitra-Varuṇa; though they were to Varuṇa singly. The animal sacrifices were all offered to single gods represented by Agni, the god of the household-fire and the sacrificial-flame, Agni-Jata-vedas, and Soma, the rain-god, born from the lightning-flash and the thunder-cloud (Eggeling, *Śat. Bráh.* iii. 6, 4, 1; S.B.E. vol. xxvi. p. 162), to the gods united in Agni Svishtakrit, and called Agni-Soma, Agni-Somáu, Indragni (*Indra* and *Agni*), Ashvináu, Vanaspati, Devá-Ajyapa, the goat (*aji*) gods, (Haug's *Ait Bráh.* vol. ii. pp. 95, 96, notes). They begin with a victim to Agni, and end with the eleventh victim to Varuṇa (Eggeling, *Śat. Bráh.* iii. 9, 1, 6-21; S.B.E. vol. xxvi. p. 218-221). They represent the theology of the age preceding that of the worship of Mitra-Varuṇa in India, and Apollo of Delos in Greece, to whom no animal victims were offered.

The historical retrospect here set forth does not bring the ritualistic national history down to the age of the year of thirteen lunar months, which is celebrated in one special Āpri hymn of thirteen stanzas (Rigveda, i. 142), belonging to the collection ascribed to Dīrgha-tamas, to which I have referred at the beginning of this Essay; while the still earlier year of twelve months, which was calculated with the year of thirteen months during the long (*dirgha*) period of darkness (*tamas*), and is celebrated in the last hymn of the collection illustrating this period, has also an Āpri hymn of twelve stanzas to itself (Rigveda, i. 13).

But in order to understand fully the time-reckonings of the Vedic age, it is necessary to consider the rules laid down for the time of celebrating the sacrifices and beginning the year. At the present day, as Bāl Gangādhar Tilak tells us, all Hindus south of the Nerbudda begin their sacrificial ceremonies in the Uttarāyana, when the sun is going north (*uttar*) at the time of the winter solstice;[1] and the yearly Sattras or series of sacrifices thus begun were, as Dr. Haug remarks in his introduction to the Artareya Brāhmaṇa, only an imitation of the sun's yearly course. This was the year of the early corn-growing races who made Orion, the antelope constellation, the patron star of their year, and of Orion's month Mārgashirsha. This is also called Āggahun, and Āggahun is the contracted form of Āgrahāyani, meaning the beginning or fore part (*agra*) of the year. This beginning of the year is fixed absolutely in all the Gṛihya Sūtras on the full moon of Mrigashirsha, that nearest to the winter solstice, except in the Sānkhāyana, which allows two alternative years, beginning with the Nakshatra Rohini before Mrigashirsha, or in the month Bhādon of the autumnal equinox.[2] It is at the winter solstice that the Devayāna division of the year begins, while that of the Pitriyāna began at the summer solstice, and the prayer at the sacrifices then

---

[1] Bāl Gangādhar Tilak, *Orion*, chap. ii. p. 31.
[2] Oldenberg, *Gṛihya Sūtra Sānkhāyana*, iv. 17 ; S.B.E. vol. xxix. p. 130.

offered is addressed to the white hind or mother-antelope.
This is the sacrifice at which the first-fruits of the harvest
are eaten; and at which, according to the Grihya Sūtra of
Āpastamba, the spit-ox to Kshetra-pati, the lord (*pati*) of the
fields (*kshetra*), was offered by those who have not set up their
household fires, that is, by the worshippers of the early gods.[1]
This was the father-bull [2] which was brought to the sacrifice
with the mother-cow and the young calf, who was to be the
father-god of the future year, after the ploughing-ox of the
old year has been sacrificed. This was the sacrifice of Prajā-
pati, who is said in the Brāhmaṇas to be the visible sacrificer
sacrificing himself, and to spring like an embryo from the
Jarayu, or afterbirth of antelope skin in which the sacrificer
is enveloped when entering the baptismal bath at the Soma
sacrifice, reproducing the yearly death of the antelope-god of
the year of Orion.[3] It was then that the tribal totem was
killed and eaten, the antelope first and the bull afterwards;
and this sacrifice belonged to the early matriarchal age, for
only relations on the mother's side could partake of the
Kshetra-pati bull.[4] This sacrifice was, in the ritual of the
Sānkhā-yana Grihya Sūtra, changed into one which is more
consonant with the national sacrifices of the aboriginal sons of
the tree, the founders of the village community, who did not,
like their Northern conquerors, sacrifice totem animals with the
fruits of the mother-earth. The sacrifice tells of the days
when the sons of the antelope were led into India by paths
passing along the sloping river banks where the short luscious
grass, the Kusha grass (*Poa cynosuroides*) dear to the antelope,

[1] Oldenberg, *Grihya Sūtra, Grihya Sūtra of Āpastamba*, 7, 19, 3-7, 20,
19; S.B.E. vol. xxx. pp. 289-291. The other *Grihya Sūtra* do not prescribe
any special time for this sacrifice.
[2] It is especially enjoined that this bull should not be gelt.—Oldenberg,
Grihya Sūtra Pāraskara Grihya Sūtra, iii. 8, 4; S.B.E. vol. xxix. p. 352.
[3] Eggeling, *Sat. Brāh.* iv. 3, 4, 3; iii. 2, 1, 11; S.B.E. vol. xxvi. pp.
341, 28.
[4] Oldenberg, *Grihya Sūtra, Grihya Sūtra of Āpastamba*, 7, 20, 18; S.B.E.
vol. xxx. p. 291.

abounds, and where they had amalgamated with the earlier cultivators of the soil. For the sacrificial offerings consisted of the leaves of Shami grass (*Panicum frumentaceum*), a grass growing on rich soil, and particularly liked by cattle; flowers of Madhūka (*Bassia latifolia*), the honey-tree, from the flowers of which the honey-drink dear to the Ashvins was distilled, and the flowers of which are still made into pounded cakes and kept for family consumption ; leaves of Shirisha (*Mimosa Sirissa*), one of the gum-arabic trees ; Udumbara (*Ficus glomerata*), the parent-tree of the Vaishya founders of the Soma sacrifice; Apamārga (*Achyranthes aspera*) plants, an antidote against witchcraft and scorpion bites ;[1] Kusha grass (*Poa cynosuroides*), jujube-fruits of the Jambu tree (*Eugenia jambolana*), the parent-tree of the central Kushika kingdom of Jamba-dvipa, and an earth-clod taken from a furrow.[2]

I have already shown that the primitive year of the Pleiades, the year of the matriarchal age divided into two periods, began with the six months consecrated to the fathers, that is, with the Pitriyāna of that age, who were the ancestors worshipped in November. It was this worship of the ancestors which was symbolised in the mother-plants, the father-antelope, and the sun-ox who died for the renewal of the year, offered at the winter solstice ; and it was this period which was changed into that called Devayāna, when the sun-god was accepted as the father of the race. It was then that the old year's sun died to make way for the new year. It was then that the period consecrated to the earthly fathers was changed ; and the change is shown in the year observances of the early Parsis, for when they made their year begin with the summer sol-stice and the month Fravashinam, sacred to the Fravashis, or spirits of the primæval mother-gods, they devoted the last five days of the old year and the first five of the new, called

---

[1] Zimmer, *Altindisches Leben*, chap. iii. pp. 66, 67.
[2] Oldenberg, *Grihya Sūtra Sānkhāyana Grihya Sūtra*, iv. 17, 3 ; S.B.E. vol. xxix. p. 130.

the Fravardigan days, to the worship of the manes of their ancestors.[1]

The ceremonies inaugurating the year beginning with the winter solstice seem, from the offerings of the sacred tribal plants and the sacrifice and eating of the tribal totem, to have been arranged as a combination of Southern offerings of fruits to the mother-earth, and of totemistic Northern tribal feasts. These were followed by those called the Ashtakas, or sacrifices of the eight, that is, of the corn-growing race who made their parent-star the eight-pointed star of the Kushikas, formed of the junction of the rain- and fire-cross, the sign of the god of seed. There were three of these, beginning with that celebrated on the eighth day of the second fortnight of Mārgasirsha, and repeated on the same day in the two following months. The last of these was called the Ekāshtika sacrifice, or the wife of the year, and was held on the eighth day of the dark fortnight of Māgha. On the first of these Ashtakas, vegetables, the sacrifices to the mother-earth, were offered to the seed-gods of the early agricultural races. On the second an animal—a goat or a cow—was killed, and on the third the offerings were baked cakes sacred to the fire-god, the wonder-working Māgha, mother of fire.[2]

It is from their arrangements of the sacrifices of Mārgasirsha and the Ashtakas, compared with the times set apart for the seasonal dances of the early Indian tribes, that we can understand the order prescribed for the Sattras, or sacrificial sessions, in the Brāhmaṇas. In two almost identical

---

[1] Dr. Gerger, *Civilisation of the Eastern Iranians*, vol. i., and Haug, *Essays on the Parsis*, p. 225, quoted by Bāl Gangādhar Tilak, *Orion*, chap. iv. pp. 92, 93.

[2] Oldenberg, *Grihya Sūtra Sānkhāyana Grihya Sūtra*, iii. 12, 13, 14; *Āshvatāyana Grihya Sūtra*, ii. 4; *Khādira Grihya Sūtra*, iii. 3, 28; also iii. 4; S.B.E. vol. xxix. pp. 102-105, 205-207, 417-420. In the *Khādira Grihya Sūtra*, iii. 4, 30, 32, p. 417, vegetables are ordered to be offered on the last, and cakes on the first Ashtaka. The arrangement given in the text is that prescribed in the *Sānkhāyana Grihya Sūtra*.

passages in the Taittirīya Sanhitā, vii. 4, 8, and the Taṇḍya
Brāhmaṇa, v. 9, intending sacrificers are directed to consecrate themselves, that is, to take the baptismal bath of initiation, on the Ekāshtika day in the beginning of February.
They should then consecrate themselves for the sacrifice to be
offered on the full moon of Phalgunī, which is called the
month of the year. This is the date on which the great
Hindu annual Saturnalia called the Huli sacrifice is held.
At the same time, this date is said to be not convenient,
because if the first annual sacrifice beginning the year is held
in March, at the vernal equinox, the Vishūvān or mid-year
sacrifice falls in the rains. Therefore, the writer who wrote
the original passage, quoted by both authors, recommends
that the sacrificer should continue to honour the Ekāshtaka,
the wife of the year, by purchasing his Soma for the sacrifice on that day, but he should not consecrate himself for the
annual sacrifice till four days before the full moon of Cheit
at the end of March, and thus he can make his Vishūvān, or
mid-year sacrifice, fall after the rains.[1]

This change in sacrificial ritual here recommended is one
which marked a departure from the successive historical
forms of the annual sacrifice observed in the popular theology.
This began with the original sacrifice to the fathers, accompanied with dances, in the beginning of November. This
was followed by the earliest year of Orion, beginning, as in
the Egyptian ritual, with the festival of the launching of
Osiris (Orion) on his year-bark on the 26th of Choiak, about
the 12th of November. He, the year-sun, was then united
to the moon-goddess, and the result of this union was the
birth of the young sun-god of the next year, the black antelope Krishṇa, on the 8th day of the light half of Bhādon,
about the 23rd August. When the sons of the mother-cow
joined the sons of the antelope, the annual festival and
national Saturnalia beginning the year was altered to the full
moon of Mārgasirsha, near the winter solstice, the date when

---

[1] Bāl Gangādhar Tilak, *Orion*, chap. iii. pp. 46, 47.

the Santals of Bengal hold their Sohrai festival and annual
tribal dance, the Madras Dravidians that called Pongol, and
the Romans held their Saturnalia. Under this arrangement
the young son-god, begotten at this festival, is born at the
autumnal equinox, the time when the barley-growing Semites
and the people of Syria, Asia Minor, Macedonia, and the
Peloponnesus began their year. But this national pairing-
time was not that of the fire-worshipping sons of the mother-
Mâgha, who dated their ritualistic history from a more
northern clime, where open-air dancing was impossible at
the winter solstice. They made the Hindu month Mâgh
(January-February), the season of our St. Valentine's Day, of
the European carnivals, and of the Greek marriage of Zeus
and Hera, the time of their great national new year's dance.
This is that celebrated by the Mundas and Ooraons of Chota
Nagpore, and this festival, which makes the young sun-god,
born in November, marriageable in February, is connected
with the original division of the Pleiades' year into two
periods of six months each, for it is this division which marks
the month of Srâvana (July-August), six months from Mâgh
(January-February), sacred to the Nâga snakes, the sons of
the bird, for it is in this month that the Hindu festival of
the Nâg-punchami or five Nâgas is celebrated ; and the offerings to the serpents, the Svâha or gods summoned as the rulers
of the rainy season, take place on the full moon of Shrâvana.[1]
It is then that the flour of fried barley is offered, and it is in
this month that the Ooraons and all Hindus of Chota Nagpore hold their great annual festival called the Kurrum, to
the Kurrum-tree (*Nauclea parvifolia*) and the barley-plant.[2]
This is a repetition of the earlier festival of the rice-growers,
still observed in this month in Chuttisgurh, and called the
Gurh-pûja, when young shoots of rice are hung up in every
house ; and this growing of the rice was one of the aids to

---

[1] Oldenberg, *Grihya Sûtra Sânkhâyana Grihya Sûtra*, iv. 15, 1 ff.; S.B.E.
vol. xxix. 127.
[2] Risley, *Tribes and Castes of Bengal*, vol. ii. pp. 145, 146.

time-measurement in early chronometry. It is in the month
before this, Asarh (June-July), that the Mundas, who offer
eggs and turmeric to the Nāga gods of the bird races,[1] celebrate their rain festival, called the Bahtauli, when each
cultivator kills a fowl, the mother-bird (*khu*) of the Kaurāvyas, born from the egg laid by Gandhārī, and inserts each
of its wings in a cleft bamboo, one of them being set up in
his field and the other in his dung-heap. In these two
festivals of the worshippers of the mother-bird and of the
ass, the parent-totem of the Ooraon barley-growers, we see
evidence that the sons of the mother-bird began their year
with the winter solstice, on the full moon of Mārgasirsha,
and held their Vishūvān, or mid-year festival, at the summer
solstice ; while the sons of the ass began their year in Māgha
(January-February), and held their Vishūvān, or mid-year rain-
festival, in July-August. It is this last year which is celebrated in Hindu epic legend as the year of the Pāṇḍavas.
In this year the two twin-seasons, autumn and winter, consecrated to the sons of Madrī, and the Ashvins, or twin-gods,
Saha-deva, the fire or driving (*saha*) god (*deva*), and Nakula,
the mungoose or snake-devouring god of winter, come last
in the order of the seasons, after the spring consecrated to
Yudishthira, the son of Dharma, the god of the immutable
order of nature, the summer to Bhima, son of the Vāyu, the
wind-god, and the rains to Arjuna, son of Indra, the rain-
god. This is the year which succeeded the earlier year of
five seasons, which are the father-gods of the five races of the
Rigveda, the sons of Yayāti, and his two wives Devayānī and
Sharmishtha. For this year begins with the two twin-
seasons, the rains and autumn, called Yadu the rainy and
Tur-vasu the autumn season, the god (*vasu*) of the Tur, the
meridian pole of the barley-growing races who began their
year in autumn ; these are the sons of Devayānī, the goddess-
mother, the daughter of Shukra, the rain-god. These are
followed by the three seasons represented by the sons of

---

[1] Risley, *Tribes and Castes of Bengal*, vol. ii. p. 103.

Sharmishtha, Yayāti's earlier wife, whose name means she who
is most protecting, the Banyan fig-tree, the Indian representative of the Syrian fig-tree, which was the parent-tree of the
barley-growing races in their mother-land of Syria. They
are Druhyu, the winter of the sons of the sorcerer (dru);
Anu, the spring, the season sacred to the village races who
worship the local tree-gods (Anu, the Anath of the Canaan-
ites); and Puru, the summer season of the Eastern (puru)
race of city-builders (pur). This last year of three seasons
was the original year of Orion, the year of the first barley-
growers.

There is not, in any of the authorities I have consulted, any
evidence showing that this year of the sons of Yayāti, which
was that of the foundation of Vedic ritualistic history, and
the worship of Soma, the rain-god, was ever brought into the
Nakshatra system. The year was, in the natural order of
Indian seasons, very clearly defined, for the rainy season on
the west coast always begins about a fortnight before the
summer solstice; while in the latitude of Benares there are
generally only a few days' difference between the first rainfall
and the summer solstice.[1] But we find in the Mahābhārata,
and in the calculations of Hindu astronomers, a great deal of
astrological evidence connected with the year of Yudishthira.
According to Garga, the rule of Yudishthira coincided with
the time when the Great Bear stood in the Nakshatra
Māgha,[2] that is, in the constellation Leo, of which the chief
star is Regulus, ruling that Nakshatra. This would make the
year-measurement of the age of the Pāṇḍavas, who, as the
sons of Prithu, were the Pārthava of the Rigveda and the
Parthians of history, to be the year of the Lion, known from
Sabaean inscriptions to be one of the years reckoned by the

[1] I know this from experience, for during the four years when I was Settlement
Officer in Chutitsgurh, and obliged to remain out in camp till the rains began,
I always arranged my tour so as to return on the 18th June. If I was later,
I had my camp deluged with rain.
[2] Sachau's Alberuni's India, vol. i. chap. xlv. p. 390.

Sabæans of Yemen.[1] This was the year of the early navigators, who, as the Yadu-Turvasu, started from Dwāraka, established themselves at Turus, the Akkadian holy island of Dilvun (*Bahrein*), in the Persian Gulf, and founded the maritime commerce of the Indian Ocean.   It was they who made the seal of the royal lion of the tribe of Judah, Yudah or Yadu, the national cognisance of the Abyssinian kings— who still build their churches as their Hindu forefathers did their village or parish temples, in the centre of the village grove.[2] This was the year which, as I have shown in *The Ruling Races of Prehistoric Times*, was measured by the revolution of the pole in the constellation Leo.[3] Its perpetual rotations were thought to give birth to the great heat of the Indian and Persian summer, symbolised in the constellation Leo.   This heat it transferred, as the creating impulse of fresh life, to the heavenly Soma-cup, the constellation Crātēr, whence it was distilled on earth in the life-giving rains by the heavenly Soma-press, the constellation Hydra, the great Nāga snake.[4] It was as the god of spring, who prepares for the rains brought by his brothers, Bhima, ruling the hot season, and Arjuna, ruling the rains, that Yudishthira ruled the year; and his birth in the Mahābhārata is said to have taken place under the eighth Muhūrta or division of the day and night called the Abhi-jit,[5] that is to say, it was ruled by the star Vega, which was the star of the Nakshatra Abhijit, which has eight Muhūrtas.[6] This is one of the stars in Lyra, which was the polar star from 10,000 to 8000 B.C.[7] This horoscope shows that the year of Yudishthira was one based on the belief that the passage of

---

[1] *Encyc. Brit.*, Ninth Edition ; Müller's ' Yemen,' vol. xxiv. p. 740.

[2] Bent, *Sacred City of the Ethiopians*, chap. viii. p. 138.

[3] *The Ruling Races of Prehistoric Times*, Essay iv. pp. 343, 369-372.

[4] *Ibid.*, Essay iv. pp. 332-334, 343.

[5] Mahābhārata Ādi (*Sombhava*) Parva, cxxiii. p. 259.

[6] Burgess, ' Notes on Hindu Astronomy,' *Journal of the Royal Asiatic Society*, Oct. 1893, p. 756.

[7] Norman Lockyer, *Dawn of Astronomy*, chap. xii. p. 128.

time was marked by the revolutions of the circumpolar stars
round the pole, and that it succeeded that in which the star
Vega was the star of the mother-bird, who as the mother
Gandhāri laid the egg when the Kauravyas, the rivals and
predecessors of Yudishthira and the Pāṇḍavas, were born. This
was the age in which the constellation called afterwards that of
the Jackal, Tortoise, and Lyre, was the constellation of the
Vulture, which was, Alberuni tells us, called Alnasr-alwāqi, or
the Falling Eagle or Vulture (nasr) by the Arabs, and he
quotes a passage from the Vishnu Dharma, stating that the
Hindu astronomer, Mārkandya, one of the legendary sages of
the Mahābhārata, also called the constellation by this name.[1]
It was this vulture, called Nasr, which was worshipped by the
Sabæan Arabs of Southern Arabia as the mother-bird, the
representative of the female principle, which ruled the
Egyptian year, and which in the hieroglyph for star represented the mother-bird, the goddess-mother of the sons of
Horus, Khu, as one of its constituent signs.[2]   It is to Vega, in
the constellation of the Vulture, called Anubis or the Jackal,
that the earliest Egyptian temples at Abydos and Luxor
were oriented,[3] and the Vulture constellation was that assigned to the goddess Ma'āt, the ruler of law and order and
of the invariable succession of natural phenomena.[4]   It is in
the astronomical myth told in the constellations Hercules,
Lyra, and Cygnus, that we see how the vulture became the
jackal, for we see in the ancient pictorial astronomy, which
has come down to us from Akkadian times, Hercules, representing the fire-god Hermes, the dog of the gods, toiling in
human form up the mountain, the mother-mountain of the
tortoise race, with bent knees, his left foot resting on the
head of the constellation Draco, and holding in his left hand

[1] Sachau's Alberuni's *India*, vol. ii. chaps. lv. lvi. pp. 66, 85.
[2] Tiele, *Outlines of the History of Religion*, 'Religion of the Sabæans,' § 46, p. 79 ; Hewitt, *Ruling Races of Prehistoric Times*, Essay iv. pp. 347, 348 : iii. 267.
[3] Norman Lockyer, *Dawn of Astronomy*, chap. xxxii. p. 238.
[4] Romieu, *Sur un Décan.* p. 39.

the snake with three heads—the year of three seasons—which
he is about to place in the constellation Lyra, while on its
other side, Cygnus, the swan, the bird (*ornis*) of Greek
astronomy, the rain-bird, is flying away with the seed of rain,
which she has taken from the mother-star.[1]   When the home
of the mother-bird was taken possession of by the fire-dog, it
became the constellation of the Jackal (*Anubis*), the heavenly
representative of Horus, the meridian-pole and fire-drill of
the tortoise earth.   The whole story appears most clearly in
the myth of Gandhārī, the sister of Shakuna, the kite, and
the wife of Dhritarāshtra, the blind house-pole and fire-drill,
the father of the life-giving heat which gave offspring to the
earth.   The earth, the home of the tortoise race, called the
Khūravyas or sons of the bird Khu, or Kaurāvyas, sons of
the tortoise Kur, was depicted in the early cosmogony of the
Akkadians as a boat turned bottom upwards, overarched by
the sky with its fixed stars, which revolved round, and with
the house-pole of the earth, the mountain of the East, Khar-
sak-kurra,[2] this was the column joining and keeping separate
the heavens and the earth.   This symbol of the revolving
mountain was that afterwards transferred in Hindu mythology
to Mount Mandara, the ancient name of Parisnath on the
Burrakur, said to be made to revolve by Vāsuki, the spring-
god, who set up the rain- or meridian-pole of the empire of
the Kushika on the Sakti mountains, the Kymore range,
immediately south of Benares.   It was from the revolutions
of this mountain that the long-eared horse or ass, Ucchaish-
ravas, the horse of Indra, the rain-god was born ; and he was
the three-legged ass of the Zendavesta, the ass-god, the totem-
father of the barley-growing races, ruling the year of three
seasons, who helps Tishtrya (*Sirius*) to bring up the rains.[3]
These always begin in the meridian of Benares, on, or very
near, the summer solstice, an epoch which has from time

---

[1] Lockyer, *Dawn of Astronomy*, chap. xii., chart on p. 127.
[2] Lenormant, *Chaldæan Magic*, chap. xii. p. 152.
[3] West, *Bundahish*, xix. 1-12 ; S.B.E. vol. v. pp. 67-69.

immemorial been consecrated to Sirius in Northern India.
The earlier mountain-home of the mother-bird Gandhārī,
meaning she who wets (dhārā) the land (gan), was on the
top of the mother-mountain of the East, the Himalayan
range, whence the waters of the mother-river of the Kushite
race, the Helmend, descend to Kandahar and the lake Kāsh-
ava, now called Zarah.[1] It was from this Eastern mountain-
summit that the year-egg, the rising sun, generating the
creating rains and mountain mists, proceeded on its yearly
and daily journeys, during which the hundred sons, the
Kushika Khūravya or Kaurāvya, who issued from Gandhārī's
egg, were born. The mother-bird who laid the egg appears
also in the bird called, in the Rāmāyana-Jatāyu, the vulture-
child of Gadura, the flying bull of light, born of the second
egg laid by Vinatā, meaning she who is bowed down, the
pregnant mother-mountain, the tenth of the thirteen wives
of Kashyapa, the father of the Kushite race. Jatāyu was
sister of Sampāti, the lord (pati) of collective life (sam), and
is said to know all things past, present, and future. She
fought with Rāvana, the god of fire and storm, killed the
asses who drew his chariot, overturned the chariot and
Rāvana's charioteer, but was finally overcome by the dis-
mounted Rāvana, who cut off her wings.[2] The epoch of
this battle, which closed the era of the Vulture and bird-
mother and inaugurated that of the god of fire and storm,
the lightning-god of the rainbow or heavenly bow, is fixed in
the Rāmāyana; as it took place when Jatāyu, who was going
to tell Rāma of the route taken by Rāvana when carrying off
Sitā, was overtaken by the ravisher of the mother-goddess,
who was first Sitā, the furrow, the wife of the ploughing-ox,
Rāma, and afterwards the crescent-moon, the mistress of
Rāvana. This change took place while Rāma was beguiled

[1] Darmesteter, Zendavesta Zamyād Yasht, 66 ; S.B.E. vol. xxiii. p. 302.
[2] Rāmāyana, iii. 20-29, quoted by De Gubernatis ; German translation
called Die Thiere in der Indo-Germanische Mythologie, part ii. chap. ii. p.
482.

from Sitā by Marīchi, the deer, whom Rāma slew, and who
went up to heaven as one of the stars in the parent-constellation of the Great Bear, called the seven Rishis or antelopes,
and became the heavenly father of Kashyapa. This was the
age when the twins Day and Night—afterwards symbolised
as the Ashvins or heavenly horsemen, the stars Gemini—
became parent-gods in place of the mother-bird; the age
of the Pāṇḍavas, who were the sons of the gods of heaven,
and of whom the two youngest, the twins Sahadeva and
Nakula, were the sons of the Ashvins. It was then that
Rāma, who from the ploughing-ox became the sun-god,
and his twin-brother Lakshman, the god of boundaries
(*Laksha*), the East and West gods, with the two ape-wind-
gods, Hanuman, son of Pāvana, the wind, and Su-griva, the
husband of Tāra, the stars, the gods of the pole, undertook
the final campaign against Rāvana, the storm-god, and
captured his fortress, the home of the winds, which became
the home of the star-god of the South Pole, the Hindu
father-god, Agastya, the star Canopus in Laṅka (*Ceylon*),
and Agastya was the father of the three races of the Dasyus,
the Cheroos, sons of the bird (*khur* or *chir*), the Cholas or
sons of the mountain (*kol* or *chol*), and the Paṇḍyas or sons
of the sun-antelope, the star deer-god, Marīchi, called Paṇḍu,
the fair (*paṇḍu*) god, and father of the Pāṇḍavas. It is this
supersession of the mother-bird by the twin-gods as rulers of
time which is alluded to in the Rigveda, where the Ashvins
are said to come in the morning as the gods of light to
drink their Soma-cup before the greedy Gridhra or vulture,
to whom the Soma-cup, called Marka or the cup of death
(*the Zend Mahrka*), consecrated to the waning moon, was
offered.[1]   It told of the supersession of the gods of darkness—
whose time-measurers were the months of gestation—by the
gods of light, the Devas or Asuras, who measured it by the
daily and yearly birth of the sun and the recurring full

[1] Rigveda, v. 77, 1; Eggeling, *Sat. Brāh.* iv. 2, 1, 1-14; S.B.E. vol.
xxvi. pp. 278, 279: Hillebrandt, *Vedische Mythologie*, pp. 224, 225.

moons, which lighted up the night. This new age was that
which reckoned time by days and nights, symbolised as the
stars Gemini, the twin stars, who twirled round as its hands
and formed part of the Hindu constellation of the fourteen
circumpolar stars called Shishu-māra, the Alligator, a
reckoning which succeeded that measured by the revolution
of the circumpolar stars of the Great Bear round the pole,
the husband of the mother-vulture, the polar star Vega, from
10,000 to 8000 B.C., and by the phases of the moon. It was
then that two seasons were added to the three seasons of the
year originally planted by the fire-god in the home of the
mother-bird. It is these three seasons of the mother-bird
which are commemorated by the name of the weaving sisters
given by the Chinese to three stars in Lyra which form a
triangle; and this name, which is a reproduction of the name
of that of the Krittakas, or spinners, given by the Hindus to
their mother-constellation, the Pleiades, together with the
figure formed by the stars, shows that they were the mother-
stars of the race who worshipped the mother-year under the
symbol of the triangle enclosed in the mother-mountain, the
Hittite sign of the goddess Ba △, for the Hittites, as the
race of the Khati, or joined, symbolised by the Hittite ideograph of the brothers joining hands, were the race born from
the heavenly twin-parent-gods.[1] In the Chinese ode in which
these stars are spoken of, they are said to pass in a day through
the seven stages of the sky,[2] showing that time was then
reckoned by weeks, each week counting for a day of the gods
of time, the seven days and seven nights making up the
number fourteen sacred to the twin-gods.

This mother-constellation of the weaving sisters, the
Norns of the North, who weave the rope of destiny, and, like
the mother-vulture Jatayu, know the past, present, and
future, was the constellation called Urakhga by the Akka-

---

[1] 'The Hittite Syllabary,' by Major C. R. Conder, No. 106 *Journal Royal
Asiatic Society*, Oct. 1893.

[2] Legge, *Shihking Decadi*, v. Ode, 9 : S.B.E. vol. iii. p. 363.

dians before that name or its counterpart, U-ga-ga, was given to the constellation Corvus, the crow, the mother-bird of the sons of the prophets. It was ruled by the monstrous stellar storm-bird, the Arabian Rukh, the bird of the breath of life (Heb. *Ruakh*), which broods over the luminous egg of the sun and rules the sparkling valley of diamonds, the starry sky,[1] the Simargh or moon (*sin*) bird (*murgh*) of the Persians and Zends.

The constellation Lyra, or the Vulture, is proved to be connected with the fire-god, not only by the astronomical myth of the constellations Hercules, Lyra, and Cygnus, which I have quoted, but also by those of Hermes and Orpheus; for it was Hermes, the wind-dog of the gods, the leader of the four dogs of the Babylonian fire-god Bel, guarding the four quarters of the heavens,[2] who placed seven strings on the shell, or constellation of the Tortoise (*Chelus*), to make it that of the lyre, and these seven strings were the seven stages of the sky in the Chinese astronomical myth, the seven days of the week. It was by the lyre that Orpheus—the Greek form of the Sanskrit word Ribhus, meaning the makers of the seasons—brought back from the realm of death his sun-bride Eurydice, who had, as the dying year, been stung in her foot by the snake of winter when dancing on the day of her marriage. She, after her death as the sun of the old year, was brought back to earth to be the bride of the new year god by the lyre of Orpheus. That the constellation of the Tortoise (*Chelus*), which became that of the Lyre, was first the constellation of the mother-bird, is proved by the fact that the mother of the tortoise was the bird Gandhārī, who laid the egg whence the sons of the tortoise were born. And that the time-measurement instituted by the rulers of the tortoise-earth and sons of the

[1] R. Brown, jun., F.S.A., *Eridanus*, p. 60. See also Hewitt, *The Ruling Races of Prehistoric Times*, Essay iv. p. 341, note 1. Baring-Gould, *Curious Myths*, Second Series, p. 146.

[2] Sayce, *Hibbert Lectures for* 1887, Lect. iv. p. 288.

mother-bird was based on the revolution of the circumpolar
stars, is proved by the Greek myth of the Harpies. These
three-winged carrion-eating birds of Lycian mythology
were a form of the mother-vulture of the year of three
seasons. They were wind-goddesses, called the dogs of Zeus
(Διὸς κύνες).[1] that is, the dogs who guarded the four quarters
of the heavens; and that they went round the pole in their
yearly watch, taking the sun with them, is shown by one of
them called the Harpy Podargē, meaning the 'bright-
footed,' said by Homer to be the mother of the horses of
Achilles,[2] the sun-god. It was this united triad of the
three mother-birds which was deposed by Hermes, the fire-
god, who, like the Harpies, was the dog of the gods, but who
also bore the wonder-working fire-drill, the caduceus or
gnomon pole, grasped and protected by the two mother-
snakes, whose meeting heads formed the summit of the
mother-mountain—the whole being another form of the
Trident. This pole was, in the Zend myth of the mother-
mountain, called Saokanta or Ushi-dhau, the *uru* or root,
the tree-trunk running through its centre, and serving as the
pipe whence the sap of life, the water distilled from the
ocean on which the tortoise-earth floated, ascended to the
summit,[3] to descend as the water of life in the life-creating
rivers which flowed from it, and which were the parent-gods
of the river-born sons of the tortoise and mother-bird.

The connection between the lyre of Orpheus and the con-
tellation of the primæval polar star Vega brings us to the
religion of the Kabiri, of which the Orphean mysteries were
an offshoot. The name Orpheus, which becomes in Sanskr
Ribhus, appears in the Teutonic mythology of the race who
changed *r* into *l*, as the Alps or Elves, the dwarf workers in
metal, who founded the fire-worshipping societies of the
smiths and artisans of the primæval world, who looked on
the measuring pole or gnomon, the fire-drill, as their sign of

[1] Ap. *Rhod.*, ii. 289.   [2] Homer, *Iliad*, xvi. 150.
[3] Darmesteter, *Zendavesta, Khōrshed Nyāyish*, S ; S.B.E. vol. xxiii. p. 352.

the divine father-god. It was they who evolved the myth of the mother-mountain as the revolving pole of the Kushite world, carrying round with it all the stars except the polar star. This last was the star worshipped by the Hindu Kushika as the 'firm star,' the pillar of the stars, and was adored by all newly married couples on the first night they spent in their own home.[1] Their chief god, the fire-god, was a dwarf, like the Hindu Vishnu, the Greek Hephaistos, and the Egyptian Ptah, meaning the opener, the Egyptian counterpart of the Hebrew Japhet. The statue of Ptah, the chief god of Memphis, is described by Herodotus as that of a dwarf-god similar to the Phœnician statues of their gods called Patāikoi which they used to place in the prows of their vessels.[2] This name Patāikoi is connected with the root 'patak' or 'patag,' to strike, which appears in the Hebrew Pat-tish, the hammer. They were the people of the hammer and the anvil, the Kabirian god Akmōn, the father of Eurytion the Centaur, the rainbow-god, who gave life and form to the dead metal by striking and fashioning it with the hammer. Hence their father-god was the god bearing the hammer or thunderbolt, the weapon of Thor, and it was the hammer or mallet which these primæval fire-worshippers used in sacrificing the victims slain by them as the tribal totems, who were, when devoured by the tribes-men, to infuse into them their divine attributes. This is proved by the statement in the Hindu Brāhamaṇas that the earliest slayers of sacrificial victims used to strike them with the hammer or mallet on the frontal bone, and this custom was followed by that in which they were slain by cutting the jugular vein behind the ear, which was again superseded by the orthodox Vedic method of strangling them.[3] This god who bore the hammer, the divining pole or sceptre, which from

---

[1] Oldenberg, *Gṛihya Sūtra, Gṛihya Sūtra of Hiraṇyakesin,* i. 7, 22, 14 ; S.B.E. vol. xxx. p. 194.

[2] Herod. iii. 37.

[3] Eggeling, *Sat. Brāh.* iii. 8, 1, 15 ; S.B.E. vol. xxvi. pp. 189, 190.

the fire-drill became the house and rain or meridian pole, was the god of the four quarters of the heavens, the Assyrian god Bel with his four dogs, and these quarters of the heavens were denoted by the upright cross of the fire-god ╈,

the form of the mythic hammer of the Kabiri, which became the trident, ╫ the Gond-god Pharsipen, the female (*pen*) trident (*pharsi*), the three weaving sisters, worshipped by the Hindu Gonds and Takkas, the artificers or makers (*tvak*). In this symbol of the divinity, the three prongs denoted the three seasons of the year, while the centre prong, the Shūla, symbolised the parent-tree trunk, the meridian-pole, said by Hindu astronomers, as stated by Śripāla, to extend from the South pole, marked by a star called Shūla,[1] or the beam of crucifixion, a little below the meridian of Canopus, which was to the Arabians the star of the South pole, and to pass through Canopus, the Hindu father-star Agastya I have spoken of above, to the North polar star at the top of Mount Mera,[2] while the seven stars of the Great Bear revolve round it. These people, who pictured the pole as perpendicular, and the east and west quarters as horizontal, placed the home of life in the east and the realm of their dead in the west, and it is to the west that the temples of Ptah at Thebes, whom the Egyptians represented as a mummy or dead god, are oriented.[3]

These people, who founded the cult of the meridian North and South pole and the twin-gods Day and Night, were the race who changed the symbol of the star from that of the Egyptian five-rayed star of Horus ✕, representing the

---

[1] The perch of the mother-bird Shu or Khu.
[2] Sachau's Alberuni's *India*, vol. i. chaps. xxii. xxx. pp. 240, 308.
[3] Norman Lockyer, *Dawn of Astronomy*, chap. xxix. p. 296.

pole as the rain-pole, the Ashera of the Jews, placed in the centre of the rain-cross ✕ and conceived as the mother-mountain surmounted by the polar star, standing in the centre of the tortoise-earth, watered by the streams flowing from it. When their pole was extended from the North polar star to the South polar star Canopus, by the conquest of the South pole by Rāma, the sun- and moon-god, they changed the symbol of the star into the six-rayed star ✳ , which is still, with the crescent-moon Sitā, the wife of their sun-god Ram. Rā, or Rāma, borne on the banners of the Turkish armies: and the sun-god then became the god who went in his boat, the ship Argo, from the east to the west through the south. starting from the star Sirius, and who, from the home of the prophet-bird Corvus in the west, climbed up the mother-mountain to the polar star, and thence descended as the rain-bird, bringing the rains of the summer solstice to Sirius. It was from these sons of the pole, the race called by the Hindus the Turvasu, or those who made their god (*vasu*) the pole (*tur*), that the Phœnicians of Byblus, the city of the Papy-rus or record (*Byblus*), the sacred city of the maritime Semites of Palestine, took their religion, for they brought it with them from the island of Tur-os, sacred to the pole Tur, the island now called Bahrein, in the Persian Gulf, and named by the Akkadians the sacred island of Dilvun, where Iā, the fish-god, first appeared. The Arabian Sea, of which the Persian Gulf is a branch, was that in which the Zend constellation Sata-vaēsa or Argo, the constellation of which Canopus or Agastya was the chief star, and it was thought not only to be the constellation of the South pole. but to rule the tides, as it is said in the Zendavesta to push the waters forward,[1] while in Hindu mythology Agastya

[1] Darmesteter, *Tir Yasht*, i. 8, 9 : *Vendîdâd Fargard*, iv. 16-19 ; West, *Bundahish*, xiii. 12 ; S.B.E. vol. xxiii. pp. 92, 95, 96 ; vol. iv. pp. 53, 54 ; vol. v. p. 44.

drank up the waters of the ocean, which was again replenished
by Gangā, the mother river.[1] These Phœnicians of Byblus,
again, were a branch of the Kabiri of Thrace, for both they
and the Thracian Kabiri called their priests Kabir, a
Hebrew name meaning great, and this name is exactly
analogous to that of the Magi, who were the Makkhu or
great (mag) ones, the high priests of the Akkadians. But the
difference in ritual of the fire-worshippers, whose priests were
Makh-khu, proves, like the difference in the doctrine of the
pole as taught by the people of the five-rayed and six-rayed
star, a difference in ethnologic history. For the Kabiri did
not, like the early fire-worshippers in Phrygia, Assyria, and
India, emasculate their priests to make them sexless like the
god of the fire-drill, but believed in the divinity of pairs,
the primæval twins, parents of the Hittite race, and looked
on life as generated from the water of life distilled by the
revolving parent of fire, originally the burning mother-
mountain Ararat, and distributed over the earth by their
parents, the rivers which supplied water for their crops and
fruit-bearing trees. They who had first been the sons of the
Syrian fig-tree became, as the race descended from the primæval
pair, the sons of the parent-tree of Babylon and Western
India, the date-palm, the male and female tree which only
bears fruit when the flowers of the female tree have been impregnated by those of the male. This race still survives
in its primaeval form as the Haranite Sabæans, described by
Dr. D'Chewolsohn. They celebrate the mysteries of their
supreme god Shemol, the god (el), Shem meaning the name
and also the left hand (semol),[2] in an underground room like
a cave, a survival of the ancient sacred caves of the mother-
mountain. This is called the house of Bogdariten, and they
call themselves his sons. This is the god Bog or Boga of
the Slavonians, the Bagha of the Persians, the Sanskrit
Bhaga, the god of the tree of edible fruit. This supreme

---

[1] Mahābhārata Vana (Tirtha Yatra) Parva, ciii.-cix. pp. 324-340.
[2] Miller, Harmoad, or Mountain of the Assembly, p. 50.

god and creator, Shemol, was imaged in the polar star; and
his chief priest, the priest of the tribe, is called Kabir,
while the mother-stars of the star-born race were the seven
stars of the Great Bear, the seven mother-Rishis or antelopes
of the Hindus. In their sacrificial sacramental ceremonies
they partook of bread prepared after the manner of the
shepherds, the Paseal cake of the Jews, and the sacred cake
of the Indian Soma sacrifice, of the Zend sacrifice to Haoma,
and of the Eleusinian mysteries. They also partook of food
and wine, the totemistic sacramental meal, on the tribal
totem, and at this meal in earlier times its blood was drunk,
and it was for this that the wine was a substitute. The
final ceremony was the preparation of cakes made of meal
kneaded with the boiled flesh of a male child offered in sacra-
fice. This was the Semitic sacrifice of the eldest son, the
earliest Passover, the sacrifice of the eldest son of the Hindu
mythical King Jantu, to make his hundred wives pregnant,[1]
the son slain to secure by his life-giving blood the fertility of
the coming year. These cakes were eaten as mystical bread
throughout the entire year.[2]

It was Sabæans who were the sons of the mother-vulture,
worshipped as their god, Nasr, by the Southern Sabæans of
Arabia.   Their religious tenets are set forth in the book of
Byblus, ascribed to Sanchoniathon, a name which means 'the
god Sakkun hath given,'[3] Sakkun being the god of the Semite
Phœnicians, who worshipped the two triads of the sun, moon,
and earth, and the rivers, waters, meadows, emblematic of
the twin parent-gods of the sons of the rivers, who were
invoked in the oath sworn by Hannibal to Philip of Macedon.[4]

[1] Mahābhārata Vana Parva, cxxvii. cxxviii. pp. 386-389.

[2] Chewolsohn, *Ssabzer und der Ssabzismus*, ii. Excursus to chap. ix. pp.
319-364, quoted in Dr. Miller's *Harmoad, or the Mountain of the Assembly*,
published by S. T. Whipple, 110 Main Street, North Adams, Massachusetts,
pp. 50, 51. See also pp. 87, 163, 186, 187, 411 of the latter work.

[3] *Encyclopædia Britannica*, Ninth Edition, Art. ' Sanchoniathon,' vol. xxi.
p. 255.

[4] Polybius, vii. 9, 2.

Their sacred city was Byblus, called by them Ge-bal,[1] the city of the stone-cutters. This name, Ge-bal, is the same as that of the Akkadian fire-god of the horizontal equator, called Gi-bil or Bil-gi. Their other god, Sak-kun, is the Akkadian Sukh-us, their name of Istar, the Hindu rain-god Suk-ra, the Sek-nāg of the Gonds, and the Pali god Sakko, the god of the wet (*sak*) land of the Euphratean Delta and North-western India, who in Buddhist mythology is the head of the thirty-three Nāga gods of the Tavatiṁsa heaven, the heaven of thirty-three (*tavatiṁsa*), the thirty-three lords of the Ritual order of the Zendavesta, the thirty-three gods of the Rigveda, and the thirty-three heavenly judges seated in the judgment-hall of the Egyptian goddess Ma'āt, who ruled the year of five seasons divided into lunar months of twenty-eight days each.[2] It was in this wet land, and in their original mother-land of Asia Minor, that they first, as the Hindu Turvasu, the people whose god (*vasu*) is the Tur or pole, learned the religion which they brought with them first to the Persian Gulf and afterwards to the Syrian coast, close to the land from which they first started on their land-wanderings in the wilderness. They there learned to revere the eight Agnis or fire-gods, the father-gods of the eight tribes of Gonds, the people who, as the sons of the wild cow, Gauri, the mother of the Hindu Gonds and of the Euphratean Gaurian race of Telloh, the ancient Gir-su, ruled the seven kingdoms of the Indian Kushika; and of these kingdoms, Jambu-dwīpa, the land of the Jambu-tree (*Eugenia jambolana*), the edible forest fruit of Central India or Gondwana, was the centre, and Sāka-dwīpa, or the wet (*sāka*) land whence Indra gets the rain, the Northern kingdom. These eight Agnis become the eight creating-gods of Egyptian mythology, who first as the six and afterwards as eight creating apes of the Eastern green land, Uetenu, called Bentet and Keftenu, or the Phœnician apes (*kaft*),

[1] *Encyclopædia Britannica*, Ninth Edition, Art. 'Jebeil,' vol. xiii. p. 613.
[2] *Ruling Races of Prehistoric Times*, Essay iii. pp. 266-267.

became in a later age the four male and four female creating-spirits, headed by Nun, the fish-god, the spirit of life, concealed in the life-giving mist.[1]   These eight creating-gods were those worshipped as the eight Cabiri in Phœnicia and Thrace, who were as stars the seven stars of the Great Bear and the polar star.   This last was the god called Eshmun, the youngest of the sons of Sydek, the just, the father of the gods of the Phœnician Kabiri; and Sydek, or Su-dik, was the revered (dag) Su, the soul or sap of life, the root of the Hindu Soma and Zend Haoma, and the Southern form of the Northern bird, Khu.   Eshmun, the pole-star, was the eight-rayed star ✳, called by the Akkadians An, meaning god or Esh-sha, the soul or bird (shu) of life (esh), the seed, and depicted as the ideograph of the seed and god in the Gir-sa inscriptions.   This was, as I have shown above, formed

by the superposition of the rain-cross ✕ on the fire-cross ✝.   It was in the eighth Mahurta, consecrated to

these eight gods, of whom the eighth was the polar star Vega. the firm star, the pillar of the stars, that Yudishthira, meaning the most steadfast (yu), was born, and it is to his birth we must now return after this long digression.   His birthday was on the fifth of the light fortnight of Kārtika, the eighth month, another appearance of the sacred number eight under the star Jyeṣṭha.[2]   This star is, in the list of the Nakshatras, Antares a Scorpio, and this is one of the stars called by the Babylonians Māsu, or the leading antelope (mas), and the star of the lord of seed, the star of the month Tisri,[3] which begins the Semite year at the autumnal equinox.   But this year-god, son of the sun-antelope Paṇḍu, born in Kārtik

---

[1] *Ruling Races of Prehistoric Times*, Essay iii. pp. 295-296.

[2] Mahābhārata Adi (*Sombhava*) Parva, cxxiii. p. 259.

[3] R. Brown, junr., F.S.A., 'Tablet of the Thirty Stars': Star xxiii.: *Proceedings of the Society of Biblical Archœology*, February 1890.

(November), under the guardian star of the year of the
Semite race, did not come to be supreme ruler of the year till
Bhishma, the uncle of the Kauravyas and Pāṇḍavas, and the
sun-god ruling the year of the Kauravya sons of Gandhārī,
the bird-mother, died, and his death, according to the
Mahābhārata, took place in the first half of the month of
Māgha.[1] So that it was with the full moon of Māgha that
Yudishthira's time of studentship ceased and his rule as year-
god began. The year that preceded that of Yudishthira,
the year of Bhishma, was that marked on his banner by the
five stars above the Tal or date-palm, the parent-tree of the
trading races of Babylonia and Western India.[2] These five
stars are the five bulls or eagles of the Rigveda, standing in
the midst of heaven marking its four quarters and the North
pole, the constellation of the mother-bird, the Vulture or
Lyra, whence Cygnus, the rain-bird, next to it, gets the seed
of life, which was to make the rain which it was bearing to
the earth fruitful. This mother-bird, the bringer of rain, is
that which is invoked in the 52nd stanza of the cosmological
hymn, Rigveda, i. 164, quoted in the beginning of this
Essay as the child of the waters and the plants, which brings
the rain. These five eagles or bull-stars, at the head of
which is the mother-eagle or vulture, or the mother-cow, are
said to guard the road against the wolf (of fire) which
troubles the young waters, another form of the hot-weather
demon which keeps back the rains; and they are said to stand
in the place of the seven rays, the seven days of the week,
the home of the supreme god Agni.[3] Thus the year of
Bhishma was one which, like that of the sons of Yayāti,
began with the rainy season. The year of Yudishthira, born
in Kārtik (November), and beginning his rule as year-god in
Māgh (February), is one reckoned on the same principle as
the Akkadian or first Semitic year of thirteen lunar months,

[1] Mahābhārata Anuṣhasana Parva, 167.
[2] Mahābhārata Bhishma (*Bhishmavadha*) Parva, xlvii. p. 165.
[3] Rigveda, i. 105, 9, 10, 11.

which began with the birth of the sun-god in Aquarius in
November, as the nursling of the moon, who passed with
the sun through the circle of the thirty stars in the three
months between November and February, and left her charge,
the young sun-god, as the full grown ram-sun in Aries in the
latter month, to pursue his path through the ten stars, of
which Regulus in Leo was the fifth, which were called the
ten kings of Babylon.   In the last of these, Skat in Aquarius,
he plunges into the Soma bath of regeneration, to emerge
again as the new-born sun-god,[1] who is again to be nursed by
the moon, who appears in the Mahābhārata as Bhishma,
the dying sun, till he is again placed in Aries in February.
This year of Yudishthira is the year of the moon-lion,
guarded by the constellation Leo, which, in the list of the
Nakshatras, is the guardian star of Māgh, the year of the
Sphinx of Assyria, Phrygia, Greece, and Egypt, the year-
goddess, who was slain by Œdipus, the god of the swollen
(*œdi*) foot (*pous*), the son of Laius or Lāüs, the stone pillar
or gnomon obelisk, and Jocasta, the crescent moon, whom
Œdipus, after he slew Laius, married as the young sun-god
parent of the coming year, the god of the year beginning
at the vernal equinox.   It is also the year of the young
Achilles, the sun-god of Greek legend, who was nursed by
his moon-mother, Thetis, and who finally died at the close of
the two years'—or ten lunar months'—war against Troy, the
city sacred to the moon, guarded by Apollo, god of the silver
bow, the lunar crescent.   This year, in which the active life
of the young sun-god began in Māgh, is that which is spoken
of in the hymn of the Rigveda telling of the marriage of
Soma, the moon-father, to the sun-maiden, in which it is said
that the oxen sacrificed to the sun-god were slain in Maghā.[2]
This was the year of the thirteen months, the thirteen wives
of Kashyapa, which is still kept by the Santals; and Varuna,
the guardian of order, is said in the Rigveda to know the

---

[1] *Ruling Races of Prehistoric Times*, Essay iv. 376-386, 391-394.
[2] Rigveda, x. 85, 13.

thirteenth of these months.[1] The mother of this year is Kadrū, the thirteenth of Kashyapa's wives, to whom the mother-bird gave the Soma or soul of life, which she brought from heaven, and which Kadrū gave to Indra and Agni.[2]

The year which succeeded, in India,[3] this year of thirteen months was that marked by the transfer of the national Saturnalia from the full moon of Māgh to the full moon of Phalgun, near the vernal equinox, when the Hindu Hūli is now held. This birth of the young sun-god took place at the full moon of Mārga-sirsha, near the winters olstice, and he celebrated his marriage at the vernal equinox in the signs of Phalgun.[4] But in the Vedic hymn telling of this event the sun-god has become the sun-maiden, and the husband to whom she is wedded is the moon-father of the Northern nations, all of whom make the moon masculine. This was the year of six seasons and 360 days, the year of the six Āditya of the Rigveda, the year of the wheel of the heavenly oil-press turned round the pole by the six year-gods, Aryaman or Aditi, Mitra-Varuṇa, representing the six months of the Devayāna, the months consecrated to the bright gods of heaven, in which the sun is going northward, and Bhaga (the fruit-tree), Daksha (the visible fire-god), and Aṅsha (the Soma stem),

[1] Rigveda, i. 25, 8.

[2] Eggeling, *Sat. Brāh.* iii. 6, 2, 8-12; S.B.E. vol. xxvi. pp. 150-151.

[3] The question of the succession of these two years cannot, as I show in Essay viii. on 'The Mythology of the Northern Races,' be absolutely determined by Indian evidence. I have shown there that the year of twelve months and 360 days was the wheel-year of the sun-ship, represented in Greek mythology by the voyage of the Argo. The year of thirteen months was the year of the fish-sun-god, the year of Perseus in Greek mythology, and this was certainly later than that of Jason and the Argo, which was the year calculated after that of the sun-horse. The year of the Scandinavians, in which the months were divided into six periods of five days each, called Fimt or Noroe, week of five days, was a year which returned to the reckoning of the year of Orion. This sanctity of the number five was based on the Hindu year of five seasons, and this wheel-year was imported from Babylonia to the North by the Phœnician traders, the crew of the Southern ship Argo.

[4] Rigveda, x. 85, 13.

those consecrated to the fathers and the gods of earth; the months of sowing and harvest, when the sun is going south, and of those Mitra, the male moon-god, Varuṇa, the god of rain (*var*), or of the dark heaven of night, and Aryaman, are called the pure Ādityā.[1] Aryaman is called in the Zendavesta Airyaman, the physician and healer,[2] like the Hindu Ashvins, the stars Gemini, and is therefore, like them, a star-god who marked the beginning of the year, as the Ashvins originally, the twins Day and Night, marked the beginnings of the diurnal computation of time. His name shows him to be a ploughing-god (*ar*, air), and he must, therefore, be connected with the black bull Puṣh, which gave its name to the first month of the Hindu year; and as the driver round the pole of the six-spoked wheel, to which the year is assimilated in the cosmological hymn I have quoted at the beginning of this Essay, he must be a charioteer; and all these qualifications are met by identifying him with a Auriga, called Capella or the Little Goat, the patron star of Babylon, which took the place of the old goat-god of time. It was this charioteer star which drove round the pole the seasons and months headed by the bull; and it was to the young bull, called Phalguni or Arjuna by the Hindus, the father-god of the bull Bhārata, as the ruling race are called in the Mahābhārata, and Marduk, the young bull of the Babylonian year, that the vernal equinox was dedicated. This year of 360 days was also the year which, as we are told in the observations of Bel, giving the astronomical records of the reign of Sargon, who ruled 3750 B.C., was the year of the Akkadians.[3] It was the bull

---

[1] Rigveda, ii. 27, 1, 2. The order in Rigveda, ii. 27, 1, is Mitra Aryaman, Bhaga, Varuna, Daksha, Anṣha, but as in v. 2 Mitra-Aryaman-Varuna are called the pure Ādityā, and as in v. 14 Aryaman is called Aditi, or the beginning, their order as gods of the seasons must be that stated by me in the text.

[2] Darmesteter, *Zendavesta Vendīdād Fargard*, xxii. 3, iv. 13-22; S.B.E. vol. iv. pp. 232-234.

[3] Sayce, 'The Astronomy and Astrology of the Babylonians,' *Transactions of the Society of Biblical Archæology*, vol. iii. p. 160.

ruling this year which took the place of the year-ox slain in Mägh, the sexless sun-god Bhishma, and he was, in the Vedic mythology, the sun- and moon-bull Vishvā-mitrā. His descendants, the king called Chakra, the wheel, born from Arjuna and Su-bhadrā, the mountain-goddess, twin-sister of Krishna, the black antelope, was the first of the traditional line of Chakra-varti, or kings of the horizontal wheel, who in Hindu legend ruled the empire of the Ikshvāku, or sons of the sugar-cane (*iksha*), from their central capital and mother-city Kushambi, the city of the tortoise (*kush*) and the Pleiades (*ambā*), which guarded the junction of the two mother-rivers of Northern India, the Jumna or Yamana, the river of the twins (*yama*), and Gañ-gu, the wetter or mother (*gu*) of the Gan, or garden of god. It is this year, beginning with the full moon of Phalgun and the vernal equinox, which is, as Bāl Gangādhar Tilak tells us, that used by all Brahmins, from the Himalayas to Cape Comorin, for fixing the times of their sacrifices.[1] This year of twelve months of thirty days each was that of the Athenians and Egyptians, both of whom divided their months not into weeks, the division used by the computers of the lunar year, but into three dekads of ten days each.

But throughout these changes in the year-reckonings, the old division of the year into two periods of six months each still held its ground, for the mid-year Soma sacrifice of the summer solstice became that which is yearly celebrated to Jaganath at Pūri and the Skirophoria festival of the Umbrella (*skiron*), the festival of the sons of the Banyan-tree (*Ficus Indica*), who made the umbrella the sign of royal dignity, which took place at Athens at the summer solstice, when the bale fires to the fathers, which used in the Pleiades' year to be lighted in November, are still lighted in the west of Ireland.

But the reckoning of time by the year of the Western Hindus, the year of the perfect circle commemorated in the

---

[1] Bāl Gangādhar Tilak, *Orion*, chap. iv. p. 70.

Kuṣṭik or girdle of the young Parsis, produced errors owing
to its being five days too short, and similar but smaller errors
arose from the calculation of the Eastern year of thirteen
lunar months or 364 days, while the adjustments of intermediate dates between the solstices and equinoxes, which
alone could be measured with the gnomon or Tur, were in
both systems difficult unless ready means of correction were
available. It was for the correction of these errors that the
circle of twenty-eight lunar mansions or stars was changed
into that of the twenty-seven Nakshatra or stations of the
field (*kshetra*) of the Nags or stars, each representing 13° 20′
of the solar circle; and the origin of the list names of these
stations is shown by the names Pūrva and Uttara, given to
the Nakshatras of Phalguni, Ashādhā, and Bhādrapada, the
months of the vernal equinox, summer solstice, and autumnal
equinox. These show that the design of the calculators who
added the Nakshrata divisions of the circle to the older forms of
ritualistic time-measurement was to reconcile the discrepancies
between the Eastern (*Purva*) year, the lunar year of thirteen
months of the eastern Kushikas, which still survives in the
Santal year, and the Northern or North-western (*Uttara*) year
of Chakravarti Ikshvāku kings of Kushambi, the year of 360
days. But in all these arrangements the priestly calculations
of national time only tried to perfect the sacrificial year, and
to make the time-honoured festivals to the mother-Pleiades,
the antelope-father Orion, the mother-antelope, and the
mother-cow (*Rohini-Aldebaran*), the ploughing but sexless ox-
father, and the bull-god, universal national festivals held on
the same day throughout the length and breadth of India.
They never dreamt of using the Nakshatras to record the position of the sun at different epochs, as they were used by the
authors of the Vedānga Jyotesha, who followed the example
set by the ecliptic observations of the Babylonian astronomers. This later innovation arose after the ecliptic lists
were completed, and the year calculated by the passage of
the sun through the ecliptic stars. The Babylonian astro-

nomers called this year the year of Anu and the year of the
horizontal circle; the wheel-year of the Rigveda they called
the year of Bil, the fire-god, the god of the right-angled

upright cross ——┼——, formed by the equinoctial and polar

stars.[1] These western traders of Babylonia and Western
India, who used the wheel-year, were the stock from which
the temple-builders of Babylon, Jerusalem, Palmyra and
Baalbec and the pyramid-builders of Egypt [2] sprang, the
masons who oriented their temples to the true east and west;
and this system dated ages before the much later orientation
of the later Egyptian and Greek temples to certain stars.

I have now shown in this sketch of the ancient Hindu
methods of reckoning time, and of some of the historical
lessons to be learnt from them, that they first divided the
year into two seasons of six months each, beginning in
November and marked by the movements of the Pleiades.
This year was followed by that of three seasons, the earliest
year of the barley-growing races, consecrated to Orion, and
beginning at the winter solstice, and in this age time was
also measured by the lunar months of gestation. This was
followed by the year of five seasons of the mother-bird, the
constellation of the polar star Vega, beginning with the
summer solstice and the Indian rainy season, ushered in by
the star Sirius. This was the year of Bhishma and of the
sons of Yayāti. Then came the year of Yudishthira of
thirteen lunar months, in which the young sun-god emerged
from his yearly bath of death and re-birth in November, was
nursed by the moon for three months till February, when he
was married to the moon-goddess, as Zeus was wedded to
Hera in Gamelion (February). The series of Vedic and
Hindu ritualistic time-measurements closes with the year of
twelve months or 360 days, in which the sun-god was born

---

[1] Norman Lockyer, *Dawn of Astronomy*, chap. xxxiv. pp. 363-365.
[2] *Ibid.* chap. ix. pp. 92-94, chap. viii. p. 85.

at the winter solstice and wedded to the moon at the vernal
equinox. This was the year of the bull, and it was to reconcile the discrepancies between this and the lunar year, and
to secure a correct measurement of time always available for
use, that the Nakshatras and five years' cycle was invented.[1]

If this view of the progress and aims of Vedic astronomy,
which I have shown to be entirely in accordance with the
popular time-measurement, be accepted as correct, we cannot
believe that it is possible to find in any Vedic references to
the heavenly bodies such notices of their position as will help
us in assigning dates to the Vedic poems, or the Brāhmaṇas
which expound the Vedic ritual, earlier than the age of the
Vedānga Jyotisha. The idea of marking dates in this way
was quite outside the purview of the Vedic Rishis, who had
been taught by the rules laid down by the primæval national
historiographers to show the sequence of events by the
order given to the signs denoting them in their national
historical myths and ritualistic observances. Both myths
and ritual started from the earliest dawn of the national
past, and noted the succession of each noteworthy minute and
hour of the clock of time by its appropriate reference to the
calendar of history or change in the ritual of the confederated
nation.

But there is still a most trustworthy source of information
as to the growth of Hindu ritualistic astronomy which I have
left unexamined, and which, as I shall now proceed to show,
gives additional and most convincing proofs of the truth
of the deductions made in this Essay, and which also enables
us to carry on the history told in Vedic beliefs and methods

---

[1] See note, p. 61, and also the Essay on the 'History of Northern Mythology,' where I show that in the true order of the lunar-solar year of thirteen
months, and that of the solar-lunar year of twelve months, their positions as
given here should be reversed. The wheel-year of twelve months was conceived before the days when the passage of the moon and sun through the
stars was traced by the Babylonian astronomers, who calculated the year of
364 days. It was this year which preceded the ecliptic year of twelve months
and 365 days.

to the later ages, when comprehensive mythic history was
superseded by that founded on annalistic records, and the
succession of events marking stages in the lives of kings and
national leaders of thought. In discussing the relation of the
history of the Buddha to the lunar-solar year of the Akkadians, in which the immortal sun dies and is reborn each
year in the baptismal bath of the constellation Pisces, filled
by the water-pouring Aquarius, I have already shown that
Buddhistic theology is based on ancient astronomical speculations, and I shall now prove that these speculations run
through every phase of their religious history, and that they
are the origin of the Buddhist belief in rebirths, and the
reproduction in each of these successive births of the condition of the reborn soul, considered by eternal justice to be
the fit outcome of the Karma, or sum of the deeds of each
individual in his former states of existence.

I have shown that the account of the conversion of
Sumedha, the sacrifice (medha) of Su, the rain-father and
soul of life, and his discovery of the Ten Perfections, the ten
steps or months of gestation of the perfect soul, is based on
the coming of the rains from the Himalayan mountains at
the summer solstice, and the ten months of gestation following their cessation at the autumnal equinox, when the seed of
immortality (su) has been infused into the earth.[1] I shall
now prove that this myth of the birth-year of the sanctified
soul is reproduced in another form in the history of the
twenty-seven Buddhas who preceded the Buddha-Gautama,
called Siddharta or grain of white mustard-seed (Seddhato),
the regenerated son of life, the great teacher, whose history,
though he was a living leader and founder of the regenerated
church, was, like that of his predecessors, told in the form of
a sun-myth. These Buddhas were the twenty-seven Nakshatra or lunar mansions, the leaders and predecessors of
Bhudda, the sun-god, by which the discrepancies of the solar
and lunar year were reconciled, and the edicts of the law of

---

[1] *Ruling Races of Prehistoric Times*, Essay iv. pp. 398, 399.

the holy life gradually brought to perfection, while their offspring, Gautama Buddha, the sun-god who went forth to the bath of regeneration from which he was to emerge as a reborn ascetic on the day of the birth of his son Rāhulo,[1] was the twenty-eighth Nakshatra, who completed the lunar month symbolising the birth and education of mankind.

But it is not only the lunar-solar year of thirteen months marked by the passage of the moon and sun through circles of stars, the ten months of gestation, and the star-circle of the lunar mansions which underlie Buddhist theology, for we find also in its foundations the earlier year of thirteen lunar months, the thirteen wives of Kashyapa, and the wheel-year of twelve months and 360 days. The mother of the Buddha Māyā, the goddess Māghā, the miraculous mother of fire, the creating-goddess who formed in her womb her future off-spring, the flame emanating from the fire-spark, died on the seventh day after she had given birth to the Buddha in the mother-grove Lumbini, under the parent Sāl-tree of the united Koliya and Sakya race.[2] His conception took place at the midsummer festival of the summer solstice in the month of Asālhi (*Assar*), when Māyā was borne in a vision to the Himalayan mountains, and placed under the great parent Sāl-tree on the crimson plain stained with the blood of the slaughtered sacrificial victims of the races who offered animal sacrifices, called Mano-silā-tal, the plain (*tal*) of the rock (*silā*) Mano,[3] meaning mind or intelligence,[4] the Buddhist form of the Hindu Manu, the father of the bull race by his daughter Idā, born out of the waters of the flood. Idā or

---

[1] The 'Nidāna Kathā,' Fausboll, *Jātaka*, vol. i. pp. 60-62 ; Rhys Davids, *Buddhist Birth Stories*, 78-82.

[2] The 'Nidāna Kathā,' Fausboll, *Jātaka*, vol. i. p. 32 ; Rhys Davids, *Buddhist Birth Stories*, p. 66.

[3] The word as a whole means crimson or vermilion, and in this sense denotes the colour of the sky when the sun is rising out of a mist, the mother of life in ancient mythology.

[4] The 'Nidāna Kathā,' Fausboll, *Jātaka*, vol. i. p. 50 ; Rhys Davids, *Buddhist Birth Stories*, p. 62.

Irā was, as I have shown, originally the sheep-mother, the
mother-goddess of the bull-race (*Gut*) of the Gautama, the
sons of the rivers called Irāvatu, their mother-river in North-
eastern India being the Irāvati or Rapti, into which Rohinī,
the river of the red cow (*Rohinī*), flows, passing between the
city of Kapila-vastu, the city of the Buddha's father, and
Koliyā, that of his mother. Idā was begotten by the sacred
sacrificial seed of clarified butter, sour milk, whey, and curds,[1]
the five ingredients of the Pāka sacrifice symbolising the year
of five seasons, and rose out of the waters of the flood as
the mother-mountain, the Phrygian goddess Ba, who is depicted as the mother-mountain with the triangular seed of
life enclosed in it △ . It was the mother-tree standing in
the plain whence the mother-mountain of the understanding
mind rose, that bore the seed of life to which the first
Buddha, in whose honour this story of the birth of the sun-
god was framed, owed his being ; and this legend of his birth
shows us that the year measured by its makers was the year
of five seasons, beginning with the summer solstice, which
I have called the year of Bhishma and of the sons of Yayāti.
On the death of the mountain mother-goddess of this year a
new chronological era began, and this is symbolised by the
sister of Māyā Mahā Pajā-pati, called Gotami, the creating
cow-mother of the bull (*gut*), the moon-goddess. It was she
who as the moon-goddess nursed the young sun-god after his
mother's death, and the thirteen months of her year, in which
the last three months were those consecrated to the birth of
the sun of the next year, the death of his mother, and the
rebirth of his moon-nurse, were symbolised by the thirteen
Buddhist Theris, or female saints, of whom she was the chief.

 The mythical stories of the former births and lives of these
thirteen Theris are given in the Manorathā Puranī, meaning
the wish-fulfiller, the name of Buddhai-gosha's commentary
on the Anguttara Nikaya, and this has been translated by
Mrs. Bode, and published, with the Pali text, in the *Journal*

---

[1] Eggeling, *Sat. Brāh.* i. S, 1, 7 ; S.B.E. vol. xii. p. 218.

of the Royal Asiatic Society.[1] In this work every one of these thirteen Theris is said to have been first born in the city of Haṃsa-vati, that is, of the moon-goose (*haṇsa*), who appears as Kaṇsa or Haṇsa, the goose-king, in the Hindu legend of the birth of Kṛishṇa, the sun-antelope, and his twin-sister Durgā or Su-bhadrā, the mountain-goddess, and she, as I shall show, is the ninth of these thirteen Theris. The time of their birth is said to have been that when Padumuttara, the northern (*uttara*) lotus (*paduma*), was Buddha. He was the thirteenth of the Buddhas, and Haṃsa-vati was his capital, and his place in the list of Buddhas at once shows him to be connected with the lunar year. The name of his father Ānanda (joy) is a reproduction of that of Nand (pleasure), a name of the god Shiva, to whom Nand-ganw, the hill-village of the husband of Rā-dhā, the giver (*dhā*) of Rā, is consecrated in the holy land of Mathura. Shiva is the shepherd-god (*sib*) of the sons of the North, and Rā-dhā is the mother of Rā, the sun-god, of whom Nand or Ānanda was father. The mother of this divine lotus is called in the Buddhist birth-story Su-jātā, meaning born of Su, the water or sap of immorality, the sacred pond or lake of the lotus consecrated to Rā-dhā as Rādhā-kund. His chief disciples were Devala and Sujātā, the latter name reproducing that of his mother in the masculine form. But the most distinctive mark of his age is that given by his sacred tree, the Sāl-tree (*Shorea robusta*), under which the Buddha was conceived. I have shown how the sons of the North, the sons of the pine-tree and the mother-bear, the seven stars of the Great Bear, adopted the Indian resinous tree, the Sāl-tree, as their parent-tree when they left the Northern pine-tree behind them in their Southern migrations, in which they married the daughters of the Southern Pleiades, the women of the Indian matriarchal races.[2] This lotus-born

---

[1] This translation and text is printed in the *Journal of the Royal Asiatic Society*, July and October 1893, pp. 517-566, 763-798.

[2] *Ruling Races of Prehistoric Times*, Essay vi. p. 511-513.

god of the Sāl-tree, the father sun-god of the Northern immigrants, the sons of the bear, was succeeded as Buddha by Sumēdha, the sacrifice (medha) of Su, the fourteenth Buddha, the eldest son, who devoted himself for the good of mankind, and discovered the Law of the Ten Perfections. His sacred tree was the Champaka (Liriodendron grandiflora), the tree whose blossoms form the sacred Hindu sacrificial garland, and which is the parent-tree, married by all Telis, or growers of oil-crops, the upland Northern crops which first competed with the Southern upland rice.[1] Sumēdha's city was Su-dassana, the city of the manifestation (dassana) of the Su, the city in which the son of god, the offspring of the divine lotus, was born; and this offspring was the fifteenth Buddha, Sujāta, born of Su in the city of Su-mangala, the auspicious (mangala) Su, and his sacred tree was the great bamboo-tree, while his female disciples were Nāgā, the rain-snake or cloud, and Nāga-samilā, the brown (samilo) snake, the cobra, the mother-goddess of the villages of the matriarchal tribes. This era was that represented in the Mahābhārata by that of Kichaka, the hill-bamboo, the son of Kai-kara, the Gond mother-goddess of the Bhārata, the commander-in-chief of the king of Virāta, where the Pāṇḍavas had taken refuge during the thirteenth year or month of their exile. Kichaka was slain and deposed by Bhima, the god of the summer solstice ending the year in the Pāṇḍava time-measurement. This age of Padu-mattura, Sumēdha, and Sujāta is called in the Nidānakathā the Maṇḍakalpa or age of water-born scum and froth (maṇḍa), from which were born the frog-sun, the imprisoned and transformed prince of Northern mythology, and the four creating-frogs, the male partners of the female snakes, who, together, made up the light-creating spirits of the Egyptian faith, headed by the Nun.[2] It was these eight gods who were represented in the Akkadian sign of the eight-rayed star, formed by placing the rain- over the fire-

[1] Ruling Races of Prehistoric Times, Essay ii. p. 87.
[2] H. Brugsch, Religion und Mythologie der alten Aegypter, pp. 158-160.

cross, which denoted both god and the seed of life, and it was from this symbol that the Buddhist doctrine of the eight-fold noble path, which leads to the destination of sorrow, was evolved. The age from which the initial conception of the eight national father-gods was evolved was that of the union of the eight tribes of Gonds, as the Bhāratas or Bhars, whose totem was the hill-bamboo, the product of the hills Jambu-dwipa, or Central India, the kingdom still called Nagpore, or that of the city (pur) of the Nags. They were the sons of the Northern migrating race, the growers of dry upland crops, worshippers of Rā, the sun-god, and the rain-god, Sukra, Sakra, or Sakko, lord of the Northern land of Sāka-dwipa, to whom the divine lotus, the child of the migrating tank or water reservoir, was sacred. It was they who, when they first entered India, made the Sāl-tree and the Champaka, the tree of the sacrificial garland, their parent-tree; and it was under their rule that the measurements of time by months, begin-ning with the ten and eleven lunar months sacred to the gods of generation, and the thirteen months of the lunar year, the thirteen wives of Kashyapa, the twenty-seventh Buddha Kassapo, and the thirteen Buddhist Theris were first adopted in place of the previous measurement of the year by seasons.

An examination of the names, functions, and histories of the thirteen Theris discloses further historical evidence. They are called (1) Gotamī Mahā Pajāpatī; (2) Khemā, the safe one; (3) Uppalavaṇṇā, the blue (uppalam) lotus, or Padumavati, the mother of the lotus (paduma); (4) Pata-cārā, she who goes (cārā) [under] a roof or canopy (paṭa); (5) Dhammadinnā, she who was given (dinna) to the law (dhamma); (6) Nandā, pleasure; (7) Soṇā, the dog or the red one; (8) Sakuta, the goddess of the Sākya or wet (saka) race, sons of the rain-god, and worshippers of the light-creative powers; (9) Bhuddā Kuṇḍalukesā, the saint (Bhuddā) with the curly (kuṇḍula) locks (kesa), called also Su-bhaddā or Su-bhadrā, the sainted (bhuddā) Su;

(10) Bhuddā Kāpilānī, the sainted mother of the yellow (*kapila*) race; (11) Bhuddā Kaccānī, the golden (*kancana*) saint-mother of Rāhulo, son of the Bhuddā; (12) Kisā Gotamī, the emaciated or ascetic (*kiso*) Gotamī; (13) Sigāla-kamātā, the mother of the jackal (*sigālo*).

The first Gotamī Mahāpajāpati, called the embodiment of experience, was the moon-goddess, who took charge of the young Buddha when he left the Tusita heaven and came to earth in the fourth age of Buddhist mythological chronology. The first is that of the Catum-maharajika-devaloko, that is, of the heaven (*devaloko*) of the bright-gods (*deva*), the hundred (*catumo*) great kings, the hundred Kushika sons of the mother-bird. The second, that of the Tavatimsa, meaning the thirty-three gods of time ruled by Sakko, the rain-god. The third, the age of the Yama-devaloko, the heaven of the twin-gods (*yama*), that is, of the race who measured time by the revolutions of the circumpolar stars, turned by the heavenly twins, the stars Gemini. The fourth was the age of the Tusita heaven of the gods of wealth (*tuso*). Of these heavens, the first covered the space between Mount Mera and Yugandharo, the mountain of the North polar star, and the land of the South pole, the land of the yoked (*yugam*) pair, the united Northern and Southern races born of the mother-bird Gandhārī, the vulture, who formed the confederacy of the allied eight tribes who made the mother-mountain of the East, Mount Meru or Khar-sak-kurra, their mother-mountain, and worshipped the eight creating-gods, parents of the eight tribes. It is above the mother-mountain that the other heavens are placed, rising one above the other. The lowest of these, the Tavatimsa heaven, is that of the thirty-three Nāga or snake-gods of time, the five seasons of the year of Northern India, and the twenty-eight days of the lunar month which measured the year of the twin-races, who looked on the gods of the eleven months of generation of the sons of the sun-horse as their parent-gods. It was these twin-races, children of the Zend Yima, the Hindu Yama, who succeeded the Nāgas as the Kathi or joined (*kath*) Hittite trading races, who elaborated the ritual of the Soma sacrifice, and made the Udumbara tree (*Ficus glomerata*) their parent-tree. This is the sacred tree of Koṇā-gamana, the twenty-sixth Buddha, representing the paired thirteen months of the lunar year, and it is this number twenty-six which is the sacred number of the Kabirpuntis, the Indian unitarian sects, who reproduce the Kabiroi of Thrace, Asia Minor, and Palestine, to which all the Karmis and Koiris, the two great cultivating tribes in North-eastern India, belong. They claim Kabir, who, they say, was a weaver living in the sixteenth century A.D., as their founder; but he was merely a reviver of the old discipline, who took the name of the consecrated priests of their faith Kabir. The chief Guru or high priest of the sect lives in Belaspur district of the Central Provinces, and, as he himself told me, it is a law of their religion that no Guru or Kabir can hold the office for more than twenty-six years. If his rule ever reaches that period it never extends beyond it, for according to their belief he then dies at once; and to a race who care so little for life in itself as the Hindus, and who look upon death merely as passing from one form of existence to another, it is no hardship to fulfil the decree of heaven by a voluntary death, if the process should become necessary. The highest of the three heavens, the third or creating season of the year of three seasons, the parent-year of the Northern fathers of the Kushika, was the Tusita heaven of the angels, the bird-born sons of Kashyapa or Kassapo, the twenty-seventh Buddha, whose parent-tree was the Nigrodha or Banyan-tree (*Ficus Indica*).

It is the three mother-seasons of this race which are commemorated in the first three Theris, as is clearly shown in the history of the third Theri, Uppalavaṇṇā, or Padumavatī, which was her first name. She, the mother of the sons of the rivers, who had been first born in the age of the Padumuttara, was reborn again as one of the seven sisters, the mothers of lunar time in the palace of Kiki, the blue jay, the bird-king of

Kāsi, the Kushika capital (*Benares*) in the city and age of
Kassapo, the twenty-seventh Buddha. After this birth she
was born again as a village maiden, who gathered a lotus in
which she counted five hundred seeds. These and the lotus-
flower she placed in the begging-bowl of a Pacceka (solitary)
Buddha, with a prayer that she might have five hundred
children, the number of the seeds of the lotus. The Buddha
took the gift to the Himalayan mountain Gandhamādano,
the mountain of intoxicating (*mādano*) odours (*gandha*), the
sacred mountain of the Pāṇḍava worshippers of incense, and of
Drupadī, their common bride, the altar of incense. He there
placed the lotus on the stairs leading to the lake Nanda
(pleasure), issuing from a cave. This was the mother-cave of
the Himalayas whence the Gond sons of the tortoise are said
in the *Song of Lingal* to have been born. Here the maiden
Padumavati was again reborn as a baby, found in the bud of
a lotus, plucked by a hermit on the Lake Nanda. He
brought up the infant, at every one of whose steps a lotus
(*paduma*) sprang up; and when she was grown she was
married to the king, who demanded her from the hermit
after he had heard of her beauty from his forester. The
other wives of the king, who were jealous of her loveliness
and miraculous gifts, persuaded the midwife who attended
her in her confinement to take away her child and replace it
by a log of wood smeared with blood, which she was to show
as the child born of Padumavati. When her child Mahā-
padumo, the great lotus, was born, 499 brothers were born
to him from the moisture of the air, the parent of life in the
mythology of the worshippers of the rain-god. They were
all taken away by the 500 wives of the king, each of whom
placed the child she had taken in a sealed box, which she
threw into the river. Padumavati, being persuaded by the
midwife that she had given birth to a log of wood, induced
her to destroy it and fling the pieces into the oven. But she
did not succeed in concealing the story from the king, who
heard it from his wives and turned her out of the palace ; and

when she went away and took refuge with a poor woman, her power of making lotus-blossoms grow under her footsteps vanished. After she had gone, the king, at the request of his wives, celebrated a river festival to the river-parents of the race called the sons of the rivers, the Indian section of the Ibai-erri or Iberian people (*erri*) of the rivers (*ibai*), named the Irāvata. As the court were crossing the water during the festival, the boxes containing Padumavati's sons were taken in their nets by fishermen, and when they were opened, by the order of the king, a child nurtured by milk miraculously provided was found in each box, with a writing of the god Sakko saying that the child was Padumavati's, and ordering that the murderers of her children were to be put into the boxes and thrown into the river. When the king, after proclaiming a reward, found Padumavati, she refused to return unless a carpet woven in colours was laid from her protector's house to the royal palace, and a canopy studded with gold stars was placed over it to shield her from the sun. Her sons, when they reached the age of sixteen, placed themselves in lotus-flowers and were miraculously conveyed away as Buddhist ascetics to Mount Nanda; and Padumavati, who died when forsaken by her children, was reborn as the daughter of a village labourer, who married a cultivator. She then feasted eight Pacceka Buddhas, who were miraculously changed into five hundred, for whom the food provided for eight sufficed; and after she had fed them she brought them eight handfuls of blue lotuses, and prayed that in her next birth she might be born with a skin of that colour. She was accordingly reborn as the daughter of the treasurer of the king, called Vi-sakha, or the two (*vi*) branches (*sakha*), the name of the Hindu month March-April.[1] This was the month especially sacred to the Buddha as the sun-god, for it was in it that the great ploughing festival of the Nagur or plough, the Akhtuj festival of the Gonds, was

---

[1] Visakha also appears as the husband of Dhammadinnā, and the father of Bhaddā Kundalakesa in their last transformation.

celebrated. This was the festival instituted by the Kushika growers of upland crops, the race who placed the king's province in the centre of the realm, just as the Vi-sakha was the central or sixth month in the Pleiades year of Western and Southern India, beginning in November and divided into two periods of six months each, of which the second began in April. It was at this festival that the infant Buddha was placed under the Jambu-tree (*Eugenia jambolana*), the central and sacred tree of Jambu-dwipa, the central kingdom of the seven Kushika kingdoms of India; and this tree, which is also called in Pali, Koli, is shown by this latter name to be the parent-tree of his mother's village of Koliya. Then the miracle took place which proclaimed the infant Buddha to be the young sun-god, for the shadow of the Jambu-tree under which he lay remained fixed, while those of the surrounding trees revolved.[1] It was to contemplate and devote herself to this sun-god that Uppalavaṇṇā, the blue lotus-flower of the dark blue sky of night, the daughter of Vi-sakha, the sacred month of the ploughing race, the sons of the plough and sun-god Rāma, refused marriage, and spent her time in gazing on the lamp of the hall of assembly of the Buddhist brotherhood, covered by the sky canopy which it was her duty, as the goddess of the night, to sweep and cleanse. She thus became the devotee of the miraculously born fire-god, who was on earth the lamp, and in heaven the sun of life.[2] It is this deification of the god of the dark blue sky which is reproduced in the Mahābhārata, where the king of the South, king of Mahish-mati, the great mother, is called Nila or the blue one, and he was conquered by the Pāṇḍava twin Sahadeva, whose name, meaning the driving-god, shows him to be the fire-god; and it must also be noticed that Uppalavaṇṇā's transformations as the lotus-mother began under the rule of Kiki, the blue jay.

[1] The 'Nidāna Kathā,' Fausböll, *Jātaka*, p. 57; Rhys Davids, *Buddhist Birth Stories*, p. 85.

[2] See the text and translation of the 'Story of Uppalavaṇṇā,' *Journal of the Royal Asiatic Society*, July 1893, pp. 532-552.

We see in her story an epitome of the history of the
Kushika race, the founders of ancient ritual, who were born
from the mountain-lake Kāsh-ava, the modern Zarah, into
which the Helmend, the Zend Haēturmaṇt, flows. This
was the lake Nanda of this story, the Rādhā-kund or tank
sacred to the Rā-dhā, the mother of the sun Rā, and wife of
Krishṇa or Rāma, the sun-antelope; and this mother-tank of
the migratory sons of the lotus reappears in the Egyptian
sacred tank of the god and goddess Min or Men, the Manu
of the Hindus, and mother of the Minæans, who appear as
the great engineering race who distributed the waters of
Lake Copais in Greece, ruled Phrygia and Crete as the
Minyan subjects of King Minos, the measurer, and made the
great Minyan reservoir of Maarit in Southern Arabia. She
was the mother-star of the corn-growing race, born from the
virgin ear of corn, the star of Virgo, called Spica Virginis,
worshipped by the followers of Amen-Rā, who opposed the
worshippers of Set,[1] the hippopotamus-god of the totem-
worshippers, and she gave her name to Mena, the first king
of Egypt who built the great barrage of the Nile, and the
name also appears in that of the Meos, Minas, or Menas of
Rajputana, who are hereditary policemen, representing the
original cultivators of that arid country. They show their
origin from the people who divided their lunar year into
fifty-two weeks by dividing themselves into fifty-two clans,[2]
while the Jāts, who represent the more aristocratic cultivators, still trace their descent to the River Helmend and
Lake Zarah. It was these sons of the lotus of the mountain-
lake who were the sons of Gandhārī, the Vulture rain-bird
who wets (*dhara*) the Gan or garden of god, the sons of the
rivers descending from the Himalayas. It was they who
changed their Northern bird-mother, the stork, to the rain-
bird who brings the rains, and who placed her in the

---

[1] Norman Lockyer, *Dawn of Astronomy*, chaps. xxix. and xxxi. pp. 297,
318, 319.

[2] Hunter, *Gazetteer of India*, s.v. 'Rajputana,' vol. ix. p. 411-414.

heavens as the mother-star, the polar star Vega. It was
under her protection, as the chief star of the star canopy,
that Padumavati returned to the palace after her sons had
been consecrated by baptism as the sons of the rivers, and
it was when they left the earth to become, in heaven, the
parent-stars, the lotus-flowers of the dark blue sky, that she
was reborn as the blue mother of the heaven of night, the
female form of the god Varuna, the god of rain (*var*) and of
the dark sky of night.

Again, in the story of the birth of her children we learn
the history of the father-god of the totemistic races who
sacrificed their totem animals by tying them by the neck to
the sacrificial stake, stained and vitalised as the father of the
race to whom the blood was the life by the blood which
flowed on to it from the jugular vein of the victims, which
was pierced by the sacrificing priest. It was this Ashēra, or
pole consecrated by the earthly blood of the sacrifices, and
by the rain, the blood of the gods, which was changed into
the god of the sacrificial flame of the star-worshippers, who
offered their sons to the fire-god Moloch, when the blood-
stained log born of Padumavati was burnt in the oven, that
is to say, when it was placed on the altar as the central fire,
the Agni Jatuvedas of the Rigveda, in place of the sacrificial
stake : and these people who placed the fire-god in the centre
of the altar were the same race who adored the sun as the
central god of day, just as the firm and constant star, the
polar or central star, was the mother-goddess of the night.
They measured time by the phases of the moon, and also by
the revolutions of the stars round the pole, and it was
this system of time-measurements, by the turning of the
heavenly oil-press, which ultimately led them to frame the
wheel-year of twelve months, the six-spoked wheel. It was
this wheel, the broad earth, which is said to have revolved as
a potter's wheel, when on the full-moon day of Asālhi (*Assar*),
the month of the summer solstice, the Buddha, who had been
conceived at this time, left on the beginning of his twenty-

ninth year the guardianship of Mahāpajā-pati, the moon-goddess, and betook himself to the baptismal bath of the ascetic life, to emerge as the sun-god of the revolving-wheel, the year of 360 days.

But before showing how this wheel-year is proved to be the year symbolised in the Buddhist wheel of the law, I must first revert to the thirteen Theris, and adduce further proofs showing that they represented the lunar year of thirteen months. In the fourth Theri, Patacārā, she who goes (*cārā*) under a roof or canopy (*pata*), we find a reproduction of the star-goddess Uppalavaṇṇā, who showed, by walking under or supporting the canopy as the polar star, that she was the queen-star of the star-worshippers, whose father was the blind revolving-pole, the king Dhritarāshtra, who holds (*dhrita*) the kingdom (*arāshtra*) of Hindu mythology, and it was she who, in the story of her life, went forth alone after she had lost her husband and two children, the three seasons of the year preceding that of four consecrated to the sacred number four of the fire-god. The fifth Theri, Dhammadinnā, she who is given to the law, was the wife of Vi-sakha, the middle month of the Pleiades year, the mother-goddess of the year of the five seasons, the year of the worshippers of the Indian rain-god, who always brought up the North Indian rains at the summer solstice. The sixth Theri, Nandā, mean-ing pleasure, the daughter of Mahāpajā-pati, reproduced the Lake Nanda, which was the birthplace of the lotus-mother Padumavati. Soṇā, the dog or red one, the seventh Theri, or seventh day of the week, was the mother of many chil-dren; while Sakulā, the eighth, the goddess of the higher vision, was the fire-seed of the sacred eight, the mother-goddess of the Sakya worshippers of the rain-god Sukra, Sakra, or Sakko. The ninth, Bhuddā Kuṇḍalakesā, the saint with the curly locks, was also called Su-bhaddā or Su-bhadrā. She, who was the daughter of the treasurer Vi-sakha, married Sattuko, meaning the enemy, the robber son of the Purohit, or family priests of the king of Rājagaha. He was the offspring of the race who had replaced the year of the mother-Pleiades by that measured by the revolutions of the father-pole of the Northern races. This father-pole, the robber of the waters of the ocean, was the conduit-pipe passing through the mother-mountain and bringing to its summit the regenerating rains. He took his bride to the mother-mountain to rob her of her jewels, the parent lotus-born stars who were sons of Uppalavaṇṇā, and this mountain was that whence thieves were cast down ; but she pushed him over the precipice, killed the doctrine of the father-pole, and became, in its place, the ruler of earth and sky, as the mother of the mountain-born Soma-plant that loves the rains, and whose sons measured their year by the thirteen lunar months. From the mountain-mother of Soma she became the mother-mountain goddess adored by the Jains, the sect developed from the Soma doctrine of the moral efficacy of the two Soma-cups, penance and consecration, and they ordained her with their highest ordination, the ceremony of plucking out the hair with the thorns of the Palmyra-tree, or date-palm (*tāla*),[1] that is to say, they were the race who made the cutting or sacrifice of the hair a preliminary ceremony, which, together with that of the regenerating bath, must be observed by all Soma-worshippers, and who substituted the date-palm, the tree of life of the Western Indian trading races and Babylonians, for the original parent-tree of the Soma-worshippers, the Udumbara-tree (*Ficus glomerata*). The tenth Theri, Bhuddā Kāpilāni, was the tenth lunar month of gestation, the mother of the yellow (*kapila*) race, the counterpart of Vinatā, meaning she who is bowed down, the pregnant mother, the tenth wife of Kashyapa, who was succeeded by the eleventh month, Kapila her son, and who was also, in Hindu legend, the mother of the undeveloped egg-born son Aruna or Araṇi, the fire-drill, the lame fire-god, and of Gadura, the bull of light, the flying

---

[1] See the ‘Story of Sa-bhaddā,’ *Journal of the Royal Asiatic Society*, Oct. 1893, pp. 777-785.

bull and sun-bird. The eleventh, Bhuddā Kaccāna, the golden saint, was the wife of the Buddha, and mother of his son Rāhulo, the little Rā-hu, or sun-god, the Kapila, or yellow one, of the Hindu lunar months. She was also called Yaso-dharā, or the renowned (*yaso*) earth (*dharā*). The twelfth Theri, Kisā Gotamī, the emaciated or ascetic (*kiso*) Gotamī, is the reproduction, under the guise of the waning moon, of the moon-mother Gotamī, who was to be the parent of future life. It was the first Gotamī who was wedded to the sun-god in the first of the eleven preceding months of genera-tion, in the last ten of which months Rāhulo, the young sun-god, was begotten, conceived, and born. When she had attained her beatitude in the birth of the young sun-god, born at the full moon, she passed away as the Northern Māyā, and became, as the waning or emaciated Gotamī, the virgin nurse of the new-born child. It was Kisā Gotamī who, in the life of the Buddha, greeted him as he passed her house with the verses telling how, on the birth of his son, his mother, father, and wife had become Nibbuto, that is to say, had passed into the stage of non-(*nib*) existence (*ba*), in which perfect peace and rest were to be found, and who had changed from being celestial luminaries who had regulated the course of events, into those who had substituted passive contemplation for active exertion. It was she who heralded the extinction of the Buddha as the young sun, for it was on seeing her that he formed his determination to end his active existence as the sun of earth in the baptismal bath of ascetic regeneration, whence he was to emerge as the spiritual sun of righteousness.[1] The thirteenth Theri was Sigāla-mātā, the mother of the jackal (*sigālo*), that is to say, the goddess who, as the moon-mother wedded to the young sun-god, was to become the mother of the next year's sun-god. The explanation of the sun as the jackal is to be found in Egyptian mythology, for it is in this that we find Horus, as

---

[1] The ' Nidāna Katha,' Fausboll, *Jātaka*, p. 60. Rhys Davids, *Buddhist Birth Stories*, pp. 79, 80.

Anubis the jackal, replacing the mother-bird as the ruler of
the pole, and it was in this capacity that the jackal Anubis
was placed in the sky as the constellation of the Little Bear.
Horus is represented as the god with two heads, of the jackal
and the hawk, and the myth of his combat with the dragon
of darkness, Set, represents him as destroying at one time
the Hippopotamus, or the constellation Draco, and at
another time that of the Thigh of Set or the Great Bear,
and these two myths clearly refer, as Professor Lockyer
has proved, to the astronomical change of the polar star
from one of the stars of Draco to one in the Great Bear.[1]
I have already shown that the year represented in the
Buddhist chronology by the incidents in the life of the
Buddha before he became a religious teacher, is one in which
the sun is led forth on his journey by the horse, or the constellation Pegasus, which was in Egyptian astronomy the
constellation of the Servant, in which the four principal stars
were the four sons of Horus, who had previously been the
stars marking the four quarters of the heavens.[2] The im-
portation of the jackal into the Buddhist year of thirteen
Theris seems to show further that the thirteen Theris were
adopted as the moon-mothers of the Jain congregation of the
saints during the astronomical period when the polar star
was one of the stars in the jackal, or Ursa Minor. This,
from the diagram in Professor Lockyer's *Dawn of Astronomy*,
would seem to be somewhere about 2500 B.C.,[3] a time long
after the original Buddhist year of thirteen months was calculated, as this denoted a time when the sun was in Aries in
February, and therefore one before 4700 B.C., when the sun
was in Aries at the vernal equinox.

It appears further from the stories of the lives of the
Theris that the year which they symbolised was one which,
like the ten Buddist Perfections or months of gestation,

[1] Norman Lockyer, *Dawn of Astronomy*, chap. xiv. pp. 144-153.
[2] *The Ruling Races of Prehistoric Times*, Essay iv. pp. 395-398.
[3] Norman Lockyer, *Dawn of Astronomy*, p. 127.

represented not only the annual succession of months, but also the growth of the world-soul in virtue and knowledge, and this growth is shown in the special virtues ascribed to each of the holy mothers. Thus, Experience is the chief characteristic of the primæval mother, the creatrix Pajāpati, the Buddhist representative of the bisexual parent-god, the antelope-star Orion, called Prajā-pati, and his daughter Rohinī, the doe-antelope or red cow, the original parents of the household-fire, Vāstosh-pati, which became Agni-jata-vedas, the perpetual-fire on the altar. Wisdom is the virtue of Khemā, the safe one, who is the second mother, Ditī, of the thirteen wives of Kashyapa, the mother Māghā ; while Uppalavaṇṇā or Padumavati, the mother of the lotus, and the mothers succeeding her up to Bhuddā-kaccānā, the eleventh, represent the gradual expansion of spiritual know-ledge, gained by their several virtues, miraculous power, Iddhi, knowledge of the Vinaya or law of righteousness, preaching, meditation, effort, higher celestial vision, swiftness of appre-hension, knowledge of former births, and the mastery of all spiritual gifts, which last was finally attained by the mother of Rāhulo, the young sun-god. These were followed by the characteristics of the new age, in which death to the world was the highest form of spiritual life ; this was symbolised in the asceticism ascribed to Kisā Gotamī, called the wearer of rough garments, who was ordained to be the mother-goddess of the self-torturing Jains, and this produced Faith, which was the distinguishing virtue of the mother of the jackal.

It was under the rule of the jackal, the dog that woke the year in the Rigveda, that the year of the revolving potter's wheel, in which the earth circled the pole in the twelve months of the year, was calculated. That this wheel-year was that which was assumed as the symbol of the Buddhist wheel of the law is proved by the picture of the symbolic wheel brought by Dr. Waddell from Thibet,[1] and the interpreta-

[1] Waddell, ' The Buddhist Wheel of Life explained through the Thibetan,' *Journal of the Royal Asiatic Society*, April 1894, pp. 367 ff.

tion of the designs drawn on it told him by the Llamas of
that country. This wheel, called the whirling wheel of life,
has, like the year-wheel of the cosmological hymn of the
Rigveda, six spokes, dividing it into six parts, and thus it
differs from the wheel with five divisions, representing the
five seasons of the year of the moon-worshippers, which the
Buddha, according to the Divyāvadana, ordered Ānanda to
make for the brotherhood.[1] In the centre nave of the revolving wheel of six spokes, the three daughters of Desire,
Rāga (Lust). Dvesa (Anger), Moha (Ignorance), are seated,
while on the rim are depicted the twelve Nidānas or causes of
existence, two presiding over each division of the wheel, just
as in the Hindu year of six *ritu* or seasons each season
covers two months.   These twelve pictorial symbols denote,
like the thirteen Theris, both months and the stages in the
growth of spiritual life.

The first stage of Ignorance, or Unconscious Will to live,
is denoted by a blind she-camel, led by her driver Karma,
the accumulated impulses inherited from former states of
existence. The second picture, of a potter modelling clay on
his wheel, is identical with the Egyptian image of the
Creator Chnum, the architect, and it depicts the creative
processes of Khemā or Wisdom, the second of the lunar
Theris, as well as the beginning of the six Sankharas or
Conformations of material life of Buddhist theology, emanating from the union between Will and Purpose.   The first
stage of new-born life is the subject of the third picture,
representing the first stirrings of conscious experience in the
undeveloped man, depicted as a monkey.   The birth of self-
consciousness, the second stage of growth, is described in the
fourth picture of a physician feeling a patient's pulse. The
third stage, the birth of conceptional knowledge, is shown in
the fifth picture of the mask of a human face, denoting the
perception of differences. In the sixth picture, which, under
the symbol of kissing, or of a man grasping a plough, marks

---

[1] Maine, *Early Law and Custom*, note A, p. 50.

the fourth stage in the growth of knowledge as attained by
the employment of the perception of differences in investi-
gating the external world. The consequences of this inquiry
are shown in the seventh picture of the fifth stage of growth
to result in the production of feeling and the growth of
thought from physical and mental sensation, depicted in the
eye pierced by an arrow. The union of perception and conception leads us to the sixth stage of growth in the desire
they excite, and this is symbolised in the eighth picture of a
man drinking wine. The highest stage of the Sankharas or
Conformations is reached in the production of desire, and the
further progress of the education of the human mind is
shown in the ninth picture, representing the indulgence of
desires under the guise of a man plucking fruit and storing it
in baskets. The indulgence of desires leads to the lessons
gained by its consequences, and the learning of these lessons
is described in the tenth picture of the married woman,
marking the completion of the time of gestation and the
approaching birth of the heir, born of Unconscious Will,
Power, and Purpose, whose spiritual life has been growing
through the months of conception. This heir of the new
divine life appears in the eleventh picture of the woman and
child, the woman being the goddess, Kisā Gotamī, of the
lunar series, who has become the nursing instead of the conceiving mother; while in the corpse, carried out to be burnt
in the twelfth picture, we see the death of the old year and
the beginning of the rule of the new. It is this series of
prophetic stars or pictures revolving round the pole, supported by the daughters of Desire, which is symbolised in the
Thibetan prayer-wheels, twirled round at the end of a long
stick or pole.

From this review of the fundamental propositions of
Buddhist theology, it is clear that the Buddhist conception
of the history of the growth of spiritual life is the product of
a long series of ages, during which the human mind was
gradually feeling its way through the wastes of Ignorance

and the mazes of Error to the perception of the essential
conditions of a higher life than that ruled by the animal
appetites, and to the conception of the possibility of making
the reign of Righteousness, first conceived as an ideal vision,
a reality on the earth.   The belief in the new birth of the
human soul which underlay this vision of a new earth
arose, as I have shown, from the assertion by the twin-races,
who founded Semitic nationality, of duality as a primary
factor in the production of continuous existence.   The
metaphysics of these dualistic philosophers began with the
pairs of animal life, generated, according to the creed of the
fire-worshippers, from the union of life-giving heat, born
from the fire-drill twirling in the socket of the mother-
earth, and, according to the creed of the rain-worshippers,
from the infusion of the divine seed brought from heaven
and infused by the rain into the womb of the same universal
mother.   This dualism was reproduced in their conception
of the Pleiades year with its two seasons, dedicated to the
gods of the dead fathers, the authors of the annual resurrection of life, and to the seed and its products, by which they
reproduced and awakened the dead or slumbering life of the
past.   These two seasons were, by their union, the parents
of life, and this dual pair became the creating triad, when
their son or daughter, the seed of future life, was added to
the creating parent-gods, and became the third season of the
year.   When among the patriarchal races this seed was worshipped as the Son of God, who descended into the earth
and died, to become the parent of future life, his annual
death was commemorated by the sacrifice of the eldest son,
the victim whose life-blood was shed to revitalise the earth-
mother.   These conceptions of the three seasons, of sowing,
growth, and seed-bearing, and of the efficacy of sacrifice as
giving fresh strength and life to the earth, gave birth to the
further conception of the necessity of propitiating the powers
who ordained these seasons, and who were originally the ghosts
or spirits of the father-gods, the gods dwelling in the air

whence the divine fire and water descended as the lightning and the rain.   As these sacrifices could only do good to the land when they brought the changes of the seasons at the right time, it was necessary to calculate the several stages in the march of time so as to insure the absence of all mistakes in selecting the dates of sacrifice.   It was this beginning which gave birth to the various methods of calculating the duration of the year, and the rings of time, formed by months, which I have described in this Essay and in the preceding Essays of *The Ruling Races of Prehistoric Times*.   These, as I have shown, culminated in the Jain and Buddhist era in the reversal to the original dualistic conception in the calculation of the year of six seasons, the double three, and of twelve months, twice the double of three, each month containing thirty days, which reproduced the original sacred three and ten dedicated to the three seasons of the year and the ten months of gestation.   This was the wheel-year of the Rigveda and of Buddhist theology, and it was this year which was used in the symbol of the wheel of the law, producing in its eternal revolutions the continual rebirth of the spiritual graces, which were ultimately to regenerate the world and to produce on earth a holy people, born from the death of their predecessors, who died in sin and sinful life on earth that they might be reborn again in their regenerated children.

ESSAY VIII

HISTORY AS TOLD IN THE MYTHOLOGY OF THE NORTHERN RACES. THE FATHERS OF THE TEMPLE-BUILDERS, THE SONS OF THE SEED OF LIFE, THE EIGHT-RAYED STAR.

I HAVE already shown in scattered observations, which I now propose to supplement by a more detailed examination of its tenets, that the mythology of the Norsemen was interfused with and altered the creeds of the Southern races. But, in considering the questions raised by an inquiry into the inter-action of national beliefs and national character, it is necessary to remember that both the Northern and Southern races who have written their earliest histories in these stories of their gods and mythic ancestors, have formed the aggregate of their beliefs as well as their national polity from an accretion of materials contributed by the successive tribes who have amalgamated to form the confederacies of the Northern and Southern nations. When this fact is once grasped, it will be seen that there runs through each religious and national system one central idea, which interpenetrates and transforms foreign incorporated additions, and marks the line of thought followed in the development of the national story by the first founders of the national faith, from whose initiative the course of the dramatic action, detailing how the destiny of the race has been worked out, has gradually grown to be a whole formed of congruent parts. This guid-ing central idea is a distinguishing mark of the historical myths framed in the three centres whence civilisation was

diffused over the world before the days of the written history
of individuals, as it characterises the Kushika mythology
of India, that of the barley-growing races of Asia Minor,
and that of the sun-worshipping sons of the North. Thus
in the Hindu historical mythology the central idea is the
growth of the kingdom formed by the union of its component provinces, and governed by a king living in the
central province, the Chakravarti or wheel-king of the
Buddhist theology; and the stages of the formation of this
conception can be traced from the growth of the original
village and province of the cultivating forest tribes up to
the confederacy of united provinces which was ultimately
organised by the statesmen of the fire- and sun-worshipping
Māghadas, and by their coadjutors and successors the
Kushika, the sons of the bird (*khu*) and of the tortoise
(*kush*), who framed the constitution and national polity of
the races who were grouped round the mother-mountain,
the traditional birth-place of the race, whence Indra, the rain-
god, got the rain. It was from the root conceptions of these
sons of the mountain that the idea of the kingdom of righteousness grew as the outgrowth of the central tree or plant
of life, whose seed contained the Su, its divine essence or
soul; and it was this idea which, when infused with Northern
individualism, developed into that of the righteous God and
Father of the theology of the Zends and Semites. In
passing from the Hindu to the Assyrian mythology we
find another development from the original type of the
Dravidian village community of the sons of the tree, for this
phase of the growth of national life received its characteristic traits from the trading shepherd and artisan races who
united with the sons of the village grove to form the first
organised population of the Euphratean countries. Among
these practical races the exigencies of trade prevented the
national thought from finally finding an outcome for its
energies in the dreamy theology which tried to turn all the
most earnest Hindus into Jain ascetics, Brahmin metaphy-

sicians, and Buddhist monks and nuns; a tendency which,
however, appeared later on in Irān in the poems of the Sufi
thinkers, Hafiz and Omar Khayyám, and in the metaphysical speculations which produced the mysticism of the
Kabbala and ultimately formed the groundwork of gnosticism.
In earlier ages the practical trading races gradually deduced
from the original traditions of the amalgamated tribes the
system of belief founded on their birth as the sons of the
fish, the god Ia of the house (*I*) of the waters (*a*), the
spirit of life which had descended from the rivers of the
mother-mountain to the ocean surrounding the land of her
sons. It was they who learnt from the immortal fish who
came to life in the dried-up inland reservoirs, as soon as the
first showers of rain filled them with water, the rules which
led to the attainment of knowledge, civilisation, and wealth,
and it was this practical tendency of the national aspirations
which made astrology first and astronomy afterwards the
sciences cultivated by the Akkadian, Hittite, and Kusho
Semite rulers of the land. It was under these influences that
the ruling Semites became the followers of Nabu the prophet, the planet Mercury, who was originally made the
leader of stars as the morning star, the herald of the dawn,
and who became, in mythological history, the Moses or
national leader of the Semite faith, after he had, in the first
beginnings of his mythic story, been the father of the kings
of the Kushite race, who was found by Uzava, the goat-god
of time, abandoned in the reeds of the Lake Kashava, the
reservoir of life-giving water filled by the Haētumaṇṭ (*Helmend*), descending from the mother-mountain of the East,
who had sent her divine son down the waters of the parent-
stream to be the god-given ruler of the sons of the mountain-
bird, or the bringer of the rain. The religion of the Hindu
Kushika, the sons of the Urœus or rain-snake, the emblem of
royal dignity in both India and Egypt, and that of the still
earlier Hindu and Egyptian worshippers of the ape and
crocodile gods, the symbols of the winds and the rain-clouds,

became, in the hands of the Egyptian philosophers and astronomers, an inquiry into the destinies of the individual soul, in its progress to perfection after it had died on earth. It is in this faith, as set forth in the Book of the Dead and its pictorial illustrations depicted on the Papyrus of Ani, that we find the individualistic impulse, which did not in the Jain and Buddhistic creeds penetrate beyond the shadowy cloudland of Nirvana and metempsychosis, developed into the realistic pictures of the regenerated life of the elected soul, who, after the appraisement of his deeds made by weighment in the scales of justice had been recorded by the moon-god Thoth, and the final verdict of approval had been pronounced by the thirty-three judges in the judgment-hall of the goddess Mā'at, was adjudged fit to participate in ' the ampler æther and diviner air ' of the Elysian fields.   The lot of the dwellers in this realm of bliss, where wars and fighting cease and abundant crops reward the toil of the cultivator, is shown in the pictures of the agricultural labours of Ani, and these tell us that, in the conceptions of the framers of the myth, the chosen soul has been looked on as one admitted to be a settler in the heavenly village regulated according to the communistic laws of the Indian matriarchal villages, so that, in this theology, the result of the individual efforts towards the attainment of goodness is, that he is elected not to be an idler but a worker in the assembly of faithful souls who people the kingdom of heaven. Thus, in all these systems of Southern historical mythology we find that the leading idea is that which lies at the root of Pantheism, and which looks on the community and all natural objects as coalescences of individuals and atoms, each impregnated with the divine impulse forcing them to perform their allotted functions, in carrying out the completion of the whole design ordained by the hidden and unrevealed lawgiver.   This design, as disclosed in social organisation, is, in the Pantheistic creed, the welfare of the community or assembly of the faithful, the Buddhist Sañga.   In this con-

ception of the social organisation the individual was an atom
who, after his reception as a component member, was quite
unthought of and disregarded as long as he performed the
duties allotted to him satisfactorily, and who only emerged
into prominence when selected as a member of the governing
body, or when, by his wrongdoing or negligence, he threw
the machine out of gear; and faults of this description, when
unatoned for by prompt repentance and reform, were soon
followed by the expulsion of the offending disorganiser. It is
as a consequence of this view of the overwhelming importance
of social prosperity, as learned from the political experiences
acquired in the communistic village, that the Egyptian philosophy, which assigns to the individual a renewal of active
life beyond the grave, only depicts him as one who has been
accepted as an inmate of the celestial village, where he has
received an allotment of land which he has to cultivate.

It is in Asia Minor and Greece that we find the individual,
the offspring of the seed sown in the mother-earth, of the
earlier village races, first deified as the father-god and hero
and the mother-goddess and protectress—the father of fire,
the fire-drill, Ixion or Axifon, the ruler and director, who
turns in the heavens the wheel round which the stars of the
mother-Artemis, the seven stars of the Great Bear, revolve.
This is the never-resting potter's wheel, from whence the
sons of Mount Pelion and Peleus, the sons of the potter's
clay (πηλός), were born. It is to them that the skill
of the heavenly potter gives the outlines of beautiful
form, which are reproduced in the masterpieces of Grecian
sculpture; while the colours, which vie with form in making
life on earth beautiful, and which inspired the art of the
Greek painters, were first deified in the flowery garland of
Koronis, the sister of Ixion, the successively recurring series
of blossoms which mark the passing phases of time as the
flower-clock of the year. It was the Greeks who made the
national mother-goddess, not the mother-grove or the tree,
or the mother-mountain of Dravidian and Finnic mythology,

but the individual mother Athene, the virgin goddess of
wisdom, who was first the flower, and Aphrodite, the beautiful mother of the adventurous race of the sons of the sea,
whence she was born. It was they who showed, in their
dramatic art, that society is not composed of a congeries of
undistinguished atoms working like ants to a common end,
but of a number of individual souls, each playing his re-
spective part in the great drama of humanity, but all moving under the impelling force of resistless necessity, against
whose decrees the noblest souls struggle in vain. For even
Hercules, the sun-god, who frees humanity from the tyranny
of lawless monsters, must end his days in the agonising
tortures caused by the blood-stained robe, the clouds of sunset, given him by Deianeira.

In all these mythologies the individual organiser and contriver, the father fire-drill, and the Egyptian god Chnum,
the architect, is looked on as the servant of the mysterious
powers who regulate, in the unseen world, the destinies of
the earth and its inhabitants. These appear as the Zeus
of Hephaistean legend, who made himself the thunder-god
of heaven and cast down Hephaistos, who was once the
thunderer, to earth, where he was to resume his primitive
position in mythologic history as the lame fire-god, the one-
legged fire-drill of the artisan races, before they were united
to the agricultural worshippers of the rain-god. The resistless fate which underlay the original myth appears in the
story telling how Nidung, the king of the Nids, or nether
powers of the earth, cut the hamstrings of Wieland, the
divine smith, the maker of the year rings, and forced him to
work for his mutilator in the cave in which he afterwards
slew the sons of Nidung, and begot, on his daughter Bathilda, the young god of the new age, Widonga, the god of
the perpetually reborn woods and meadows. It is to the
mythology of the North that we must look for the history of
the birth of the idea of the perpetual struggle between the
authors of law and the powers of nature, deemed to be law-

less, which developed into the conception of the personal god, the judge, father, and divine ruler of the sons of Danu the judge. He is the living and moving divine leader of the armies of the righteous strivers against wrong, a being, in the original conception of the Northern theologists, totally different from the god of the Pantheistic creed, who is the remote and dimly realisable promulgator of a law laid down in the first throes of creative organisation. It is this living god who is, in Greek mythology, Zeus, who casts the lawless Titans out of heaven, and who, in the dualistic Zend theology, is Ahura or Asura Mazdeo, the god of righteousness. But it is in the religion of the Northern sons of Odin that we find the most distinct realisation of the conception of the ruling will, as told in the story of the victory of the man-god, the son of the seed of the tree of life and immortality, the victor trained in the observance of the laws which hedge in the course of the righteous ruler, who insists on working out his own purpose, and who shows his right to rule by his successful eradication of lawlessness and wrong.

This idea of the rightful rule of the divine will and purpose grew from the conceptions of the palæolithic hunters, the race who in the long freezing winters and short summers of the Glacial Period could only maintain life by killing wild animals for food, and who were also obliged to defend themselves from the attacks of beasts of prey, whose claws, strong limbs, and teeth provided them with natural weapons superior to those given to man. Thus human life was in those days a constant contest between the cunning, perseverance, bravery, and fortitude of the hunter, and the brute strength, swiftness, and wiles of the prey he sought to kill, or which tried to kill him. In this mutual struggle the victors learned that the possession of mental gifts was not confined to the human animal, and that those who sought to be the most successful hunters must be nearly allied to the wild animals, not only in strength but also in craft and tenacity ; and hence the right to rule and direct the annual

96 THE RULING RACES OF PREHISTORIC TIMES

campaigns against their animal foes was conceded to those who could cope with the difficulties of snaring, trapping, tracking, and slaying the animals wanted for food, and who could defy, conquer, and kill those who sought to hunt down and devour the human hunter. It was among these people that the belief in father-totems and the transmission of ancestral qualities by descent arose. This faith caused the tribes who excelled their neighbours in courage, strength, and cunning to take the names of the sons of the bear, and those in whom craft, guile, and endurance were the most conspicuous qualities to call themselves the sons of the wolf. The incorporation into their bodies of the qualities to which they aspired was, in their belief, aided by draughts of the blood of the totem, which mingled itself with and gave fresh strength and vigour to their own; and hence in the Saga which tells how Hjalte (the sword-hilt), helped by Bjarke (the birch-tree), succeeded in marrying Rute, the daughter of Rolf, who was the son of Ursa, the bear-mother, and of Helgi, the holy one (German *heilige*), the son of Hālfdan, the half of the father Dan, the judge, we are told how Hjalte gained the strength which made him worthy of his royal destiny by drinking, by Bjarke's advice, the blood of a bear he had slain.[1] Again, in the Saga of Hadding, the hairy king (*hadd-r*) of the North, the bearded sun-god who went down to the under-world as the Southern sun of winter, after he had wedded Ragnhild, the daughter of the winter

---

[1] Elton and Powell, *Saxo Grammaticus*, book i. pp. 17, 18; book ii. pp. 62, 63, 68, 69; Preface, pp. cxvi. cxvii. Halfdan was, as Dr. Rydberg, *Teutonic Mythology*, pp. 98, 101, has shown, the equivalent of Gram, the son of Skiold, the father of the Scioldings, born of Skiold, Scyld or Secaf, the slayer of the bear, the miraculously born child who, in the poem of Beowulf, drifted to the land of the Danes and became the fish-sun-god Gram, who married Gröa, meaning growth, the spirit of life, and is identified by Mr. York Powell with Örwandil, or the year-star Orion. Hence his son Helgi, the holy one, who is in the *Niebelungen Lied* the son of Sigmund, the master-smith, the king of the Volsungs, the woodland sons of the tree, and of Borghild, the earth-goddess (*hilda*) of the mountain-castle (*burgh*), was the sun-god, the ruler of the year of Orion, the father of life.

twilight (*Ragna*), to get the angelica, or plant of life, he, when entrapped in his earlier career by Loki, the fire-god, and exposed to wild beasts, freed himself, by Odin's advice, by slaying the wolf who attacked him and drinking his blood.[1] It is under the pine-tree, born from the hair of the wolf, that the bear-father was nursed, in Finnic legend, by his maiden mother, the stars of the Bear.[2] This bear-father became in Finnic mythology the eternal forger, the master-smith Il-marinen, the second god of the triad of Vaïnämöinen, the god of moisture, Il-marinen, the hammerer, and their son Ukko, the thunder-god.[3] It was while sleeping after eating bear's flesh that Wieland, the master-smith, the German Il-marinen, was surprised and bound as the sun of winter by Nidung, the king of the Nids, or people of the nether-world, who maimed him and made him work for him in an underground cave.[4] It was the divine frenzy of their father-god which inspired the warriors called the Bersarker, or men of the bear (*ber*) shirt (*sark*), who led the hosts of the North to victory. It was these sons of the wolf-mother and the bear-father who became the twin races who called the gods of day and night, the children of the wolf-mother Leto, Apollo, the wolf, and Artemis, the bear, and made them their national gods. When these twin races went southward and joined themselves with the agricultural village races, the sons of the tree-grove, they incorporated their custom of eating their parent-totem and drinking its blood into the sacrificial ritual ; and when they had made the earth their mother, they transferred to her the sacrificial draughts of totem-blood by pouring it out on the ground at the foot of the altar, and thus making blood-brotherhood between the father-gods of the North and the mother-goddess of the South.

---

[1] Elton and Powell, *Saxo Grammaticus*, pp. lxviii. lxix. book i. 29, 37.

[2] Abercromby, *Magic Songs of the Finns*, iii. (*i*) *Folk Lore*, vol. i. No. 1, March 1890, pp. 27, 28.

[3] Lenormant, *Chaldæan Magic*, chap. xvi. pp. 246, 247 ; De Gubernatis, *Die Thiere*, German translation by Hartmann, pp. 113, 114.

[4] *The Edda*, by Hans von Volzogen ; ' The Wieland Saga,' pp. 210 ff.

But in this mythic genealogy it is the wolf-god, the god of fire, the Loki of the Edda, afterwards cast out from the number of the supreme gods by Odin, the god of wisdom of the star-worshippers, who is father of the Northern conquering tribes, the hunting race, who made the dog their divine guide.   It is in the wolf's valley, by the wolf's lake, that, in the legend bearing his name, Wieland, or Völundr, the crafty master-smith, son of the Finn king, takes up his abode with his two brethren, Egil, the archer, and Schlagfeder, the cutter (*schlag*) of feathers (*feder*), he who puts the directing feathers on the arrows shot from Egil's bow.   There they are visited by the Valkyr maidens, the wind-goddesses, the swan-white snow-maiden Hludh-gund, she who strives (*gunr*) with the earth (*hlödh*), and Her-vor, the foreseeing, called also Alvitr, the all-knowing daughters of king Hludh-vig, he who conquers (*vig*) the earth.   The third was Ael-run or Ol-run, the old knowledge (*run*), daughter of King Kjär of the Welland or lowlands.   Egil, the god of the bow, that is, the rainbow, married Ol-run, and thus represents the race who brought the lessons of experience learned by the farmers of the South to the mountain-glens of the North; while Schlag-feder, he who cuts the feathers of the snow-bird, wedded Hludh-gund, the winter-maiden, the mother-bird, who is to be the mother of the future year.   Wieland, or Völundr, he of the master-will, married Hervor, the all-wise. They lived eight winters together, and in the ninth, the number sacred to the gods of heaven, his two brothers and the three wives fled away to the South, and left Wieland alone to become the god of the races who succeeded the bird race, and who believed in one god and father.   He in his solitude plied his craft of the master-smith, and forged the seven hundred year-rings which he hung up on the walls of his house, just as the timber-tree stores up within its bark its year-rings of annual growth.   This story is precisely similar to that of Puru-ravas, the Eastern (*puru*) roarer (*ravas*), the thunder-god, who was forsaken by Urvasi after she had seen

him naked, and had learned that her husband, the master-
workman, was the god of heaven revealed by the lightning
flash.   Like Pururavas, Wieland remained and worked alone
as the unseen and unknown father-god hidden in the inner
chambers of the tree and in the clouds of heaven.   And he
is shown to be the god who directs the course of the year
and the path of the sun by the English form of the Wieland
legend, in which, as Weyland the smith, he is concealed in
his underground cave to make the shoes of the white horse,
the sun of the year.   When dwelling in his hidden solitude
as the giver of fire, the master of the metal-working elves,
the dwarfs who know the secrets of wealth, he, as the lightning-god, made the trees break out into flames as he passed
through the forest to hunt the bear ; and when he returned
and ate the prey he had slain, he was attacked by Nidung,
the king of the Nids of the nether-world, who wore a
sword shaped like the snake-mother of the Southern races,
the sword of the crescent-moon, and bound in the bonds
of the wintry cold.   He was thus not allowed to wander
through the world as the lawless and unrestrained storm-god,
but was hamstrung by Nidung and made into the lame god,
the unseen director of events, forced to do the bidding of his
captor, the god of the law of natural order.   But when
Nidung's two sons came to his prison-cave to learn his craft
he slew them, and sent their skulls to Nidung to represent
the two solstices of the year of four seasons, the turning-
points of the sun-god in his yearly voyage, while the equinoxes were represented by the eyes of his two sons, which he
sent to Nidung's wife, their mother, the spying-goddess, the
moon-mother.   From their teeth he made the year-necklace,
which he sent to Both-vildr, Nidung's daughter, meaning,
she who delights in strife (*both*), the German earth-goddess,
Bathilda, made fertile by the storms.   Both-vildr came to him
with the year-ring which her father had carried off from
Völundr's forest home, and which she had broken, and asked
him to mend it.   He intoxicated her with ale, and begot on

her their son Widonga, meaning woods and meadows, who
became the god of the twin races who measured time by the
mended ring of the months of gestation, the Roman annus
or year-ring. When he had begotten his son, Völundr made
himself wings and became the bird-king who brings rain to
the mother-mountain, whence it is to descend to the plains
as the fertilising water of the rivers. It was on these wings
that he flew to Nidung, and made him swear to protect
his son. This phase of religious evolution, in which the
thunder-god became the parent-bird,[1] is represented in the
Pururavas story by the meeting of Pururavas with Urvaśī as
the moon-goose, swimming on the Plaksha lake, a meeting
which results in the birth of Āyu, the father-god of recorded
time. This birth-story of Widonga is also the Norse counterpart of the Greek story of the birth of Erichthonius, the son
of Hephaistos, who was, like Völundr, the lame fire-god, and
of Atthis, the daughter of Kronaos, or, in other words, of
Attika, the land of Athene, the flower-goddess, called both
Atthis and Kronaos, or the rocky land. Erichthonius ($\dot{\epsilon}\rho\iota\chi\theta\acute{\omega}\nu$)
means the very fertile earth ($\chi\theta\acute{\omega}\nu$), just as $\dot{\epsilon}\rho\iota\beta\hat{\omega}\lambda\alpha\xi$ means
the land with large clods ($\beta\hat{\omega}\lambda\alpha\xi$), and both stories tell how
the very fertile and deep-soiled earth, which had taken its
substance from the hills above it, and is watered by the streams
descending from their sides, is the son of the fire-god and the
pregnant earth-mother, who was once the burning mountain.
This interpretation is confirmed by the further legend of
Erichthonius, which depicts him as a snake-god hidden by
Athene in a chest buried underground, whose existence she
revealed to Agraulos, the dweller in the fields ($\dot{\alpha}\gamma\rho\acute{o}\varsigma$),
Pandrosus, the dew-fall ($\delta\rho\acute{o}\sigma o\varsigma$), and Herse, the spring dew
($\ddot{\epsilon}\alpha\rho$). In this story the meaning of the myth is unmistakable, for it tells us how the fertility of the earth is revealed
by the goddess of wisdom to the watchful dweller in the fields,
who marks how the dews of the warm spring bring to life the

---

[1] Just as Ukko, the Finn thunder-god, was the parent of Linda, the
Esthonian mother-bird : Essay ix.

seeds hidden in the earth, which are to become the parents
of future life as they grow up, blossom, and bear ripe fruit.
The whole story of Wieland tells us of the union of the
Northern hunters and smiths with the Southern races, the
farmers of the South, whose supreme god was the rain-god
of the heavenly bow, and his minister the arrow-maker, the
priest of the hunting-tribes of the South.[1]   It is they who
wed with the wind-maidens, the rain-clouds of the South,
and it is they who looked to the thunder-god as their father-
god, and worshipped him as the heavenly reproduction of
the Northern fire-god, who first evoked fire from the mountain flints.   It is he who became in the woodlands of the
South the fire-drill and the god of life-giving heat and rain-
bringing lightning, who is the father of the children of the
year of three seasons, the three united pairs of the Wieland
story, who, as the fire-drill and the socket, create the heat
which fosters life.   It is they who form the six creating-gods
or double pairs, worshipped by the confederacy of tribes
gathered round the mother-mountain of the East.   Of these
six, Völundr alone has offspring in his son Widonga, the
growing-seed born of the earth-mother, who, again, belongs
to a confederacy of five gods—Nidung and his wife, two
sons, and Bothvildr, the four quarters of the heavens, and the
pregnant mother-mountain in the centre, depicted in the

star-rain-cross of Horus $\times$, the sacred sign of the

earliest star-gazers, who looked on the stars as the best
guides to mark the passage of time.   It is this god of the
pole fixed on the top of the mother-mountain who takes, in
the legend of Wieland, the place occupied in Hindu genealogical mythology by Kashyapa, said to be the father of men,
for his name is derived from the Finnic Khu, meaning bird,
while, in the mythology of the Edda, Wieland, who is the soul
or essence of life in the tree-trunk and bird, becomes Bör, the

[1] *Ruling Races of Prehistoric Times*, Essay ii. p. 54.

first man, whose name means the tree (*bör*), and he is said to
have been born from the stones covered with salt and hoar-
frost which had been licked by the mother-cow Audhumla,
the void (*aud*) darkness (*hum*). The sons of the tree (*bör*),
and the bear (*bessi* or *bersi*) mother Besla were Odin, the
god of wisdom (*Odh*), the Northern counterpart of the Greek
Athene, the wise goddess, Vili (will), and Ve (justice).[1] Of
these three brethren, Odin was the father-god of the sons of
the horse, whose eleven months of gestation were probably
the first origin of the worship of the eleven gods of generation, the number of the united families of Völundr and his
brethren, and of Nidung. These sons of the twin races
invariably worshipped these eleven gods, who appear, as I
have shown, in the mythologies of the Hindus, Akkadians,
Semites, and Egyptians.[2] In the Edda these eleven gods
are the eleven horses of the gods led by Sleipnir, Odin's
horse with eight legs,[3] the eight-rayed star, the sign of seed
and god in Akkadian and ancient Chinese mythology, the
eight Agnis of the Hindus, and the eight creating-gods of
the Egyptians. It is this mythology of the eleven gods of
generation which marks the union of the confederacy of the
yellow gardening and dwarf mining races of the North-east,
the white Northern sons of the fire-wolf and the dark farming races of the South ; and it was they who traced their
descent from the eleven horses who drew the car of time, and
were the nursing-mothers of the son of the divine seed, the
eight-rayed sun-star, who was born to rule the confederated
corn-growers, gardeners, shepherds, artisans, traders, and
hunters, the dwellers in the land girding the mother-mountain
of the East, the Khar-sak-khurra of the Akkadians, the
Mount Ushi-dhau of the Zends, Meru of the Hindus, and
the Asgard of the Edda, the garden of the sons of the
mountain (*as* or *isa*) overarched by the Himinbjorg, the

---

[1] Prose Edda, chap. vi. ; Mallet, *Northern Antiquities*, p. 403.
[2] *Ruling Races of Prehistoric Times*, Essay iii. pp. 265-267.
[3] Prose Edda, chap. v. ; Mallet, *Northern Antiquities*, p. 411.

heavenly (*himin*) mountain (*bjorg*) at the summit of the rainbow bridge Bifrost, the home of Heimdall, the sun-god, the
son of nine virgins.[1] In the mythology of the Edda we find
the process of intermixture between Northern, Eastern, and
Southern national combinations still more clearly defined
than in the Wieland myth ; and here also, as in the Wieland
myth, the supremacy of the father-god of the North, the
divine creating will, is unhesitatingly asserted. The Edda
tells us of the three ages of the world's growth when it was
ruled—(1) By the giant races of Jötunheim, the name of the
devourers (*iotunn, coten*, from *eta, itan*, to eat) of time, the yet
un-understood and uncontrolled powers of nature sojourn-
ing in the land of the East. (2) By their successors, the
dwarfs of Muspellheim, the home of fire (*muspell*), the
Southern world. This was the age in which Wieland was
master of the elves, before he was captured by Nidung. (3)
The age of the warrior Æsir, ruled by Odin, the god of
wisdom, dwelling on his mountain-throne called Hlidskjalf,
or the sloping shelf (*skalf*).[2] He received his knowledge
and wisdom from the prophet-ravens Hugin (prophetic insight) and Manin (thought), who sit on his shoulders. These
are the birds who represent the last transformation of Wieland into the bird, and it is in these ages that we find a reproduction of the year of three seasons with its wintry cold,
the work of the frost-giants, the summer heat brought from the
south by the sun on its northward journey, and the genial
middle climate of harvest when the fruits of the victory of the
fire-god, the sun, over the frost-giants are gathered in ; and it
tells also of the earlier age of two seasons, when the year was
divided into the cold season of winter and a warm summer.

In the mythology of the gods of the Edda we find further
proof of the ethnology of the component sections of the
Northern confederated tribes, for we find that they are
divided into the gods of the Æsir or Āsa, the gods of the

[1] Prose Edda, chap. xxvii. ; Mallet, *Northern Antiquities,* p. 421.
[2] *Ibid.* chap. xvii. p. 414.

Vanir of Vanirheim and the fire-gods, the gods of the dwarf races, Loki, the fire-god, with his sons, Fenrir, the wolf, and the Midgard-serpent. It is this serpent who encircles the habitable world, in the midst of which stands Yggdrasil, the ash-tree, the tree of life of the Northern nations. These divisions tell us of the earliest ages of civilisation, when the primæval village was encircled by the ring of cultivated land, which became the ocean-serpent, the parent-god of life of the nations, whose home was bounded by the Indian Ocean, the Caspian and Black Seas, the Mediterranean, and the Red Sea. It tells also of the earlier days when the confederated culti- vating tribes of Asia Minor were congregated round the burning mountain of the East, Mount Ararat, the home of fire and of the rule of the gods of fire, who were, in the mythology of the cultivating races, governed by the thunder- god, who distributed the rains in the uncertain proportions common to the temperate zone, where the climate alternates between periods of drought and of seasonable and excessive rains. This was the reason which made the Northern races represent Loki as the god of trickery and caprice, and it was the period of his rule of the heavens which was called that of Mid-Odin or the Middle Odin. The first period was that of the worshippers of the mother-snake, when Odin had for- saken the world owing to the adulteries of his wife Frigga, the seed- (*frio*) mother, or, in other words, the age when the mother-earth was the mother-goddess of the sons of the village begotten in her consecrated grove, her sylvan temple, when their parents were not united in matrimony but only in the temporary unions formed at the village seasonal dances between the women of one village and the men of another. It was the midland country of the Southern section of the temperate zone and the tropical forests of India, where the village originated, that was the home of the matriarchal races, called in Norse mythology the Vanir. Their name shows them to be the worshippers of the goddess of love and desire, the Sanskrit Vena, the golden-winged

mother-bird of the children of Prishni, the mother-earth in
the Rigveda;[1] the Latin Venus, whom they called Vanadis,
the goddess (*dis*) Vana, and also Freya, the goddess of seed
(*frio*). It was in the land of the Vanir that the Priapic
cult arose, the worship of the signs of sex, the Linga and the
Yoni, the cult which was born out of that of the mother-
earth when the Southern sons of the tree met with the
Northern Finns, who looked on the wedded pair who lived
in their own home as the national unit, the offspring of the
primæval twins. It was they who brought into the Northern
mythology the gods Njord, the god of the North, the ruler
of the winds, the god of the pole-star, the Ashēra, or husband
of the land, who was worshipped as the gnomon or rain-
predicting pole and father of life by the people of Asia Minor.
He and his children, the god Frey and the goddess Freya,
the male and female seed (*frio*), were taken by the Āsir from
the Vanir in exchange for the horse Hœnir, the sun-horse of
light,[2] who brought the sun from the North to the South in
the autumn season of harvest, and whose career, as I shall
show presently, is described in the account of Sigurd and his
horse Grāni in the *Niebelungen Lied*. In the Edda, Hœnir
was the companion of Odin and Loki when they went on the
expedition which ended in Loki recovering the apples of
Iduna, the seeds of life, the apples of the Hesperidæ of
Greek mythology, from the giants, by borrowing the falcon
plumage of Freya, the hawk-mother-goddess of the mining
races, and by taking Iduna from Jötunheim in the form of a
prolific sparrow. It was then, at the time of the summer
solstice before the fruit harvest, that Thjassi, the frost-giant,
who appears in this myth as the vulture-bird, the devourer
of dead time, was burnt by Loki on the walls of Asgard;
and it was when seeking compensation for his death from
the gods that Skadi, Thjassi's daughter, the goddess of the
mountains, who became the ominous magpie (*skadi*), came

[1] Rigveda, x. 123, 1, 5.
[2] Prose Edda, chap. xxviii. ; Mallet, *Northern Antiquities*, p. 418.

to Asgard. When there, she was, by her own consent, wedded to Njord, whom she chose by seeing his feet;[1] for it was only the feet, which marked his track through the fields of time, which she was allowed to see by the conditions of the compact which permitted her to choose a husband among the gods.[2] It was by the same test Ragnhild, the daughter of the twilight (*ragna*), the autumn-sun-goddess, chose for her husband Hadding, the hairy (*hadd-r*) god, a counterpart of the bearded Njord, the Northern sun-god, having recognised him by the ring, the year-ring, which she had sown into his feet when tending his wounds. That this ring was the ring of the year beginning with the autumnal equinox, the year of the people of Syria, Asia Minor, Macedonia, Sparta, and the Peloponnesus, is proved by the further story telling how—after his wedding to Ragnhild—he, as the seed-god of the winter sun, went to the under-world to get the plant angelica.[3] It was from this union that Frey and Freya, the gods of seed, the twin-children of Njord and the mountain-goddess, Skadi or Ragnhild, were born; and the whole story tells of the formation round the mother-mountain of the confederacy of the farming and artisan races, worshippers of the phallus and the fire in Asia Minor, the land of the snowy peak, the mountain consecrated to the great snow-mother-goddess Niobe. Its top, whence the waters of life descended in the rivers flowing from Ararat, was inaccessible to men, and could only be reached by the mother-bird, the eater of dead time and dead flesh, the carrion-eating mountain-eagle; and it was she who became in lowland mythology the magpie or raven, the bird of speech, the prolific sparrow and the hawk-plumaged Freya, the mother-goddess of the mining races. It was in this age of the rule of the mother-bird that Frey, the son of Njord, gave his sword of light to Skirnir,

---

[1] That is, the stars.

[2] Prose Edda, part ii., *Bragi-rædur*; also part i. chap. xxiii.; Mallet, *Northern Antiquities*, pp. 459-461, 418-419.

[3] Elton and Powell, *Saxo Grammaticus*, book i. p. 37; Preface, pp. lxviii. lxix.

the bright (*skir*) one, the god of light, who ruled the year, and who was also called Swip-dag, the hurrying (*svipa*) day, or Orwandil, the constellation Orion,[1] the wandering sun-god. He received the sword of light for his success in wooing for Frey, Gerda, the greedy one, the mother-vulture bird, raven, or wolf (*geri*), the mountain-goddess Geri ; and it was on parting with his sword that Frey took for his arms stag's antlers,[2] and thus became the father of the sons of the antelope or deer, the race whose parent-star was the constellation Orion, called in Hindu astronomical mythology Mriga-shiras, the deer (*mriga*) star. This was the star which ruled the year of the ploughing-race, the sons of Rāma or Rā, the sun-god, who made the deer which showed them where the best grass- and corn-land lay, their totem-god and guiding star.

It was these gods of the race of the sons of the seed-grain, the god of the house-pole, pointing to the North, the mother-goddess of the sons of the egg and the seed, the bird-goddess Freya, and the god who imparted to the seed the life it transferred to its offspring, the father-god Frey, who became the gods of the three seasons of seed-sowing, growth, and ripening. These were transferred from the South to the North when the Northern sun-horse, Hœnir, was brought southward as the leader of the sons of the horse, and these three gods replaced in the Northern pantheon the earlier triad of the fire-worshipping artisans, the god Loki, the fire-father, the thunder-god, with his children the Midgard-serpent, the mother-snake, and Fenrir, the wolf of light, the lightning-flash. It was their successor, the sun-horse, who led the sun through the course of the year, in his journeys round the points of the compass, from the summer solstice in the north-east to the winter solstice in the south-east, passing on his way through the due east and west of the equinoxes, when he gathered in the autumnal equinox the apples of Iduna from the gardens of the Hesperides in the west.

---

[1] Elton and Powell, *Saxo Grammaticus*, Introduction, p. cxviii.
[2] Prose Edda, chap. xxxvii. ; Mallet, *Northern Antiquities*, pp. 428, 429.

With Njord and the bisexual gods Freya and Frey, there were united in the earliest mythology Thor, the thunder-god, the bearer of the grinding-hammer Mjölni, the Hebrew Patash, the root of which appears in the Hebrew Japhet and the Egyptian Ptāh, the dwarf-god. These were the opening (*patah*) gods, variant forms of the divine smith, whose car is driven by the two goats Tonngnistr and Tanngrisnir, the rulers of time to the shepherd sons of the goat-god Pan; and these two goat-steeds show by their names, meaning he of the gnashing (*gnistr*) and he of the divided (*grisnir*) teeth (*tann*, *tönn*), that they were descendants of the devouring wolf, who became the carrion-bird and the thunder-god. It was Thor who, in his journey to Jötunheim, first attacked the power of the giants, and thus marked the beginning of the first attempts to learn the secrets of nature, a knowledge which must precede their conquest. In this journey, Thor took as his helpers and companions the peasant children Thjalfi and Röska, and thus showed that science began with the knowledge gained by and from the tillers of the ground.[1] This contest between human intellect and perseverance and the untamed and un-understood forces of nature, begun by the

thunder-god, armed with the double-hammer, the cross $\dagger$,[2]

culminated in the victory of the Æsir sons of As or Is, the mother-mountain, the abode of Odin, the god of victorious knowledge.

It is the accomplishment of the supremacy of Odin which is told in the Hadding Saga, which describes how Hadding, the hairy (*hadd-r*), the bearded god of the pole, was brought up under the guardianship of Wagnhofde, the wagon- or

---

[1] Prose Edda, chap. xliv. ; Mallet, *Northern Antiquities*, p. 436.

[2] Count Goblet d'Alviella, *The Migration of Symbols*, chap. i. pp. 15, 16, shows that M. de Harlez has proved that the sign $\dagger$ is the simplest form of the hieroglyph representing the hammer of Ptāh, the two-headed mallet of the Celts and Germans.

wain-head (*hofile*), the polar constellation of Charles's Wain, or the Great Bear, and how while living under Loki, or Mid-Odin, he ate the wolf's heart, and freed himself from Loki's toils. He then defeated and killed Svip-dag, the hurrying day, the son of Orwandil,[1] the constellation Orion. This victory tells of the age when the computation of time by the movements of the wandering sun-god, as shown in the changes of the seasons and the lunar phases, was superseded by the reckoning based on the observation of the circumpolar stars and the movements of the sun from north to south and south to north at the solstices. It was the authors of this more scientific method of interpreting nature who first recognised the unvarying regularity of the succession of natural pheno-mena, and who looked on the laws thus ascertained as more satisfactory guides than the haphazard predictions of the witch-doctors and medicine-men of the age of magic called in the myth the age of Mid-Odin. The victory of the wise father Odin put an end to the rule of the matriarchal worship-pers of the mother-earth, his consort Frigga, and also of that of Loki, the capricious god of fire; and it was then that the conquering father-god, the upholder and executor of the divine law, became supreme lord of the sons of the mother-mountain and of the sun-horse, which runs its daily and yearly courses in the paths traced for it by the unerring knowledge of the divine smith. It was the gods of Odin's heaven who fed daily on the flesh of the divine boar, the sun-god called Sæhrimnir, the father-god of the fire-worshipping sons of the forest, cooked by Audhrimnir, the breath (*hrimnir*) of thought (*aund*), in the kettle Eldhrimnir, the ancient (*eld*) mist or breath; and this boar sun-god of the woodland race, the forest (*baso*) Basques or Vasques, is the Northern counterpart of the boar sun-god Vishnu of the Rigveda, whose votaries also worshipped the forest-god Vāsuki. The gods of these people, both in the ritual of the Edda and of the Rigveda, drank the mead of poetical

[1] Elton and Powell, *Saxo Grammaticus*, Introduction, pp. cxix. cxxiii.

inspiration made in the Edda by the dwarfs of the blood of Kvāsir, the leaven (*kvas*), who prepared it in the kettle Odhrœrir, that which stores (*hrœra*) the mind (*ō-dhr*). It was stored in the casks Bodn (or command) and Sön (forgiveness),[1] which may be compared with two caskets of Soma, the water of life, of Hindu ritual, called Dīksha (consecration) and Tapas (penance). These were brought to earth and given to Kadrū, the tree (*dru*) mother, by the snow-bird Shyena, who brings the rain, and entrusted by Kadrū to the guardianship of Indra, the rain-god, and Agni, the fire-god. The worship of the Soma stored in the caskets belonged to the ritual of the Ashvins, the heavenly twin horsemen (*ashva*), the great drinkers of mead in the Rigveda; and it was before the Ashvins were the star-horses of the sun, and when they were the primæval twins, Day and Night, that the expiatory bath of forgiveness (Sön), which in Phrygian and Greek ritual cleansed the sinner from his sin, was the bath of the blood of the divine totem-parents of the twin and woodland races. In the Hadding Saga this victory of the gods Odin and Njord is ascribed to the agency of the birds, the birds who, in the Vedic ritual, brought the Soma. It was they who helped Hadding to take the hill (*dun*) fort of Duna, the mother-mountain, only accessible to birds. And this, in the Saga, is said to be, like the Mysian Olympus, situated on the Hellespont.[2]

This series of historical myths, beginning with the solitary rule of Völundr or Wieland, the master-smith, is repeated in the *Niebelungen Lied* in a still more striking form than that disclosed in the separate legends I have tried to piece together. This story opens in the land of the Volsungs,[3] or

---

[1] Prose Edda, chap. xxxviii., *Bragi-rœdur*; Pfeiffer, *Alt Nordische Lesebuch*, p. 43 ; Mallet, *Northern Antiquities*, pp. 429, 462, 463.

[2] Elton and Powell, *Saxo Grammaticus*, book i. pp. 29-32.

[3] In this abstract of the *Niebelungen Lied*, I have taken the incidents almost entirely from W. Morris's poem entitled *Story of Sigurd the Volsung, or The Fall of the Niblungs*. In this poem, as well as in his other poetical versions of Greek and Northern myths, he follows with the most scrupulous accuracy the details given in the original poetic stories.

the united people, sons of the pole (*volr*). Their king, like
the Hindu Kushika or Chakrivarti king, dwelt in his palace
built in the centre of his woodland realm. In the midst
thereof grew the parent-tree, the house-pole of the race,
rising above the rafters fixed into its trunk, and spreading
its branches, wherein the mother-hawks, the mother-birds of
the mining races, built their nests high over the raftered roof.
It was to this palace that the ambassadors of Siggeir, king
of the Goths, came in the month of May to ask the Volsung
king to give Signy his daughter in marriage to their master
Siggeir. Siggeir, the king of the Goths, the sons of the bull
(*gud* or *gut*), is the counterpart of Swip-dag, Orwandil, the
wandering sun-god, and his ally Guthorm, the son of Gröa,
the growing (*gröa*) mother. Signy, whom he married in the
Niblung legend, is the same mythical mother who appears in
the Hadding Saga as Signy, the mother of Hadding and the
daughter of Sumble, the king of the Finns.[1] Thus the story of
the Volsungs begins with the account of the alliance between
the sons of the Finns, the united metal-working sons of the
father house-pole and mother-tree, and the wandering sons
of the bull, or, in other words, between the woodland agriculturists and the pastoral herdsmen.

On the day of the wedding of Signy—who is, by her name
as Sigyn, the wife of Loki, clearly shown to be the fire-socket,
the mother of fire—an aged but bright-eyed man, clothed in
a hood blue as the sky and a kirtle of cloudy grey, strode
into the hall of the marriage-feast. He there drew from
underneath his cloud-cloak a gleaming sword, the sword of
light, which he fixed by one stroke into the trunk of the
Volsung parent-tree growing in the centre of the hall. He
said that he who could draw out the buried sunshine should
gain mastery as the first of men. Siggeir and his Gothic
earls, King Volsung, and nine of his sons all tried in vain to
draw out the sword, and the feat was at last accomplished by
Sigmund, the tenth of the Volsung princes. Siggeir begged

---

[1] Elton and Powell, *Saxo Grammaticus*, book i. pp. 23-25.

that he might be allowed to hold this sword of Odin at his wedding, but Sigmund refused his request, and it was after this refusal that Sigurd gave his treacherous invitation to the Volsungs to visit his country. This was gladly accepted by King Volsung, who promised to undertake the voyage in two months' time. Next morning Signy, the wise woman, warned her father that Siggeir meant to kill the Volsungs when they came to Gothland, but King Volsung declared that as his word was given he must go as he promised, and he still persevered in his determination, and refused to return when Signy met him on his landing and told him that all the Volsungs were to be slain. When the doomed men arrived at the top of the hill overlooking Siggeir's palace they saw the whole valley filled with armed warriors; and though they united themselves in the wedge-shaped, shield-guarded phalanx, they were unable to withstand the onset of the Goths. All, including King Volsung, were slain, except the ten Volsung princes, who were taken prisoners. Signy begged for their lives, but Siggeir, who now rejoiced in the possession of the sun-sword of light, taken in the battle from the captured Sigmund, refused to do more than grant them a respite, and ordered them to be bound and left in a glade of the forest as a prey to the wolves. Nine were slain, Sigi being the last who was slain by the he-wolf, who, with his mate the she-wolf, attacked and devoured the prisoners. Sigmund, when attacked by the she-wolf, fixed his teeth in her, drinking her blood, like Hadding, and, aided by the strength thus gained, he in his struggles burst the bonds which bound his hands, seized her by the throat and killed her. He then remained as the wolf of fire, who had, by drinking the blood of the mother-wolf, imbued himself with her spirit and power. He then, after he had met Signy, who came to see how her brethren fared, lived in secret in a cave known only to Signy, working alone, like Völundr, the master-smith. In this phase of Sigmund's life we see him as the wolf fire-god, the offspring of the tenth month of gestation, the divine son of the Finnic miners and cultivators, the Greek wolf-god Apollo, the god of day, the son of the mother-wolf Leto. He, like Wieland the smith, who lived in the wolf's valley, had brought to the sons of the bull-race the knowledge of the craft of the smith, the master-mind who thought out and solved the problems brought before him by the changes observed in the metals and materials used in his work, discovered the best patterns for the tools he wanted, and fashioned them in the best way with the aid of the fire and the hammer. After ten years of solitude, a period which, like the number of the sons of the Volsung king, represented a time of gestation to be followed by a new birth, Signy sent to Sigmund her ten years old son to learn and help him in his craft; but when the boy was left at home to bake while Sigmund went to search for game, he left the meal untouched through fear of the snake which guarded it, and Sigmund, disgusted with his craven hesitation to face difficulties, sent him back to his mother. It was then that, in order to procure for Sigmund a helper who equalled him in courage and skill, Signy, taught by the spells of a Finnish witch-wife, exchanged her shape with her teacher and went into the forest, not as the fair blue-eyed queen of the North, but as a black-haired and black-eyed, but beautiful, Finn. She appeared before Sigmund's cave as a woman who had been lost in the forest, and who craved shelter. She won his love by her beauty and blandishments, and passed the night with him, and it was from this meeting that her son Sinfiötli, meaning the sinew (*sin*) chain (*fiotr*), the man of nerves and strong sinews, was born. He when he was ten years old was sent, like his half-brother, to Sigmund, and was set to bake bread. But instead of fearing the snake that guarded the meal, he slew it and mingled its body with the bread. It was then that the rule of the sons of the snake and of the single potter and smith, the hidden architect and contriver, passed away, to be succeeded by the rule of the twins, the wolves of light

who marked the changes of night and day, the wolf-born
children of the Greek Leto, who was originally the Czech
goddess Leto and the Wend goddess Lada, mother of the
stars and spouse of the moon-god, the Wend counterpart of
Freya,[1] and the mother-goddess of the star-worshipping races,
of the sun-god Apollo, and the star virgin-mother Artemis.
In the Niblung myth the brother and sister twins had become
the twin-brethren, the father and his son. This transformation is set forth in the story which tells how, when Sinfiötli
had reached the age of manhood, the pair came upon two
men sleeping alone in a house who had hung above their
heads the wolf-skins in which they used to appear as were-
wolves in their waking hours. They took the skins, and
thenceforth they used to prey on mankind in the guise of
ravening wolves, the fire- and sun-god, till at last Sigmund in
his increasing frenzy killed Sinfiötli, the sun-god, his foster-
son. While struck with remorse for the deed he had done
he noticed two weasels who fought, one of whom killed the
other. The slayer at once went into the wood and found the
three-leaved herb of life, which developed into the Soma-
plant of Hindu ritual and, as I shall also show, into the
Holy Grail of the Arthurian legend, and with this he restored
his companion to life. Sigmund did the same, and when
Sinfiötli was made alive again he and Sigmund were disgusted
with their lives as wolves, and resumed their human shape.
It is this part of the story which tells how the year of three
seasons, the three-leaved holy shamrock, was made the
measurer of time by the wolf-race of Northern Europe, sons
of the fire-god, born from the wooden fire-drill and socket of
the sons of the tree; and it also tells how these hordes of
ravening wolves, who made the sun and moon their gods,
came down from the North and conquered the lands of the
temperate zone.

The sons of the woodland fire-wolf, in order to complete

---

[1] Tiele, *Outlines of the History of Ancient Religions*, iv.: ' Religion among
the Wends or Latto-Slavs,' § 113, p. 185.

their conquest, determined to slay Siggeir, the magician-
husband of the witch-wife, the murderer of the Volsungs.
They made their way to the wine-cellar adjoining Siggeir's
banqueting-hall, and hid themselves among the wine-casks.
They were seen by the two youngest children of Siggeir and
Signy, who warned the feasters of the presence of the two
sons of light, 'whose hats were wide and white, and whose
garments tinkle and shine.' They were captured after a
fierce struggle by the Gothic warriors, and after Sinfiötli had
broken the bones of the betraying children and thrown their
corpses at Siggeir's feet. The father and son were bound
and buried alive under a mound divided into two chambers,
but before the thatch covering them in was finished, Signy
cast down to Sinfiötli through the uncovered space, the
wonder-working sword of Odin, the sword of light, taken by
Siggeir from his father and fellow-prisoner, Sigmund. Its
possession made the conquering twins rulers of the world, for
with its help Sigmund and Sinfiötli cut their bonds and
escaped. After slaying the guard, they went to the fire-
wood stacks, where winter wood was stored, and took thence
logs, which they piled up round Signy's palace and set a
light to them. In the conflagration both Signy and Siggeir
were burnt, for Signy, after escaping from the flames to tell
Sigmund that Sinfiötli was their son, returned to die with
her lord.

This conflagration tells us of the completion of the conquest of the world by the twin brethren, and the close of
the rule of the magical fire-god, the age of witchcraft and
sorcery. The age which succeeded was that called in the Rigveda and Brāhmaṇas the age of the Ashvins, the twin-race,
when the worship of Soma, the sap of life, the life-giving
drink and the water of immortality, was introduced. This
was the age of the rule of the confederated or united twin-
races, the joined (khat) Khati or Hittites, who made the
mother-mountain of the East their mother-mountain. This
mother-mountain appears in the Sigmund legend as Borghild,

the earth-goddess (*hilda*) of the Burg or hill capital, whom
Sigmund married after he had recovered the sword of light,
and had emerged from his obscurity as the master-smith to
become the ruler of the united Volsungs and Goths. But the
alliance between her kin—Helgi the holy one,[1] her son by
Sigmund; Gudrod, her brother, the greedy trader; and
Sinfiötli, the fighting and avenging sun-god—was broken
after the death of Helgi by a quarrel between Gudrod and
Sinfiötli over the division of the spoil, the best part of
which Gudrod wanted to keep for himself. In this quarrel
Gudrod was slain, and in revenge for his death Borghild
poisoned Sinfiötli, whose body was carried by his father to
the boat of the dead, which carried them to the West, the
land of the dying sun, and thus it was taken over the sea as
the body of the dead god of time who ruled the age when
the storm-sun-god of the mountain-bred tempests, and Sigmund, the conquering (*sig*) moon (*mond*), were kings of
heaven. In this mythology, Sigmund, the male moon-god of
the North, was, as I have shown, the son born as the tenth
of the sons of the father-house and gnomon-pole, the parent-
tree, thus showing his birth in the tenth lunar month of
gestation as the father-god of the woodland and barley-
growing sons of the tree.

Borghild was driven away after Sinfiötli's death, and the
sons of the mountain-moon and storm-god descended into
the plains and fertile valleys, a migration attested in Greek
mythology by the change in the race parent-tree from the
mountain-pine, sacred to Cybele, to the laurel of Apollo, the
sun-god. It is this change which is depicted in the Sigmund Saga by his marriage of Hjordis, the mother of the

---

[1] In the Sagas used by Saxo Grammaticus, Helgi is the son of Halfdan,
the son of Orwandil, the star Orion, and therefore represents the Southern
race of barley-growers, who measured time by the year of Orion, while Sinfiötli is an equation of Hadding, the hairy (*had*) sun-god of the North, who
was, like Sinfiötli, a son of Signy. Helgi was the husband of his own
daughter, Ursé, the bear-mother, mother of the bear race. Elton and Powell,
Saxo Grammaticus, Preface, pp. cxvi. cxvii. book i. p. 24 ; book ii. pp. 61-63.

herds (*hjord*), the cow-mother of the lunar race, who had
been the mountain-mother, a transformation shown in the
change of Isis, the mountain-goddess, into Isis, the holy cow,
mother of the bull Apis. Hjordis was the daughter of
Eylimi, king of the islands (*ey*), the ruler of the seafaring
traders, but her marriage had offended Lyngi, the king of
the mountain heath (*lyng*), who was also a suitor for Hjordis'
hand. Lyngi's forces landed in Eylimi's country, attacked
the army led by Sigmund, and in the battle Sigmund and
his hosts were defeated, and Sigmund left for dead. When
Lyngi and his men had gone, Hjordis, who had watched the
battle, came down and found Sigmund, though dying, still
alive. He told her to gather up the shards of the sword of
Odin, the lunar phases, which showed the stages of the life
of the moon-god, and which had been broken by his slayers,
and to keep them for forging the sword of the young sun-god,
whom she was to bear. On the death of the moon-god, born
in the evolution of scientific and astronomical knowledge from
the fire-drill, which became the heat-engendering world-pole,
which was the father of life, whose revolutions were measured
by the lunar phases, Hjordis dwelt under the protection of
King Elf, the king of the mining elves, who had come to
the land of the Islands after the battle, and had buried Sigmund ; and after the birth of Sigurd, the young sun-god of
the new age, she married Elf.

In determining the history of the myth of Sigurd, the
rider of the sun-horse, we are helped by the genealogy of
King Elf, his foster-father, who was grandson of the brother of
Grip or Gripir, the giant who, as we shall see presently, kept
the horses of light. This descent shows that Elf belonged to
the kin of the two dogs Gialp and Greip, Yelp and Grip,
the sons, or according to another myth the daughters, of
Geir-rod, the god of the lance (*geir*), or in other words of
the gnomon rain-pole. In short, they were the two star-
dogs, the star Sirius, the dog of Orion, and the star Procyon,
the fore (*pro*) dog (κύων), the constellations Canis Major and

Canis Minor, who kept the gates of the Milky-way, the bridge running North and South between the East, the home of life and light, and the West, the realms of death and darkness. It was Grip, the seizing-dog, the dog of the hunter Orion, who kept the horses of the sun in the East, while the Elf-king, the foster-father of the yet unborn sun, was king of the dark West. He was the foster-son of Regin, the rain-god (*Regn*) of the race of the mining elves, who was, like Völundr and Sigmund, 'the master of masters of the smithying craft.' Regin became the tutor and foster-father of the sun-god Sig-urdr, the conquering (*sig*) stone (*urdr*), the sacred stone-pillar to the east of the temple erected by the worshippers of the sun-horse, which, as I shall show, received on its summit the first rays of the sun rising at the solstices. By Regin's advice, Sigurd went with a token, given him by King Elf, to Grip, and obtained from him the gift of his best horse, the grey sun-horse, called Gräni, the grey one. But before he started on his career as the conquering sun-knight he had to obtain the sword of brightness, which was forged for him by Regin, the rain-god of the showers and sun-gleams of spring, from the shards of the sword of his father, Sigmund, the moon-god, which had been treasured by Hjordis, the goddess of the race who reckoned time by the year-ring of the months of gestation. Armed with this gleaming sword, the sun's rays issuing from the grey clouds, the horse on which he rode, Sigurd, after visiting and obtaining the blessing of Grip, the star Sirius, the guardian of the unrisen sun of the year beginning with the summer solstice, set out, followed by Regin, the spring rains, the master-smith, who shoes the white horse of the sun-god. They went to seek the realms of darkness, the mountain-cave of the Glittering Heath, where Fafnir, Regni's brother, the black cloud-snake, who will not give up the rain it encloses, guarded the Sacred Treasure, the Golden Light. This was the treasure which Loki, the fire-god, had taken from the dwarf-guardian of light, Andvari, the wary (*vari*) spirit (*anda*), and given to Hreidmar, the keeper of the nest (*hreid-r*) of the mother-bird, the father of Regin and Fafnir, as a ransom for the release of himself, Odin, and Hœnir, the horse of light. They had been imprisoned by Hreidmar as a punishment for causing the death of Otr, the otter-fish sun-god, born of the waters, one of the three brother-gods of the race of the elves, who was slain by Loki. This story of the imprisonment of Loki, Odin, and Hœnir is a counterpart of that in the Prose Edda, which tells how these three gods were met in their wanderings by Thjassi, the winter-eagle, who wanted to take Iduna's apples, the seeds of light and life, out of Asgard, and place them in Jötunheim, the home of the giants. Thjassi, when he came upon the encampment of the gods, seized the leg and two shoulders of the ox, the father-god of the ploughing race, which they were trying to cook but could not succeed in boiling. When Loki struck him, the stick with which he struck remained fixed on Thjassi's back, while the other end clung to the hand of the striker. Thjassi, the winter-eagle and mother-bird who brings the rain, who thus conquered the fire-god, dragged his victim over the rocks and forests till he was nearly killed. He only obtained his release by promising to bring Iduna and her three apples out of Asgard, and to give them to Thjassi. He then enticed Iduna into a forest near Asgard, where she, with her apples, was seized by Thjassi the eagle. Loki then borrowed the falcon plumage of Freya, the bird-mother of the mining race, and flew to Jötunheim, from whence, in Thjassi's absence, he brought out Iduna in the guise of a sparrow. When Thjassi, the eagle, returned and followed the fugitives, he all but caught them up as they were entering Asgard, when the gods lighted a fire of wood chips on the battlements and burnt off his wings as he tried to fly over.[1] This story tells how the falcon of the mining races brought back to Asgard the treasured seeds of light

---

[1] The Prose Edda, *Bragi-rœdur*, chap. ii.; Mallet, *Northern Antiquities*, pp. 459-461.

and life which had been stolen by the winter eagle, and this treasure was, in the Sigurd myth, that which was guarded by the dwarf Andvari, the mining son of the falcon bird-mother. This treasure was no longer the life-giving food and drink, the seed and sap of life, but the glittering wealth brought to earth by the sun-god, comprising (1) the helm of Awing, the darkness whence the morning-sun is born; (2) the glistening hauberk, his golden coat of mail; and (3) the golden ring, the recurring days, months, and seasons of the sun's year. These were the seeds of life found by the fish-sun-god, the fisherman who finds the ring of the year of con-jugal union in all legends of its loss by the mother of the future race; and it was these weapons of the sun-god which took the place of the apples of youth in the evolution of solar mythology. It is these three weapons which Sigurd must win before he becomes the sun-god ruling the year. To gain these he had to kill Fafnir, the dragon-snake of dark-ness, who guarded them after they had been brought by Loki to the nest of the mother-bird, the star-mother Vega, the vulture mother-star, whose treasures were guarded by the circumpolar stars of the constellation Draco. Aided by Regin, the rain-god, Sigurd went westward to the glittering heath, the home of the sleeping sun, and then he left Regin behind and went on foot, followed by Gráni, the sun-horse, through the twilight till he met an ancient man, who showed him the path down which Fafnir came each morning to drink the water of life from the mother-lake, the home of the fish-sun-god. He told him that Fafnir was only vulner-able in one place, like all sun-gods, who invariably run their destined course and then die. He must, therefore, dig a hole in the path down which Fafnir came, where he could stand and drive the sword of light into the dragon's entrails. Thus instructed, he waited for and slew the dragon, and was consecrated in his streaming blood as the knight of the sun-horse, who succeeded the stars of night, the constellation Draco, as the ruler of the year. After the death of Fafnir,

Regin came up and drank his brother's blood, and then, in
revenge for his brother's death, declared that Sigurd must be
his Thrall and cook for him Fafnir's heart, which, when
eaten, will make him master of all wisdom. Sigurd did his
bidding, and cooked the heart while Regin slept; but, while
he was cooking it, he tasted it, and thus learned the speech
of birds and the meaning of the words spoken by the seven
eagles, the mother-birds of time, reckoned by the weeks of
seven days. They told him that if Regin woke and ate the
heart of the dragon as he had drunk his blood, he would be
master of light and darkness, and would, as the untamed fire-
and rain-god, disseminate evil and strife through the world.
Sigurd, therefore, slew Regin, and after eating Fafnir's heart, as
Hadding ate that of the fire-wolf, and thus becoming the guardian-god of the treasures stored by the wisdom of the past, he
went up to the cave where, besides the helmet of darkness, the
golden mail and the golden ring of Andvari, the red rings forming the mass of the treasure, were kept. These red rings were
the seven hundred year-rings, made by Völundr during his
solitude, and which he hung up on the wall of his cave.
Out of these, Nidung, when he captured Völundr, took one,
the ring that Bodhvildr broke, and which is the golden ring
of Andvari in the Sigurd myth. When he had clothed himself in the garb of the conquering sun-god, and had loaded
Gráni with the red rings of the treasure, the lore gathered in
the rings of time, the days, months, and years of the bygone
ages, Sigurd, mounted on his cloud-horse, rode to the top of
the mountain called Hinda-fjall, the hill of the deer (*hinda*).
We find in this name a distinct chronological index of the
mythological place of the myth of the sun-horse. It clearly
succeeded, in the computations of time, that which reckoned
the year by the movements of Orion, the deer-star and deer-
hunter, and by the lunar changes. This lunar year was the
year anciently reckoned in Northern Germany, which, when
compared with that measured by the movements of the
Pleiades in November and April, and by the solstices indi-

cated by the gnomon, was found to be twelve days too short.
It was during these twelve intercalary days preceding the
winter solstice, that in North Germany, as stated by Professor
Kuhn,[1] the traditional narrative-play, depicting the hunting
of the hind, the moon, by the star-stag Orion, was acted. In
this the actor of the part of the deer-star follows the hind,
who sings unchaste songs, till he is shot by the wild hunter.
The weapon with which he was killed is, as I have shown in
the Essay on the astronomy of the Veda,[2] the arrow of three
seasons, the three seasons in Orion's belt, which puts an end
to the yearly course of Orion when the year is completed at
the end of the twelve days occupied with the final chase of
the dying year. These twelve days ended with the winter
solstice, which, in the story of Sigurd, is represented by his
appearance on the mountain of the deer to inaugurate a new
reckoning of time which, instead of dividing the year into
three seasons, as in the year of Orion, measured it by four,
through the solstices and equinoxes; and this year is, as I
shall prove presently when describing the plan of the
temples of the sun-horse, that which was introduced by the
race who called themselves the sons of the horse and worshipped Odin, the god of knowledge. When Sigurd, the sun-
god of the pillar (*urdr*) of the rising sun, reached the top of
the mountain of the deer, he found there the divine mound,
built by the giants as the home of the mother of life. She,
in the mythology of the days when the mother-mountain
was an inaccessible snowy peak, was the mother-bird, the
mountain-eagle; but in the age of the rule of the pastoral
ploughing race, the sons of the twin brethren, born of the
mother-cow Hjordis, the terrible mother of the storm-clouds,
the carrion-eating bird, had become the hawk-mother of the
sons of the fish-god,[3] born in the rivers flowing from the consecrated mother-hill. This is, in Hindu mythology, the hill

---

[1] Letter from Professor Kuhn to Dr. Rajendralal Mitra, *Indo-Aryans*,
vol. ii. pp. 300-302. Bāl Gangādhar Tilak, *Orion*, chap. vi. pp. 138-140.

[2] *Ruling Races of Prehistoric Times*, part ii., Essay vii. pp. 18, 22.

[3] *Ibid.*, Preface, p. xxxix.

Barsāna, the mother-hill of the sons of the rain-god Bar or Var, the god Varuṇa or Ouranos, father-god of the sheep-race, whose totem was the ram, the Bharata of Indian history. This was the hill consecrated to Rā-dhā, the giver or mother (*dhā*) of Rā, the sun-god, the spouse of Krishṇa, the black antelope, the deer-god of the deer-hill.[1] It was on the giant-built mound on the top of this hill that Sigurd found Brunhilda, the earth-goddess of the springs (*brunnen*), sleeping, the sleeping beauty who is to be wakened to fresh life by the kiss of the spring sun. He wakened her by taking off the helmet of darkness, thus appearing as the risen sun, and by ripping open her clothing with the shining sword of the sun's rays. The goddess of the springs who was thus awakened is, in the Hindu legend of the birth of the Kaurāvya or Khurāvya—the sons of the bird (*khu*), who are called as sons of their totem, the hawk, the Cheroos—the bird-mother Gaṇḍharī, she who wets (*dhārā*) the land (*gan*), and she is the wife of Dhritarāshtra, the blind king who, like Sigurd, represents the gnomon stone lit by the rays of the rising sun. In the Sigurd legend Brunhilda is a Valkyr or wind-bird-goddess, the victory-wafter of Odin, and it was with her, the mother-hawk, that Sigurd plighted his troth on the top of the mountain of the deer by placing on her finger Andvari's ring, and thus becoming the god of the sons of Varuṇa, the rain-god who first consecrated in the South the conjugal union which had been the source whence the Northern family born in one house and owning its own property took its rise. The plighted pair swore to meet in Lymdale, the fertile land at the mountain-foot, watered by the rivers which flowed from the springs consecrated to Brunhilda. This land of Lymdale is, in Hindu mythology, the temple-grove of Rādhā-Raman, called Sanket, 'the place of assignation,' between the hills of Bar-sāna, sacred to Rādhā and Nand-gānw, the hill of the deer-god Krishṇa, where, in the annual historical drama giving the mythical history of the

---

[1] *Ruling Races of Prehistoric Times*, Essay iv. pp. 451, 452.

race, the men still fight with deers' horns,[1] like Frey, the
father-god of the sons of the hawk-bird Freya, in the Edda.
Sigurd, after he had made Brunhilda his betrothed bride and
given her the year-ring, left her on the mountain-top and
came down to Lymdale, where he met in the forest Heimir, the
earth (*heimr*), king of the country, whose wife was the sister of
Brunhilda. Sigurd, the sun-god, remained with him through
the spring, and was in his wanderings led by his falcon, the
consecrated bird of Brunhilda, to her home, where, at the
vernal equinox, they once more met and renewed their pro-
mises. He returned from Brunhilda's house to the palace of
Heimir, whence, towards the summer solstice, he departed to
the land of the Niblungs, the sons of the cloud, the Greek
Nephele, the German Nebel, the mist who dwelt in the west,
the land of the evening sun. This was the land ruled by
king Giūki, his wife Grimhild, the witch-mother of the dark
(*grim*) night, and their black-haired sons Gunnar, the fighter
(*gunnr*), and Högni (*hagr*), the wise, with their daughter
Gudrūn, god's (*gud*) knowledge (*rūn*), the divine prophetess-
mother of the sun of the future year; and, besides these, there
was Grimhild's son by another husband, Guthorm.[2] Guthorm,
who, in one form of the Sigurd myth, is, as we shall see, the
slayer of Sigurd, is in the Hadding Saga Hadding's half-
brother, son of Grōa, the growing (*grōa*) mother, the night-
goddess, brought up by Hafli, the goat (*hafr*) god, the
primæval god of time, the rival and predecessor in time-
computation of Wagnhofde, the wain-head, the stars of the
Great Bear, the guardian of Hadding. Guthorm was the
ally of Swip-dag, the star Orion,[3] the deer-god, and he thus
belonged to the kin of Brunhilda, the dog-goddess of the
rising sun of the East, while Gudrūn was, as we shall now see,
to become the wife of the setting sun of the West, the trans-

---

[1] *The Ruling Races of Prehistoric Times*, Essay iv. pp. 452, 453.

[2] In Morris's story of the 'Fall of the Niblungs,' Guthorm is represented
as son to Gyūki and Grimhild, but in the 'Skáldskaparmál' he is called
Gyūki's stepson. Pfeiffer, *Alt Nordische Lesebuch*, p. 52.

[3] Elton and Powell, *Saxo Grammaticus*, book i. pp. 24, 25.

formed Sigurd. The Niblungs received him with a warm
welcome, and he led their troops to victory, but he still
remembered Brunhilda, and failed to return the love of
Gudrūn. But this mood was changed when he drank of
the magic cup prepared by the witch Grimhild. After drinking it Sigurd went forth, not on his own horse Grāni, but on
a horse of the Niblungs, and rode round the house of Brunhilda till the dawn of day, when he returned in a dream to
the Niblung capital, and Brunhilda's house was thenceforth
surrounded with a ring of fire. In the hall of the Niblungs he
forgot her, was betrothed to Gudrūn, and asked that Giūki
would adopt him as his son. At his wedding to Gudrūn he
drank the cup of brotherhood, the magic cup of the dwarf
race, and ate of the flesh of the boar, the father-god of the
sons of the forest, the race of the Volsung. It was over the
Volsung totem that Sigurd, Gunnar, and Högni united their
blood in the loving-cup and swore blood-brotherhood. But
according to another variant form of the myth it was only
Sigurd and Gunnar who became sworn brethren, while Högni
stood apart from the compact. In the form of the story in
which Högni is one of the united three, Guthorm is left out,
as he is said to have been absent from the wedding because
he was fighting in the East. After his wedding Sigurd went
with Gunnar and Högni to woo the forgotten Brunhilda as
the bride of Gunnar, whom his mother Grimhild wanted him
to marry. But when they came to Brunhilda's house encircled with the fire kindled by the defection of Sigurd, the
departing sun-god, Gunnar's horse refused to pass through
the flames, and when Sigurd gave him Grāni, the sun-horse
refused to carry Gunnar through the ring of fire. It was then
that, under Högni's spells, Sigurd and Gunnar exchanged
shapes, and Sigurd became the black-haired Gunnar clad in
the black mail of the Niblungs, while Gunnar became the
fair-haired Sigurd, with the golden mail of the sun-god. In
this guise Grāni still recognised his master, and carried him
through the flames. When he met Brunhilda he wooed her

as Gunnar, and the two plighted their troth. He shared the bridal bed with her, placing the sword of light between them, and in the morning, when he left, Brunhilda placed on his finger as her morning gift Andvari's wedding-ring, and promised to come in ten days to the Niblung palace and live with Gunnar as his bride. Sigurd on his return gave Gudrūn the ring he had got from Brunhilda, and when Brunhilda came as the bride of Gunnar she greeted Sigurd coldly, and took no notice of Gudrūn, the bride of the autumn sun of storms and tempests. After this the two queens quarrelled, when Brunhilda recognised on Gudrūn's finger the ring she thought she had given to Gunnar. When Gudrūn told her that Sigurd had given her the ring, Brunhilda was wroth with him for his guile and treachery in wooing her under the semblance of Gunnar, and stirred Gunnar and Högni to slay Sigurd. They having sworn blood-brotherhood with him, were unable to obey her behests, and the murder was finally completed by Guthorm, who had returned from the war in the East, and who slew Sigurd while sleeping by the side of Gudrūn unprotected by his impenetrable gold corselet. In the variant form of the myth in the Götterdammerung it is Högni, called Hagen, who strikes the fatal blow by stabbing him in the back, his one vulnerable part, when he was reclining after hunting, and had cast aside his protecting shield. In the mourning for the dead it was Brunhilda who finally asserted her right to be burnt on the pyre with her first love, while Gudrūn departed in wrath from the company of those who had compassed Sigurd's death. The story of the death of the sun-god in the tempests and storms of winter is continued in the further episode of the coming of ambassadors from Atli, the brother of Brunhilda, called the Budlason, or son of the booths (*bud*), the tent-dwelling king of the East. They asked the Niblungs for Gudrūn's hand for their master. Gudrūn, who had fled, was found by Gunnar and Högni in the house of Thora, the queen of the woodland, who in other myths is the wife

of Helgi, the son of Halfdan or Orion, the daughter of the
king of the Finns and Permian Sclavonians, and the mother
of Urse, the bear-mother.[1] Gudrūn, the wearer of the year-
ring of Andvari, consented to become the bride of Atli, but
only with the intention of consummating her vengeance on
the slayers of Sigurd. Her opportunity came when her
brethren the Niblungs consented to leave the cloudy land of
the West, and when Gunnar, the dying sun of the expiring
year, went to the East, the home of the rising sun of a new
era. He and his brother Högni were, as the gods of the
night and the cloudy winter, slain by Atli, and it was after
their slaughter that Gudrūn, the wearer of the year-ring of
Andvari, fired the palace of the sun-god of the water-springs
and the rain-clouds, and plunged into the sea, whence she
emerged as the sea-goddess worshipped by the sons of the
star-ship Argo, whose fortunes are told in the Greek myth
of Jason. In the North her story survived in those of her
children by Sigurd, Sigmund, and Svanhild or Svanhvīt.[2]
It was the witch Gudrūn who charmed the armour of the
brothers of Svanhild so that it could not be cut through,[3]
and this made them sons of the sun-god; and it was Svanhvīt,
the daughter of Hadding, who was, as I have shown, the sun-
god, the counterpart of Sigurd, who wedded the royal shepherd youth Ragnar, who became, like the Hebrew David or
Dod, the father of Hadad-Rimmon, the father-god of the
shepherd race, and king of the people whose totem was the
ram of Varuṇa. Dr. Sayce has proved that the god Dāda of
Aleppo was one of the forms of the god Rimmon or Ramānu,
the god Ram, the sun-god of Damascus, called in the list of
Edomite kings in Genesis, Hadad, the son of Bedad,[4] that is
to say, Ramānu, the son (be or ben) of the beloved one (dad),
the sun-god. This god Dod, Dodo, or David was, as we

---

[1] Elton and Powell, *Saxo Grammaticus*, book ii. p. 62, 63 ; book iii. p. 87.
[2] 'Skáldskaparmāl,' 46 ; Pfeiffer, *Alt Nordische Lesebuch*, p. 52.
[3] Elton and Powell, *Saxo Grammaticus*, Preface, p. lxxviii.
[4] Gen. xxxvi. 35.

know from the Moabite Stone where the Arel or altar of
Dodo is spoken of, worshipped as the sun-god of Southern
Palestine.[1] This was the god who became in Northern
mythology Ragnar, a name meaning the son of the gods
(Gothic, *Ragin*), and connected with Regn, rain, and Ragna,
twilight. It is, in short, a form of the name of the Lithuanian sky and sun-god Rai or Rojas,[2] who became in India
Rā-hu, in Egypt Rā, and who subsequently developed into
the Hindu god Rāma, the god of darkness (*rāma*), answering
to the northern twilight (*ragna*). He became the Assyrian
Ramānu, and the Hebrew Ab-ram, or the father (*ab*) Ram.
Ragnar was found by Svanhvīt, the swan-white moon-goddess,
as Samuel's messenger found David, tending sheep. She
recognised his kingly nature and made him her husband, and
gave him the sword of light, the disperser of goblins and
evil spirits.[3] It was Ragnar, the father of Erik, the shrewd-
spoken,[4] the sea-trader, who became the hero of the Vikings
or sea-kings, the sea-rovers, who succeeded the sons of the
horse as the rulers of North Europe, the race who are represented in Greek mythology by the Minyan crew of the Argo,
led by Jason, the pupil of Chiron, the healing physician and
teacher, sprung from the horse-race, the Centaur-sons of the
horse, who drove the bull-stars round the turning-pole of the
heavens (*tūr*, *taur*), and who, like Sigurd, the rider on the sun-
horse, went about the world doing war with wrong, protecting
the oppressed, and healing the sick. It is in the details of this
myth of Sigurd that we find the clearest explanation of the
widespread story of the sun-horse, the white horse of the

---

[1] Sayce, *Hibbert Lectures for* 1887, Lect. I. pp. 55-57.

[2] Tiele, *Outlines of the History of Ancient Religions* : ' Religion among
the Wends,' § iii. p. 182. The name Ragin or Regin is said by Grimm to
mean counsel or deliberation. The word is certainly connected with the
Hindu Rājā, born of Rā, the Lat. *rex*, the counselling king or ruler ; and
the root *ra* is probably connected with that of Reason, the Lat. *ratio*, from
reor, to think, whence the German *rath*, counsel. Hence Rā, the creating
sun-god, became Rā, the thinker.

[3] Elton and Powell, *Saxo Grammaticus*, book ii. pp. 50-54.

[4] *Ibid.*, book v. p. 156.

English chalk country, shod by Weyland the smith. The
white horse, with Weyland Smith's cave, still survives in the
Berkshire vale of the white horse, and it is in this white
horse of the sun, which was also drawn on the hill-sides at
Westbury near Bath, on Alton Hill, also near Frome, and at
Weymouth, though the rider on the white horse, who was
originally the sun, has in this last representation been changed
into George III. mounted on a white horse. This white horse
of the sun depicted on the hillsides, the god Epona of the
Britons, is evidently an outcome of the mythology of the
race who were the sons of the ash-trees, the Ash Yggdrasil
of the Edda. This tree is indigenous in limestone and flint
districts where the flint weapons used by the earliest hunting
and warrior tribes were made. Their home was the conti-
nental and island area in North Germany and the Baltic Sea
belonging to the great chalk and limestone formation which
stretches from the Wiltshire Downs on the west, to Pomerania and the land of the Lithuanians in the east. The sons
of the horse and the ash-tree were the race represented in the
story of Sigurd as the subjects of King Eylimi, king of the
Islands (*ey*) whose mother-goddess was Hjördis, the mother of
the herds (*hjörd*), the mother of Sigurd, the cow-mother Isis
of Egyptian, and Go or Gos of Hindu and Zend mythology.
It is in the Wiltshire, Berkshire, and Hampshire valleys of
the geological formation distinguishing the home of the sons
of the white horse that we find everywhere traces of the
old terraced cultivation that marked the husbandry of the
earlier Iberian sons of the rivers.[1] They were the Basque
emigrants from Asia Minor who introduced corn cultivation
into Europe in the Neolithic Age, and who are represented
in British ethnology by the Silurian races, the Silures of
South Wales, called by Tacitus Iberi,[2] who were the ruling
tribes in Wales when the Romans conquered Britain. These

---

[1] See the list of some of the localities in this part of the country where traces
of terrace cultivation are still visible in Gomme's *Village Community*, pp. 73-7.

[2] Tacitus, *Agricola*, xl.

terrace-makers were the first growers of dry crops requiring
a well-watered and well-drained soil, and when they found
that the soil on the hillsides was much cut up after violent
storms by the descending waters, which escaped too quickly
to benefit the crops, they formed the hillsides into terraces,
which would catch the water as it came down, and by which
they could regulate its flow, retaining it as long as it was
wanted for the fertilisation of the soil, and letting the over-
plus flow down on to the terraces lower down. These people
were the Basque sons of the wild boar, the wood-swine, the
totem-god of the Volsung race, who, when they were, as in
the Sigmund and Wieland myths, united with the sons of
the hammer, the metal-working artisans who first evoked
fire from flints, became the votaries of the fire-drill, the
creator of life-giving heat taken from the mother-tree.
These united sons of the village grove, the parent-tree, the
house-pole and the fire-parents, the sacred flint and the
wooden fire-drill and socket, became, when they reached
the German land of the fair-haired and blue-eyed Belgæ of
British ethnology, as the sons of the cow-mother Hjördis, the
worshippers of the white horse, the parent-god of the sons of
the sun-god. They then, in accordance with the customs of
this North-western race, made the family the national unit instead of the village of the cultivating races, and the tribe,
or national craft guild, of the hunters and fire-worshipping
artisans. In this revolution of ancient customs, the influence of
the cultivating and artisan races appears in the round graves of
the Bronze Age, reproducing the circular huts of the Southern
nations. These were substituted for the long barrows of the
Northern Neolithic hunters, which reproduced the long and
gabled houses, built in the form of a parallelogram, to contain the united generations of the family. It was on the
gables of these houses that, in the Lithuanian and Gothic
fatherlands of Mecklenburg, Pomerania, Lüneburg, and
Holstein, they used to place horses' skulls, and thus consecrate the house to their totem-god. It was these sons of

the horse who looked on the twin-doorposts as sacred representations of the father and mother of the house united in
wedlock by the lintel that joined them ; and it was as a sign
of this belief that the doorposts were, in Roman weddings,
anointed by the newly-married bride with the fat of the
wolf-mother, of the Guelph or wolf-race, whose mother-land
was North-west Germany, which afterwards became the
head-quarters of the horse-race. They also made the father
the head of the house till he gave up his rights, when
weakened by old age, to the son he chose to succeed him,
and gave to each family its own plots of land in the tribal
territory, separated from that of their neighbours by definite
boundaries. These people, when united as the pastoral
agricultural races with the metal-working artisans, made
Odin, the god of knowledge (*odh*), their national god, and
called their sun-god Rai, Raj, or Roj, the name given to him
by the Wend Lithuanians of Pomerania and the Vistula
valley. When they went southward and became members
of the confederacy of the sons of Kuṣh, they brought, as their
contribution to the national theology, the worship of their
sun-god Rā-hu, the god of the Māghadas, served by priests
called Ojhas, or men of knowledge (*odh* or *ojh*), that is,
inspired by Odin. It was then that, first as the sons of the
cow (*go*, the Zend *gos*), they took the leading place in the
national hierarchy, and superseded the Dravido-Iranian
priests, the sons of Idā or Irā, the sheep-mother. Their
progress southward from Pomeranian Gothland can be traced
in the Getæ of the Balkans and Thrace ; the Massagetæ, or
greater (*massa*) Getæ, living in the Kur and Araxes valleys
on the slopes of Ararat ; the Gaurians, or sons of the mother-
cow (*gauri*), of the Euphratean countries ; and the Jāts of
Northern India. As the Garans of Kurdistan, they were
the leading cultivators and ruling tribe in the lands of the
fire-worshippers of the petroleum country of Baku, on the
shores of the Caspian Sea ; and it was these people, who worshipped the sun as their father-god in the North, who became

worshippers of the golden calf, the sun of the coming year. This was the yellow bull Mnevis of Heliopolis, the Egyptian city of the sun, whose worship was introduced by King Kakau of the second dynasty, who brought the use of the solar year into Egypt. This sun-bull was the successor and unsuccessful rival of the earlier and more universally popular star and moon-bull, the black-spotted bull Apis. The combination of these two sacred bulls, the yellow and black bull, is found in the Hindu spit ox sacrificed to Rudra, the red (*rud*) god, who afterwards became the god Īshāna, the god (*āna*) of the mountain (*ish*). This bull-god Īshāna must not be speckled, or, if it has spots, the spots must be black, while if black, its colour must incline to copper colour. It was sacrificed at the close of the year to make way for the rule of the year-calf, its son, called Jayanta, or the conqueror, and before being sacrificed must be sprinkled with water mixed with rice and barley, so that it is a sacrifice of the barley-growers, the race who began their year with the autumnal equinox, which is one of the seasons prescribed for the sacrifice in the Hindu Gṛihya Sūtra.[1] It was these people who used as their indicator of the points of the compass the upright right-angled cross, and who looked on the true east and west as the quarters of the heavens sacred to the rising and setting sun of the equinoxes. It was to the true east and west that they oriented their temples in Palmyra, Baalbec, and Jerusalem; and it is this orientation which is repeated in the temples of the Egyptian pyramid builders.[2] It was at the headquarters of this barley-growing race, in North Palestine, that the death of the old year was mourned and the new welcomed, in the Tammuz festival held there at the autumnal equinox, when the year of Mace-

---

[1] Oldenberg, *Gṛihya Sūtra Āsvalāyana, Gṛihya Sūtra*, iv. 8, 1-7; *Gṛihya Sūtra* of Āpastamba, vii. 19, 20; *Gṛihya Sūtra* of Hiranyakesin, ii. 3, 8; S.B.E. vol. xxix. p. 255; vol. xxx. pp. 291, 292, 220-222.

[2] Norman Lockyer, *Dawn of Astronomy*, chap. ix. pp. 92-94; chap. viii. p. 85.

donia, Asia Minor, Syria, and Sparta began,[1] the year still used by the Jews, who begin their year on the 1st Tishri (Sept.-Oct.) with the Feast of Trumpets, when the year is welcomed by the music of the ram's-horn trumpets, made from the horns of the ram sacred to Varuna, the god of rain (*var*), and the dark night, the ram of the Golden Fleece, the starry heaven made golden in the dawn by the rising sun. This agricultural people, who worshipped the yellow sun-calf, the golden calf made by Aaron when Moses was hidden in the mount,[2] the golden calf of Dan,[3] are now represented by the Kurdish Druses of the Lebanon, the race who keep their creed secret, but who are said to worship the calf El Ejil, the calf of Dan. They live in the highlands of Bashan, the ancient land of the tribe of Dan, the Danaoi of Greece, the Dānava of India, the Turanian sons of Danu of the Zenda-vesta, who is, in the Rigveda, the cloud-god Danu, son of the atmospheric vault, his mother, who was slain by Indra, the rain-god.[4] It is his descendants, the worshippers of the rain-god of Syria, the gods of the upright right-angled cross of St. George, who were the central stock round which the future confederacy of the Semites crystallised. Their land of Bashan is the land of the bull, or Bœotian race ; and it is there that the Jordan, the mother-river of the sons of the bull of Bashan, rises at Hasbeya, and it is there also that the sources, venerated as sacred at Banias, still issue from the temple, called by the Greeks the temple of Pan, the goat-god of time.[5] Their priests are called Akals, or those possessing understanding (*akl*), corresponding to the Hindu Ojhas, the possessors of wisdom (*ojh*), and they are all carefully trained and initiated into their doctrines, which are allied to those which have emanated from the theology of the Northern

---

[1] Lewis, *Astronomy of the Ancients*, chap. i. § 6, p. 29 ; Sayce, *Hibbert Lectures for* 1887, Lect. iv. p. 231.

[2] Exodus xxxii. 4-6.   [3] 1 Kings xii. 28-30.

[4] Rigveda, i. 32, 5, 9 ; *The Ruling Races of Prehistoric Times*, Essay I. p. 10, note 4.

[5] *Life of Sir R. Burton*, by Lady Burton, vol. i. pp. 507, 509.

god of knowledge. They worship one god who has made himself known to men in ten successive incarnations of Ali, Albar, Alya, Moill, Kaim, Moezz, Aziz, Abu Zechariah, Mansur, and Hakim. These ten incarnations, reproducing the ten lunar months of gestation and ending in Hakim, the physician, point to a theology belonging to the era of thought which, in Greece, made Æsculapius, the divine physician, one of the chiefs of the gods, and made Jason, the healer, the leader of the crew of the star-ship Argo, that brought the Golden Fleece to Greece; which described the Ashvins, the twin-horsemen, the stars Gemini, who, in the Rigveda, bore the daughter of the sun in their chariot [1] as the physicians of the gods, and made Aryaman, who is, in the Rigveda, associated with Mitra, the moon-god, and Varuna, the rain-god, the god of the dark heaven, one of the triad of gods who ruled the sky. For Aryaman is, in the Zendavesta, called Airyaman, the holy bull, who heals all the diseases caused by the evil spirits, and makes in the earth the nine furrows which testify to the descent of the gods of heaven, whose sacred number is nine.[2] He is, as I have shown, probably the star Capella *a* Auriga, the patron star of Babylon,[3] where the temple of Bel was oriented to the true east and west; and it was under the influence of the theology which made the sun-god the golden bull-calf, that the seven stars of the Great Bear, who had been the seven antelopes, became the Haptoiringas, or seven father-bull (*iring*) stars of the Zendavesta. The Cyclopean architecture, which is the distinguishing mark of the age which, in Greece, produced the cult of the divine physician Æsculapius, Jason the healer, and Cheiron the centaur, their teacher, the possessor of the healing hand (*cheir*), is also found in the cities of Bashan and the ruins which abound in the Druse country.[4] Also

[1] Rigveda, x. 85, 8, 9.
[2] Darmesteter, *Zendavesta Vendīdād Fargard*, xxi. 1, 20, 21; xxii. 20; S.B.E. vol. iv. pp. 224-229, 234.
[3] *The Ruling Races of Prehistoric Times*, Essay v. p. 419 note 2.
[4] *Life of Sir R. Burton*, by Lady Burton, vol. i. p. 515.

the Northern affinities of the Druse dwellers in this land are shown in the constitution of their society, which is based on the family, in their monogamist marriages, and in the reverence they pay to women, who are not secluded like the majority of the women in the East, but who have, in national belief, the same right to rule as men, for when Sir R. Burton was consul at Damascus, the Jumblatts, the most powerful family among the Lebanon Druses, were ruled by a chieftainess, the Sitt Jumblatt. Their ancient connection with Egypt is also attested by the existence, near Cairo, of a body of Crypto-Druses. That their religion is based on a lunar-solar reckoning of time is proved not only by the ten incarnations of their supreme prophet, but by the further story that Hakim assigned a period of twenty-six years,[1] the number of lunar phases in a lunar year of fifty-two weeks or 364 days as the time during which the message of mercy should be open to mankind. This period of twenty-six years, as I have shown in the Essay on the astronomy of the Veda, is the period assigned in India for the duration of the apostle-ship of each successive Guru, or teacher of the Unitarian creed of the Kabir Pantis, the creed professed by the members of the great agricultural castes, the Koormis and Koiris of North-Eastern India.

This analysis of the deductions which we may justifiably make as to the history and mythological age of the people who called themselves the sons of the cow, and worked out the first germs of their monotheistic creed, proves that they were almost certainly the race who used as their sacred symbol of the god who rules the four quarters of the heavens

the upright right-angled cross ─┼─ of St. George, the

worker (*ourgos*) of the earth (*ge*), the plough and ram-god of Cappadocia and the Euphrates valley, who is worshipped everywhere in Syria as the holy George whose central shrine

[1] *Encyclopædia Britannica*, Ninth Edition, Art. 'Druses,' vol. vii. pp. 483-486.

is at Lydda, the city of the holy George, known in Arabia as El Khudr, the water-god, and in Greece as the prophet Elias, or Zeus Huetios, the showery Zeus. This was the mother-land of the growers of fruit trees, the sons of the olive and the fig-tree, who first divided the year into four seasons. The right-angled cross of St. George is, as I have shown, one of the component parts of the eight-rayed star, the sign of god and seed in the earliest Akkadian and Chinese script. In completing the history of the eight-rayed star, we must now proceed from the story of the sons of the cow to that of their successors the sons of the horse, who, as I shall show, completed the eight-rayed star by the addition

of the transverse cross ✕ which I have hitherto called

the rain-cross, but which is, as I shall show, the cross indicating the path of the sun as the god who brings the rain, the god who creates the seed of life. When we trace the migration of these sons of the horse to India we find that they there learnt to connect the victory of the sun over the frost giants culminating in the summer solstice with the rising of Sirius at the opening of the Indian rains, and it is this sun-god of the summer rains who is depicted in the Egyptian star of Horus with its five points, indicating the five seasons of the

Indian year ✕ . But before their emigration southward

they had learnt to use the measuring rod of the builders, the fire-drill of the fire-worshippers, as the gnomon which marked the sun's path by the length and direction of the shadows it cast, and worshipped the deer which browsed on the short, sweet grass of the chalk and limestone downs as the symbol of the god which led the corn-growing races to the well-drained lands best suited for the growth of their national crops. It was in their mythology that the stars of the Great Bear, which they worshipped as Artemis, the bear-

mother, the guiding star of the wanderings of the Mountain
Finns, the sons of the pine-tree, became, to the sons of the
deer, Artemis Elaphela, that is, Artemis the deer (ἔλαφος)
goddess, and the rounded hills of the chalk and limestone
country were imitated in their ritual by the artificial
mounds such as that of Avebury,[1] on which they built their
temples.   In these mounds, and their round barrow tombs,
they reproduced both the rounded house of the artisan races
and the burning mother-mountain, the home of the fire-
stone, the Shu stone of Indian and Akkadian mythology,
the mother and father of life of the sons of the bird Khu
or Shu, and this stone, the father of fire, was first the flint of
the limestone and chalk country.   It was on the sacred mound
at the top of Hindafjall, the mountain of the deer, that
Sigurd, the gnomon sun-pillar (urdr), found and awakened
from her winter sleep Brunhilda the mother of the springs
(brunnen), the Gandhāri, or wetter of the land, the bird-mother
of Hindu mythology.   The deer worshippers were the mixed
race formed from the union of the sons of the mother-tree,
the mother-bear and wolf, the lordly boar and the prolific
sow, the mother-cow, the mother-mountain, and father fire-
stone, the people who looked on the sun-god of the
equinoxes and solstices as the god who made their crops to
grow, and who ripened their barley, the seed of life (zi), the
Zeus of the Greeks, which gave its name to the Deus of the
Latin and the Theos of the Greeks, the Manx god Ji.
This father sun-god was the god on the grey-white horse,
the clouds, the white horse in Zend mythology of Tishtrya,
the star of the summer solstice who succeeded the golden-
horned bull of the bull race, as the adversary and conqueror
of the black horse,[2] and the black bull or dragon, the cloud
which will not give up its rain, which was in Northern
mythology the winter frost giant.[3]

[1] The Mound at Avebury is 130 feet high.
[2] Darmesteter, *Zendavesta Tir Yast*, vi. 16 ; S.B.E. vol. xxiii. p. 98.
[3] The White Sun-horse is still worshipped and fed daily at Kobe, in Japan.
Moseley, *Notes of a Naturalist on H.M.S. 'Challenger,'* chap. xix. p. 417.

It was this white horse, the sun-god of the limestone, flint, and chalk country, which was the god of Stonehenge, the temple whose ruins still remain to set before us with absolute certainty of the correctness of the deduction in its main details, the complete ritual of this primæval worship. The worshippers of the sun-god who built this temple are proved to have belonged to the Bronze Age, by the number of round barrow tombs found immediately round it; for Sir R. Hoare counted three hundred of these tombs within twelve miles of it, and Stukeley (A.D. 1723) counted one hundred and twenty-eight as visible from a hill close by.[1] The inner shrine of the temple is formed by an outer circle of thirty Wiltshire sarsen[2] stones topped by coping-stones joining their summits. As I shall show presently that the temple certainly depicts the annual course of the sun, there can be no doubt that these thirty stones mark the thirty days of the month of the solar-lunar year of 360 days, the wheel-year of the chronometry of the Rigveda.[3] Within this circle of thirty stones, there was before the temple had been allowed to fall into ruin, an inner circle of forty smaller stones, all but four being made of syenite, which must have been brought from

[1] Boyd Dawkins, *Early Man in Britain*, chap. x. p. 376; *Archæologia*, vol. xliii. p. 305.

[2] Sarsen stones are the boulder stones left by the ice sheet of the glacial period on the Wiltshire Downs.

[3] This is proved conclusively by the fact that the Scandinavian fathers of the sons of the horse divided their months into six periods of five days each, which they called Fimt, meaning the week of five days. (Vigfusson, *Icelandic Dictionary*, s.v. Fimt.) This year was also that of the Iranian sons of the bull and the horned-horse Keresāspa, who measured their solar-lunar year by twelve months of thirty days each. The first five days, Panchak Fartum, were consecrated to the new moon, or Antare Maungha, meaning the moon within; the next five days, Panchak Datīgar, formed the Perenō-maungha, the moon full; the next five days, Panchak Sitīgar, were those which lasted up to full moon, and were called the Vīshaptatha. The names of the last three Panchaks have not been preserved. (Darmesteter, *Zendavesta Māh Yasht*, 4; S.B.E. vol. xxiii. p. 90, note 5). This was also the year of the Hindu Karaṇas. It is this year which still survives in the Shan States of Burmah, and in Western China and Yunan, where the weekly markets are always held at intervals of five days. (Sir D. Forsyth, *Autobiography and*

Dartmoor. This must have been the original circle of Ia, the god of the house of the waters, the god Yah of the Jews, and the Ya-du or holy (*du*) Ya of the Hindus, whose sacred number is forty. While the thirty-six syenite stones in this circle reproduce the thirty-six steps of the Hindu Vishnu, the year-god of the royal race of the Ikshvāku, the sons of the sugar-cane (*iksha*), who are represented in the *prastara* or rain-wand of Ashva-vāla, or horse-tail grass, used in their Soma festival as the sons of the horse. These steps are said in the ritual to have been made from the middle peg to the eastern end of the great Soma earth-altar, and are said in the Brāhmaṇas to be reproduced in the thirty-six syllables of the Brihati metre.[1] They symbolise the thrice begotten year of twelve months, a re-echo of the sacred nine, the three times three, the number of the gods of heaven, the Igigi of the Akkadians. But these three gods represent the gods ruling the year of three seasons, whereas the year of the fruit-growers is one of four, hence the sanctity of the four times nine making thirty-six; and as the year is a solar-lunar year, the four or fundamental number has to be added to the

Reminiscences, p. 188.) This was also the year of Numa Pompilius, who added 57 days to the ten-months year of 304 days of Romulus. (Lewis, *Astronomy of the Ancients*, pp. 35, 55 ; *Censorinus*, c. 20 ; Macrob. *Sat.* i. 12, § 3, 58, 1·13 ; § 1·7 ; Solinus, 1·37, 38.) This was the year which is alluded to in the passage in the Kalevala, which tells how Vaïnämoïnen made the first sacred Kantele or harp from the jawbones of the pike, the river fish-god, and how on the teeth he fixed five strings taken from the hairs of the tail of a magic steed, the Finnic counterpart of the Indian Soma prastara or magic rain-wand made of the Ashva-vāla or horse-tail grass (*Saccharum spontaneum*). This was the year of the sun-horse, as distinguished from the lunar-solar year founded on the week of seven days, and divided into 13 months of 28 days each, making the whole year 364. This was the year of the astronomical sun-fish of the sea, the Sallimanu, or wise-fish, the year of the Semites, which is, as I have shown in the Essay on the Astronomy of the Veda, the year of the Buddhist Theris which preceded, in the Buddhist ritualistic development, the wheel-year of 360 days. It is also the year of the Assyrian fish and seed-gods Ia and Anu, worshipped on the twenty-sixth, twenty-eighth, and thirtieth of the month in which every day is dedicated to a god. (Sayce, *Hibbert Lectures for* 1887. Lect. i. p. 70.)

[1] *Sat. Brāh.* iii. 5, 1. 4, 9 ; S.B.E. vol. xxvi. p. 112, 113.

thirty-six to make forty, the four times ten, or the ten
months of lunar generation proved in the divine fire represented by four, the sacred number of the fire-god.

Next to these two circles, there are round the west end of
the altar, inside the inner circle, the two sacred horse-shoes
representing the shoes of the sun-horse, and also the crescent
moon, whose changes, like the horse-shoes, mark the steps of
its progress in its annual journey. The innermost of these
two horse-shoes, that nearest to the altar, was formed of
nineteen stones of Dartmoor syenite, rising in height from
six feet at the north-east end to nine feet behind the altar.
Here we have again the ten lunar months of gestation and
the nine gods of heaven who, in the Egyptian hierarchy, are
named Shu and Tufnit, Geb or Seb and Nut, Set, Osiris,
Nepthys, Isis, and Horus. It is this number nineteen which,
as we learn from the teachings of the Vajrāsun or thunderbolt (*vajra*) stone forming the floor of the Holy of Holies in
the Buddhist Mahābodhi temple, was especially sacred to the
worshippers of the sun, among whom the Buddhists hold a
conspicuous place. For they regarded the Buddha as an
incarnation of the sun-god led forth by his horse Kanthika,
the thorned (*kantha*) or rayed horse, to the baptismal bath
of ascetic regeneration which was to make him the example
and guide to perfect righteousness for all mankind. He was
the mythical successor of the year-horse Karṇa, the horned
(*keren*) son of Ashva, the river-horse, the sun-horse of Hindu
mythology who was the ruling god and mythical king of
Añga, the burnt (*añga*) land, the volcanic regions of the
modern Behar. He had, like Sigurd, impenetrable golden
armour, of which he was beguiled by Indra with the gift of the
dart which could not be baffled; the dart is the revolving pole
of time with which each year kills the old year and brings
the new year to fill its place. The border of the Vajrāsun
or thunderbolt throne of the sun-god shows the connection
between the supreme god of light and the number nineteen
in the double rows of nine squares at its sides, and of ten

squares at its ends, each square being filled with a representation of the double Vajra or thunderbolt , the sacred symbol of the heavenly father of the race who worshipped the thunder and rain-god. The four border rows of nine squares and the four of ten contain seventy-six squares, and the religious significance of the nineteen squares, each containing the transverse cross as the symbol of the double thunderbolt, is shown by the treasures buried under the foundation stone of this holy seat. These consist of nineteen jewels and seventy-six discs. This number nineteen is repeated again in the number of circular markings made to denote the Buddha's steps in his walk at the side of the temple.[1] From the evidence I have adduced in the Essay on the astronomy of the Veda, proving that the Buddhists used two years, one of thirteen lunar months, and the wheel-year of twelve solar months, it would seem that, in their mythology at all events, the nineteen meant the ten lunar and the nine solar months of gestation, and this was probably their meaning in the Stonehenge temple, where the thirty stones of the outer circle evidently point to the use by the builders of a solar year of twelve months of thirty days each, answering to the wheel-year of the Buddhists.

Within this inner horse-shoe is the altar of Derby micaceous sandstone, which was placed north-east and south-west along the line traced by the rays of the rising sun on the day of the summer solstice. At the back of the inner horse-shoe,

---

[1] F. Pincott, 'The Vajrāsun, or Thunderbolt-seat of the Mahābodhi Temple,' *Transactions of the Ninth Congress of Orientalists*, vol. i. pp. 246, 247, 251. That this number nineteen was one sacred to the sun-worshippers and circle stone-builders of South-west England is proved by the existence of four hundred circles each of nineteen stones at Boscawen and adjacent places in Cornwall, Thurnam on *Megalithic Circles*, Decade iv. ; Lubbock, *Prehistoric Times*, chap. v. p. 117. This theology told in stones, and common to India and Southern England, as is conclusively proved by the nineteen stones and the thirty-six spoken of in page 139, certainly did not originate in the stoneless plains of Northern India. It is most probably a product of the great country of stone circles, Bashan, the land of the bull race, the Gautuma of India.

and between it and the circle of forty stones, stood the outer horse-shoe, formed of seven pairs of sarsen stones, each pair crowned and united by a lintel stone at the top. These fourteen wedded stones were placed in the form of a horse-shoe, the pairs at the two north-east ends being lowest, and rising in height to the great pair behind the altar, which must have measured, as Dr Stukeley says (A.D. 1723), about thirty feet high. The broken pieces of one of them now measure twenty-six feet, eleven inches.[1] These fourteen paired stones, each pair united by its lintel coping-stone, represent the fourteen days of the lunar phases, and the two wedded weeks of each phase as conceived by the race who believed in the divinity of pairs and the sanctity of marriage; and they also represent the lunar-solar year of thirteen months of twenty-eight days each, as distinguished from the solar-lunar year of the outer circle, composed of twelve months of thirty days each, divided into six weeks of five days. But apparently, at Stonehenge, the full lunar year of thirteen months is only represented in its inchoate form of ten lunar months of gestation which appear in the horse-shoe of nineteen stones. Thus the chronological order of the year, as reckoned by months, appears to be (1) the ten lunar months of gestation, (2) the twelve months of the wheel solar year, which still preserved the sacred ten of the first reckoning in their month of thirty days and its divisions into six periods of five days. This was the year of the sun-horse and the river-fish, the pike father-god of the Kalevala, the year of the oil-press with its four seasons, (3) the lunar-solar year of thirteen months of twenty-eight days each, founded on the week of seven days, the seven strings of the lyre fixed on the tortoise by Hermes, while the former year was founded on the five strings of the tail of the magic horse fixed on the harp called the Kantele, formed by Väinämöinen, as recorded in the

---

[1] These and all other details relative to the former construction of the Stonehenge temple are taken from the official handbook prepared by Mr. Judd, a local antiquary living at Maddington, near the temple. See Part I. pp. 9-22, especially pp. 10, 13, 14, 15, 20, 21 ; also Part II. pp. 34.

Kalevala, from the jawbones and teeth of the father-pike. This last year, which is not apparently represented in the theology of Stonehenge, is the year of the sun-fish of the sea, Salli-manu, the fish who, as I have shown, emerged from Aquarius in November, remained three months under the guardianship of the moon, and in the ten months of his active career, passed through the ten stars called by the Babylonian astronomers the ten kings of Babylon. This year of thirteen months and 364 days was the first year founded on astronomical observations showing the track of the moon and sun through the stars.[1] This, as I shall show in the story of the myth of Jason, was the year of the heavenly dove, sacred to the sea-fish, sent forth by Noah, the last of the ten months of the sun's active life, to fetch the leaf of the sacred olive-tree, and also the year represented in the Greek myth of Perseus, the sea-fish of the Euphrates, and ancestor of the Persian kings. The proof of the general correctness of these deductions is completed by the sacred stones placed outside the centre circle, four of which mark the course of the sun on the days of the summer and winter solstices. The principal of these stones, placed to catch on its summit the rays of the rising sun at the summer solstice, is the stone called the Friar's Heel. It stands about 110 feet to the north-east of the circle, and midway between it and the circle is the slaughter-stone, on which the victims offered to the sun-god were slain. This slaughter-stone, and the animal victims slain on it, prove that the ritual at Stonehenge was very similar to that of the Indian Soma-sacrifice of the days of the Ashvins or heavenly twins, for it was in their honour that animal sacrifices were offered at the summer and winter solstices, called the Turāyana, or times (*ayana*) of the Tur, the revolving pole; and these sacrifices were continued during the succeeding age of the purer ritual of the milk libations and the Soma-cups. The course of this revolving pole indicated by the sun, which was supposed to turn round with it, as the oil-mill is turned round by its

---

[1] *The Ruling Races of Prehistoric Times*, Essay iv., pp. 382-386.

directing beam, is shown by the other three sun-stones placed to catch the rays of the setting sun of the summer, and those of the rising and setting sun of the winter solstice, and the significance of the connecting lines between the east and west points is shown by the slaughter-stone and the altar, both of which are placed upon the line between the north-east and south-west points, marking the rising sun of the summer and the setting sun of the winter solstice. Thus the four stones were clearly intended by those who set them up to mark not only the solstitial points but also the terminals of a transverse cross exactly similar to the Egyptian

cross of Horus ✕ in the hieroglyph for a star. This, called St. Andrew's cross, is the cross on the Hindu altar, where the diagonal lines run, like those of the rays of the rising and setting sun at Stonehenge, from north-east to south-west, the path by which Indra, the rain god, brings up the rains with the south-west monsoon ; and from south-east to north-west, the path across India, traversed by the Northern Māghadas, the worshippers of the household fire, who entered it from the north-west. But while the Indian cross has, according to the statements of the Brāhmaṇas, a meteorological and historical meaning, it is clear from the evidence of the Stonehenge sun-stones that it, like the right-angled cross of St. George, was originally of Northern and solar origin, and that it was brought to India by Northern immigrants. For the four stones at Stonehenge showed, as indicated in the accompanying diagram, the path of the sun at the two solstices.

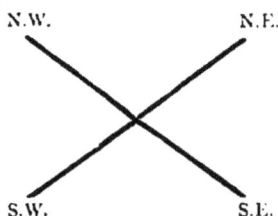

Here the dotted lines between N.E. and N.W. and S.E. and S.W. show the track of the sun across the heavens on the days of the summer and winter solstices.

The cross thus formed is the same as that depicted in the Svastika inscribed in the triangle drawn on the figure of the goddess-mother, found in the second city from the bottom of the six cities on the site of Troy, a city which belonged to the Bronze Age. The figure of the mother-goddess reproduces, as I have shown, the Hindu altar made, according to the Brāhmaṇas, in the form of a woman, and the Agni or Svastika in the centre of the triangle on the altar representing the year of three seasons, is clearly a symbolic picture of the sun rising at midsummer in the N.E. and setting in the N.W., and at the winter solstice rising in the S.E. and setting in the S.W.;[1] and this interpretation is corroborated by the Hindu ritualistic rule requiring the Samidhs, or kindling sticks, which lighted the Hindu midsummer bale-fire sacrifice to the sun-god, to be placed on the lines drawn on

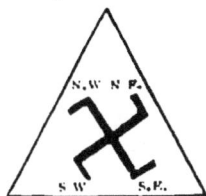

the altar to represent the ╳ transverse sun-cross, marking

the path by which Indra brought up the S.W. monsoon rains at the summer solstice. It was this cross which was the first germ of the conception which, in the progress of astronomy, led the Euphratean students of the stars to conceive the ecliptic equator of Anu, the god of seed, as opposed to the horizontal equator of Bil, the fire-god, and it was when the diagonal cross was placed on the right-angled cross of the observers of the horizontal equator that the sign of the eight-rayed star ✳, formed by the two united crosses became that which denoted, in the Akkadian writing of Gir-su, the divine essence, called both Anu and Esh-shu, both of which names mean both god and seed, while Esh-shu is composed of Ish, which

---

[1] See the figure of the goddess as drawn on p. 170 of *The Kuling Races of Prehistoric Times,* and on p. 67 of Schuchhardt's Schliemann's *Excavations,* also p. 167, and Eggeling, *Sat. Brāh.* i. 3, 4, 5; S.B.E. vol. xii. p. 91 note 1.

appears in the Hindu Īsh-āna, and the Akkadian Is-tar, a name which originally meant the mountain, and thence came to mean the swelling seed, the egg of the Shu or Khu, the mother-bird named in the second syllable of Esh-shu.

The four stones, marking the sun-cross of the solstices, as well as the slaughter-stone, the fourteen paired stones forming the outer horse-shoe between the inner ring of forty stones, and the inner horse-shoe of nineteen stones, and the thirty stones of the outer circle are all of sarsen stone, and they reproduce the time-measurement of those races who made the moon describe in each month a circle of thirty days, the number calculated in the Hindu Karaṇas.[1] These sarsen stones, and perhaps also the inner altar, were probably erected by the later founders of the ritual of the Bronze Age, while the earlier cult of the rain and fire-god and of the sun-horse is represented by the syenite stones of the inner circle, and of the inner horse-shoe, and their great antiquity is shown by the fragments of syenite found by Sir R. C. Hoare, in the bell-shaped barrow to the south-east of Stonehenge, proving that this foreign stone was regarded as hallowed by the makers of the tomb, and its presence in the tomb, therefore, in those very conservative days, prove the great antiquity of the syenite monumental stones.[2] The mythology of the people who set up these stones was that of the votaries of Ia, the fish-god, whose sacred number was forty, and who looked on the ocean, the home of the mother-fish, the sea-dolphin who bore the prophet priests to Delphi, and of the Midgard serpent of the Edda, as the mother of all life ; the source whence the waters were drawn upward by the pipe supposed to run through the mother-mountain, to become the clouds and mist wreathing its summit, and

---

[1] Sachau's Alberuni's *India*, vol. ii. chap. lxxviii. pp. 194 ff. They divide the month into two halves, the light half and the dark half, the half of the waxing and that of the waning moon, and they make the 15th and 30th days of rest, leaving, as in the fourteen paired stones at Stonehenge, fourteen days for each of the two lunar phases.

[2] Judd, *Stonehenge*, Part IV., p. 52.

descending thence to the earth as the rains which filled the irrigating rivers with life-giving water.

A further indication of the age to which the Stonehenge mythology belongs is given by the hippodrome which can still be traced about half-a-mile north of the temple, with which it is connected by an avenue about forty cubits wide. This is divided into two branches, about 1700 feet from the temple, the eastern hand going eastward to Radfin, a ford on the Avon, and the western curving round to the ancient chariot-course. It was here that the seasonal games took place, said by Macrobius to have been celebrated by the Druids, when sacrifices were offered to the gods.[1] This ancient Campus Martius, running east and west, is about 10,000 feet or 6000 Druidical cubits long, and 350 feet or 200 Druidical cubits wide, and on the east side is a long bank, extending nearly its whole length, which must have served as a place for spectators, while on the west side is a curve to allow for the turning of the competing chariots.[2] There can be no doubt whatsoever that this racecourse represents the ancient site of the national games, instituted by the sons of the horse, which are said in Greek tradition to have been founded by Akastus, king of Iolcus, after he had driven out Jason and Medea the sorceress. This was the age when, as I have shown, surgery by the healing-knife (ἄκη), symbolised, under the name of Akastus, meaning he who heals by the knife, superseded the cures made by cautery, oil, and magical incantations and charms;[3] the age when scientific investigation began to work more by actual experiment than by random guesses. This was the age which is represented in Indian historical ritual by the Vājapeya sacrifice, and its accompanying chariot races and national games. These, like the games instituted by Akastus, were celebrated in honour of deceased ancestors, the Pitaro Barishadah, or Fathers seated on the Barhis of Kuṣha grass, and the Pitaro

---

[1] Macrobius, *Satur.* i.   [2] Judd, *Stonehenge*, Part III., pp. 39-42.
[3] *The Ruling Races of Prehistoric Times*, Essay vi., pp. 524-526.

'Gnishvāttāḥ, who burned their dead, and to whom Kuṣha
grass, ears of fresh barley, roasted barley, and the milk of a
cow suckling an adopted calf were given. This sacrifice was,
as we are told in the ritual, the sacrifice of a long-haired, and
therefore of a Northern race, like that of the sons of the
horse, and in the ritual a cup of mead, the honey drink of the
North, and of the people who made the Ashvins or heavenly
horsemen their parent-gods, was given in exchange for the
seventeen cups of Surā, or distilled spirits prepared by the
Neshtri priests of the Southern matriarchal races.[1] This festival to the dead of the Bronze Age, was held on the ground
near Stonehenge, which, as the sacred temple of the British
worshippers of the sun-horse, is surrounded by the tombs
of the chiefs and nobles who successively ruled the country :
and it was originally like the festival of the Hindu Piṭaro
Barishadah held at the autumnal equinox, the time when, as
I shall now prove conclusively, the year of the barley-
growing races began in the countries of South-western Asia,
where the cross of St. George was the sacred sign of the rain-
god, and this annual New Year's festival was altered to the
time of the summer solstice, when the sons of the horse
reached India, and found that it was at that time that the
rain-god brought the annual rains to fertilise the lands of
Northern India. And it was the maritime traders of India
who started from its western coasts, where they found ample
supplies of shipbuilding timber, and who founded the maritime
commerce of the Persian Gulf, Southern Arabia, and Egypt,
where no such timber is grown. It was these people, the great
migrating race of the Minyæ, whose history I shall set forth
presently in telling of the myth of Jason, who used, as the
sign of the seed which preserved and continued life on earth,
the two united sun-crosses, marking both the equinoxes
sacred to the growers of barley, the seed of life, the dwellers

[1] *Kātyayana* xiv. I. 1 ; *Ṣat. Brāh.* v., 4. 1, 2 ; Hillebrandt, *Vedische
Mythologie*, pp. 247-249 ; *The Ruling Races of Prehistoric Times*, Essay iii.,
pp. 207, 208.

in the land of the upright right-angled cross of St. George, the plougher of the earth, and also the solstices, the coming of which had from time immemorial been announced in the North by the bale-fires, still lighted at midsummer in all Western lands where the primitive cult survives, and by the Saturnalia held in Asia Minor and Italy at the winter solstice, to celebrate the birth of the young sun-god, an annual event commemorated in the more Northern lands by the eating of the father-boar. These indiscriminate national dances and intoxicated orgies of both sexes were survivals from the ritual of the matriarchal village races, who provided by the institution of their seasonal dances, for the meeting between the men of one village and the women of another, and this secured the observance of the national law, which forbade any man to become the father of the child of a woman of his own village. They were, when adopted by the races sprung from the union of Northern immigrant men with Southern women, changed into dances of the men and women of the same village, for these people had, like the Massagetæ described by Herodotus,[1] and the earlier Jats of India, adopted the custom of communal marriages between all the men and women of the same village, and this formed a link between the matriarchal races of the South, and the monogamists of the North. These dances produced, as their natural consequences, the birth of children ten lunar months afterwards, and it was to assert their rights in the parents of these children thus begotten that the Basque fathers introduced the custom of the couvade.

Thus the temple of Stonehenge is one which reproduces in its symbolism the whole mythology of the race of the Druids, or sons of the tree (dru), the sacred oak of Dodona, from the days when the first cultivators of Southern forest clearings were united, first to the shepherd races, who ate their father totems, and thus introduced animal sacrifices, and secondly, to the fire-worshipping artisans, up to the time when these

[1] Herod. i. 219.

confederated tribes made their measuring-rod and fire-drill
into the gnomon-pillar, the symbol of the creating-god Tur,
and when they exchanged the small forest clearings of the
earlier matriarchal tribes for the wider horizon of the hill
downs, where herbage took the place of the timber of the
woodlands. It was there that they learned the use of the
gnomon as the announcer of the changing seasons of the
year, and began to look on the sun-god as the creating
father Rā, who ripened their grain. It was this worship of
Rā which first led to the use of the place consecrated by the
prophetic gnomon as the site of the national altar ; and the
first altar thus used was that appropriated to the worship of
the Greek Hekate, the mother of a hundred (ἕκατον) sons,
the three-formed goddess, the Greek counterpart of the
Hindu Gandhārī, the bird-mother of the hundred Kaurāvyas,
and also to the worship of the Hindu three mother-goddesses
of the three seasons, and their mate Rudra, the god of the
gnomon, or sacrificial stake ruling the fourth season, called
Rudra-Triambika, or Rudra with the three (tri) wives.
Sacrifices to them were offered, both in India and Greece, at
the place where four cross roads meet, or, in other words, in
the centre of the right-angled cross of St. George, the rain-
god, denoting the four quarters of the heavens, and the totem
animals offered to Hekate were dogs, black ewe lambs, and
honey, showing that the cross-road goddess was the mother
of the united fire-worshipping and shepherd races, who called
themselves first sons of the dog, and of Idā, the sheep-
mother, and subsequently, sons of Varuna, the rain-god,
to whom ewes and rams, but more especially the ram, were
sacred.

This was the stage in ritualistic evolution which is still, in
company with other primæval rites belonging to the same
line of religious thought, found surviving in the ritual of the
worship of the polar star. This is the star worshipped as the
Olma d'nhoora, the world of light, by the Sabæan Mandaites
of Euphratean Mesopotamia, known locally as the Saibi

or Sabium, meaning the washers or baptists,[1] dwelling in the land enclosed between the Euphrates and Tigris, the twin rivers, parents of the twin races, who are known in history as the Khati or Hittites. The time consecrated to the adoration of the parent of light is the last evening of the old and the first morning of their new year, which, like that of the Syrian worshippers of St. George, begins at the autumnal equinox. This is the season answering in the Jewish calendar to the first of the month Tisri, the day of the feast of ram's trumpets, falling on the first new moon after the autumnal equinox. It is, in the cosmogony of the Akkadian star-worshippers consecrated to the star Antares *a* Scorpionis, which is called in the Tablet of the Thirty Stars 'the lord of seed in the month Tisri.'[2] This is the star of Dan, the judge, and therefore of his sons, the Hindu Dānava, the Greek Danaoi, the Zend and Akkadian Turanian sons of Danu ; for it is he who in ancient mythology was the goat-god Uz or Esau, who watched the revolutions of the solar disc on the Babylonian monuments.[3] He was the father of the Hittites, who holds the staff of empire, the goat-headed sceptre of Osiris, the goat and ram-god, the rod (*rhodon*) of Rhadamanthus, the Greek judge and diviner (*mantha*) by the rod, the axial pole of the cross of St. George, which was also the road running from South to North in the cross roads of the mother-goddess Hekate. This was the cardo, the pivot axis of the world, with which the Roman agri mensores, the guardians of the traditions handed down by the first measurers of landed property, measured their templum or standard field, and this cardo was cut by the east and west line called Decumanus, so that the whole formed a right-

---

[1] Sir G. Birdwood, Introduction to Comte Goblet d'Alviella's *Migration of Symbols*, p. xv.

[2] R. Brown, junr., F.S.A., 'Remarks on the Tablet of the Thirty Stars,' Star No. xxiii., *Proceedings of the Society of Biblical Archæology*, February 1891.

[3] Sayce, *Hibbert Lectures for* 1887, Lect. iv., p. 285.

angled cross.[1] Its great and ubiquitous antiquity is also shown by the ancient Chinese sign for the earth and the popular Chinese saying that God made the earth in the form of a cross,[2] which is also preserved in the rectangular field of the Zend Varena, the maker of the garden (*vara*) of god. That the star of Dan, the judge, the holder of the gnomon rod, was the star of the Scorpion, is proved by the arrangement of the Jewish camp as told in the Book of Numbers, for there the camp is said to have been pitched towards the four points of the compass. These are marked by the standards of Dan, Reuben, Judah, and Ephraim guarding the north, south, east, and west, and the cognisance on the northern banner of Dan was the scorpion, on that of Reuben a man, while the east and west were watched by the lion of Judah and the bull of Ephraim.[3] These united sons of the seed were also the sons of the rivers, and they chose for their parent-rivers those which, like the Euphrates and Tigris, descended from the mountains of their mother-range of hills dominated by Mount Ararat, and which ran from north to south. The section of the confederated sons of the bull and cow, who descended into Bashan as the tribe of Gad, followed this example of their predecessors by taking for their parent-river Jordan, which flowed parallel to the Euphrates. These people also worshipped the polar star, the guardian star of the Kushite race, and it is this

[1] Smith, *Dictionary of Greek and Roman Antiquities*, Art. ' Agri men-sores' and 'Cardo.'

[2] D'Alviella, *Migration of Symbols*, English Translation, chap. i. p. 14 ; *The Indian Antiquary*, 1880, p. 67 ff.

[3] Numbers ii.; Blake, *Mythological Myths*, chap. iv. p. 106. In Jacob's dying blessing (Gen. xlix. 16-17) Dan, the judge, is called ' The serpent in the way, a horned snake (*adder*) in the path, which biteth the horses' heels.' This is an allusion to the constellation Draco, the constellation of the serpent, which, in the earlier polar astronomy, was made the guardian of the pole of Dan. It is on this serpent's head that the foot of Hercules, the fire and sun constella-tion, rests in pictorial astronomy, and hence it was looked on as biting the heels of the sun, (that is, as regulating and measuring its course.

primæval faith which looked on the north star as the parent
of light, and on Antares as the star of the west, which is
preserved by the surviving remnant in Mesopotamia, who
call themselves the Mandaites. The ritual of their new
year's festival has been most carefully described by an eye-
witness, whose account was published in the *Standard* of
19th October 1894, and the attention shown by the narrator
to the minute details which are necessary for a perfect
understanding of the meaning of the rites shows that it is
most accurate and reliable. From these details we can glean
a complete history of the growth of this primitive church,
and are able to connect these people directly with the early
eaders of religious thought in other countries, and especially
in Persia and India.

I have already, in the Essay on the astronomy of the
Vedas, shown that the Haranite Sabæans dwelling in the
mountains of the Northern Euphrates valley near Haran,
worship the polar star, which they call Shemol, the star of
the left hand (Heb. *Semol*). They call themselves Bogdariten,
or the sons of the Slavonian god Bog, the Phrygian Zeis
Bagaios, the Sanskrit Bhaga, the Persian Baga, the god of
the garden (*bagh*) where the tree with edible fruit grows, and
they are therefore the gardening race. The name of the
star of the left hand, given by them to the polar star, marks
them as a people who, like the Jews in their traditional
camp in the wilderness, looked to the East, the place of the
rising sun, the lion of the tribe of Judah, as their mother-
land. They call the cave in which they celebrate their
mysteries, the 'house of Bogdariten,'[1] and they thus trace
their national birth back to the days when they were the
sons of the mountain-caves of their mother-mountain Ararat.

When they emigrated southward into the plains of Mesopotamia, they became the sons of the rivers which gave water
to their crops and cattle, and have been long known as the

---

[1] Chwolsen, *Ssabier und der Sabaismus*, chap. ix.; *Excursus*, pp. 319-363 :
Miller, *Harmoad, or the Mountain of the Assembly*, p. 50.

Nabathæans, whose agricultural knowledge has been transmitted to us by the Mandaite Kuthāmi, whose Arabic
translator, Ibn Wahshiyah, identifies St George with the
Assyrian Tammuz, the Akkadian Dumu-zi, or the son of
life, while another Arabian historian, Masudi, tells how the
national patron saint of Syria and the Euphratean countries,
Ghergis, the knight of the right-angled cross, was sacrificed
three times, reviving after each execution like the Hindu
Kacha, the tortoise slain three times by the Dānava, or sons
of Dan.[1] This is a mythical method of describing the three-
fold resurrection of the indwelling spirit of life which made
the three seasons of the year revive again after their annual
death. It was these same people who worshipped the mother-
bird, called El-Nasr, the eagle or vulture, by the section of
the race who ruled in Southern Arabia as the Sabæans; and
this mother-bird must have been, to a people who, like the
Sabæans and Kushites, looked on the polar-star as their
parent-star, the star Vega of the constellation of the Vulture,
which was the polar star from 10,000 to 8000 B.C. This was
the star called by the Kabiri of Byblus, the town of the
Papyrus (*Byblos*), their sacred town in North Palestine,
Eshmun, and it is this name which, as I shall show presently,
is still preserved in the Mandaite liturgy. The name
Mandaite denotes the sons of Manda, meaning the Word of
God, the creator of light, who said, 'Let there be light,'
and there was light. It was this god, whose sign is the
pole-star, who, we are told in the Ku'rān (vi. 77), was adored
by Abram as the god who made heaven and earth, the
invisible and ever-working creator; and the whole history of
the origin of this early worship of the one god is to be found
in Finnish and Phrygian mythology. The supreme father-
god of the Phrygians was Attis, the son of Cybele, also
called Pappas. His name shows him to be the god also
called the Old One in the Kalevala, for Pappas or Pappos
means grandfather, and this proves that the name Attis is

---

[1] *The Ruling Races of Prehistoric Times*, Essay i., pp. 12-14.

connected with the Sanskrit Attā and the Norse Edda, meaning grandmother, the Gothic and Frisian Atta, and the German Otto, meaning father:[1] and he is thus proved to be one of those primaeval creating powers who were first called mothers in matriarchal times, and afterwards fathers. Attis is said to have unsexed himself under the pine-tree, the tree-mother of the Phrygians and Finns; and this parent pine-tree was, like that placed on the roof of each newly-built German house, believed to grow on the top of the mother-mountain. He was called ὕψιστος, the highest, and is proved by his starry cap, Ἀστέρωπος πῖλος,[2] to be the god that rules the stars. This identifies him with the god called by the Finns and Esthonians Taara, the star-god, and Ukko.[3] Ukko is called Taivahan napanen, meaning the navel of heaven, and this is called in the Kalevala Tähtela, the place of Tähti, the pole-star,[4] the star at the top of the heavenly mountain. His daughter Linda (from lind, a bird) is born of an egg, and she and her sister are queens of the birds, that is to say, Ukko's daughter is the mother-vulture, the polar star Lyra in Vega. This mythical genealogy tells us how the worship of the father and mother bird became the adoration of the unseen creator, whose sign is the unsetting polar star, who shows his creative power as the heavenly fire-drill making life and heat, and shown to be continually revolving by the perpetual revolution of the circumpolar stars. It was from this conception that the custom of emasculating the priests of Cybele and Istar arose among the Phrygians and Ak-kadians, and it was this degrading custom which was denounced by the successors of the fire-worshippers, the people who adored Rā, the sun-god, and worshipped the pole-star as the left-hand star, the Iranian sons of Idā, or Irā, the sheep and cow-mother, and of Yima, the twins who

---

[1] J. O'Neill, *The Night of the Gods* (Quaritch, 1893), p. 373.
[2] *Ibid.* p. 488.
[3] Kirby, *Hero of Esthonia*, Introduction, p. xxvii.
[4] Schoefer Castren, *Finnish Mythology*, pp. 32, 33.

would not allow an impotent or diseased person to enter the
Vara, or garden of god. It was to this god Rā that the sun
ram of Varuṇa, the ram offered by Abram, was sacrificed :
but, as we shall see presently, the Puritan Mandaites, who
offered a wether to the sun-god, preserved, while adopting
the sacrifice of the new totem, the earlier reverence for the
unsexed fire-god. These Euphratean Mandaites are not
only a pastoral race celebrated for the excellence of their
dairy produce, and, therefore, like the Hindu Gotama, the sons
of the cow, the Hindu star-goddess Rohinī, the star Aldebaran, which was also a mother-star of the Sabæans of
Southern Arabia, but in their principal settlement at Mardin
in the Bagdad district, and in other towns, they are noted
for their skill as metal-workers and goldsmiths, so that they
are a mixed race, formed by the union of shepherds and
cattle-herdsmen, gardeners, corn-growers, and workers in
metal.

Their year reckoning, starting from the autumnal equinox,
is one that, in the historical order of the methods of calculating time, coincides with that in Egyptian mythology which
began with the birth of Horus in the Papyrus Marsh at
Buto, to which Isis was, at the command of Dhuti, the
moon-god, led by the seven scorpions, Tefne, Bene, Mastet,
Mastetef, Petet, Thetet, and Matet, the scorpion stars of
Dan.[1] This is the year which chronologically followed that
beginning, as the year of Orion did, with the winter solstice,
which was preceded by the year of the Pleiades, beginning in
November. It was this year of the barley-growers, beginning
at the autumnal equinox, which was that of the Hindu
Piṭaro Bariṣhadah, the Kushite barley-growing fathers,
seated on the Barhis of Kuṣha grass, to whom barley was
given at the autumn festival, in which they were commemorated ; and it was as the herald of this year that Kṛiṣhṇa,
the god of the black antelope (kṛiṣhṇa), who succeeded the

[1] H. Brugsch, *Religion und Mythologie der Alten Ægypter*, pp. 392,
402-404.

sun and moon-god, Rāma, the son of Rohinī, the red cow-star Aldebaran, as year-god, was taken by his father Vāsudeva across the floods of the river Yamuna to Gokul, the cow-pen, the home of Rāma, on the 8th Bhadon, the 23rd August, when the rains are beginning to moderate. Their approaching cessation, which was to be consummated at the autumnal equinox, is indicated by the birth of Durga, the mountain-goddess, the daughter of Jasodā, the 'exhausted or super-seded' moon-goddess of Gokul, who was the foster-mother of Rāma. Vāsudeva left Krishṇa with Jasodā, and brought back with him her daughter Durgā, whom he placed in the bed of his wife Devaki. When the infant awoke with her cries the guards placed by Kansa, the moon-goose (*kans*), to slay Devaki's children, the new-born mountain-goddess, raised, like the mother-mountain in Noah's flood, from the waters, went up to heaven as the mother-mountain, born at the autumnal equinox.[1] These united sons of the barley, the seed of life of the sons of the antelope, always, as in the settlement of Sook-es-Shukh, where the ceremonies I am about to describe were witnessed, chose as a site for their villages some place near rivers, for an ample supply of water is necessary not only for their cattle and cultivated lands, but also for their religious observances. Sook-es-Shukh is a name which recalls that of Suk-us, the Akkadian name of Istar, the mountain (*is*) goddess of rain. These people also show their dependence on the goddess of seed by holding their markets on a Friday, the day of Frio, the Northern seed-goddess, called Sukra-bar, or the day of Sukra, the rain-god, by the Hindus and Persians. These people have no permanent temples, showing that their creed dates from an age before brick and stone sacred buildings were built by the sons of the fish-god, Ia. Hence all meetings for public worship on a consecrated spot must be preceded by the erection of a house for the national god and his priests. This is placed close to the river Euphrates; and early on the

---

[1] *The Ruling Races of Prehistoric Times*, Essay v., pp. 467, 468.

last day of the year, called the Kaushio Zahlo, or day of
renunciation, the elders of the village, under a Shkando or
deacon, go down to the river to prepare their Mishkna or
outdoor temple.   This is done by cutting reeds, like those
which form the basket or birth-house in which the father of
the Kushite race, the sons of the rivers, is hidden in the
tribal legendary history, the reeds of the papyrus marsh of
the Egyptian Buto, the birthplace of Horus. These some
weave into basket-work screens, while others mark out, by
fixing stout reeds firmly in the ground, an oblong space running due north and south, about sixteen feet long and twelve
broad, for the temple enclosure. These reed-posts are joined
by strong cords, to which the basket-work screens are
attached so as to form an enclosure, as depicted in the
annexed plan, about seven feet high, with two openings in
the centre like windows, looking west and east, and on the
southern side of the enclosure there is space left for a door—

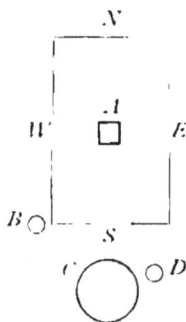

A Altar. *B* High Priest's hut. *C* Baptismal pool. *D* Bather's hut.

An altar of beaten earth (*A*) is raised in the centre of
the enclosure, and the interstices of the walls are daubed
with wet clay, which speedily hardens. On one side of the
altar a small furnace of dark earthenware with a little
charcoal, and on the other a small hand-mill is placed. A
circular basin (*C*), about eight feet across, is made close to
the southern wall, answering to the Rādhā-kund or pool
(*kund*) of Rā-dhā, the wife of Krishṇa, at Gobardhan, 'the

keeper of cows' in the sacred land of Mathura in India.[1] Water is led into this by a channel cut between it and the river. Two small wicker-work huts, just large enough to hold one person, are made, and one of these (*D*) is placed by the site of the baptismal pool, and the other (*B*) close to the southern wall beyond the entrance. This last is consecrated to the Ganzivro, or high-priest, and no layman may touch it after it has been set up. The doorways and windows are then hung with white curtains.

Towards midnight the people of the village, both men and women, begin to come down to the Mishkna, and, as each arrives, he or she goes into the wattled huts by the sacred pool, and bathes in it after undressing in the hut. A tarmido, or priest, stands as they bathe, and pronounces over each the blessing, 'Eshmo d'haï Eshmo d'manda haï madhkar elakh.' 'The name of the living one (*Eshmo*, that is, the Eshmun of the Kabiri), the name of the living word (*Manda*), be remembered upon thee.' After bathing each baptized disciple puts on the 'Rasta,' the white garments worn at religious festivals by the star-worshippers, in which they are buried after death. The garments included in the specific name Rasta are—(1) The long Sadro, or white shirt, the Sadarah, or sacred shirt of the Zendavesta,[2] which became the Ephod of the Jews; (2) the Nassifo, or stole, round the neck falling to the knees; (3) the Hiniamo, or woollen girdle, the Kōsti of the Zendavesta, the sacred girdle of the Parsis, descended from the Brahmin girdle, or sacrificial thread, the belt sacred to Orion, the deer-god;[3] (4) the Shalooal, or white over-mantle; (5) the Gabooa, or square head-piece, reaching to the eyebrows, over which is worn (6) the Kanzolo, or turban, one end of which hangs over the shoulder. Each disciple when clothed sits down on the

---

[1] *The Ruling Races of Prehistoric Times*, Essay iv., p. 460.
[2] Darmesteter, *Zendavesta Fargard*, xviii. 54; S.B.E. vol. iv. p. 199 note 3; 191 note 4.
[3] *The Ruling Races of Prehistoric Times*, Essay iv., pp. 402-408.

ground in front of and facing the south door of the Mishkna, saluting before he sits his or her predecessors by saying, 'Sood havi-lakh,' 'Blessing be with thee,' to which all reply, 'Assootah d'haï havi-lakh,' 'The blessing of the living one be with thee.' Two Tarmidos (priests), holding lamps, guard the door, and keep their eyes fixed on the pointers of the Great Bear, the guiding stars of the barley-growing worshippers of the polar-star, which they called successively the stars of the Great Bear of the Northern Finns, the seven Rishis or antelopes of the Hindus, and the seven bulls of the Zendavesta. When these by their position show that midnight is near, the guardian priests wave their lamps, and in a few minutes after this signal the priests march down in procession to the consecrated ground. The procession is led by four Shkandos, or deacons, wearing the Rasta, and also a Tagha, or silk cap, under their turban. After them come four Tarmidos (priests), whose names apparently reproduce the root of the Hindu Dharma, divine justice and law, father of Yudishthira, the king and spring season of the Pāṇḍava year. These Tarmidos have undergone the baptism of the dead, and each wears a gold ring on the little finger of the right hand and carries a tau-shaped cross of olive-wood, showing that they are priests of the god of the sacred olive-tree. Behind them comes the Ganzivro, the high-priest, who has completely renounced the world, and who has been chosen to his office by his colleagues as the Pope is chosen by his cardinals. After him come four Shkandos, or deacons. The first of these carries the large olive-wood cross, called Derashvod Zivo; the second the sacred scriptures, called Sidra Rabba, the Great Order; the third a cage with two live pigeons in it; and the fourth a measure of barley-meal and sesame seeds. They march through the rows of worshippers, who leave a passage for them and kiss the garments of the Ganzivro as he passes. After the two guardian Tarmidos have drawn back the curtains covering the doorway the procession files into the Mishkna, where the

Shkandos, or deacons, range themselves on the right or east, and the Tarmidos on the left or west side. They thus leave the Ganzivro standing alone and facing the polar star.

He goes up to the altar on which the deacon who carried it has laid the Sidra Rabba, folded back at the division which separates the liturgy of the living from the ritual of the dead. The high-priest then takes one of the live pigeons, the sacred bird of Is-tar, and of the fish-god of Nineveh, the fish-city, and the ancient capital of Mesopotamia, the bird which has been adopted by the people who sanctified conjugal union as the successor of the Nasr, the vulture of the Haranites, Hindus, Zends, Egyptians, and South Arabian Sabæans. Holding the pigeon, he extends his hands towards the polar star, upon which he fixes his eyes, and lets the bird fly, saying, 'Bshmo d'haï rabba nishabbah Zivo kadmaya Elaha Edmen Nafshi Eprah.' 'In the name of the Great Living One, blessed be the primitive light, the ancient light, the Divinity self-created.' As these words are spoken all the outside worshippers rise and prostrate themselves in adoration of the North Star. When the Ganzivro steps aside they all rise and reseat themselves, while the senior Tarmido takes the place of the Ganzivro to the south of the altar, and begins to read the Shomhotto, or confession of the sect, ending each section with the words, 'Mshobbo havi eshmakhyo Mandad'haï.' 'Blessed be thy name, O word of life,' which is repeated by the outside congregation.

While this reading is going on, two of the Tarmidos prepare the Peto Elayut, the high mystery, the sacramental bread answering to the Drōna or sacred cake of the Zend Parsis. One lights a charcoal fire on the earthenware stove standing at one side of the altar, while the other grinds some of the barley-meal brought by the deacon appointed to carry it and the sesame-seed. When the meal is ground he presses on it some oil from the sesame-seed, and prepares with it a mass of dough, which he kneads and divides into small cakes the size of a two-shilling piece. These, which differ from the Parsi Drōn [1] in being made of barley-meal and oil instead of wheaten-meal and clarified butter,[2] are baked in the oven of the stove, and are then taken out and handed to the fourth Tarmido, who has taken the remaining pigeon from the cage and cut its throat, taking care that no blood is lost. He places the neck of the bird over the cakes as they are given to him, and causes four drops of blood in the shape of a tau cross $\circ\,{}^{\circ}_{\circ}\,\circ$ to fall on each cake. They are then taken outside and placed in the mouths of the worshippers by the Tarmidos who prepared them, the priests while giving the cake saying, 'Rshimot bereshm d'haï.' 'Marked be thou with the mark of the living one.'[3] After

---

[1] Drōn, Zend Draona, the equivalent of the Sanskrit Droṇa, the sacred cask containing the Soma or sap of life of the barley-growing races, which is called in the Brāhmaṇas Prajāpati, the Supreme God. Drōna is in the Mahābhārata the tutor of the Kaurāvya and Pāṇḍava princes, and the father of Ashvatthaman, the sacred fig-tree, the Ficus religiosa (aṣhvattha), the parent tree of the Ikshvākus, or sons of the sugar cane (iksha), who were originally the sons of the sun-horse, whose parentage is indicated by their use at the Soma sacrifice of the Ashva-vāla or horse-tail grass (Saccharum spontanæum), instead of the Kuṣha grass, the parent-grass of the Piṭaro Barishadaḫ, the Kuṣhite race seated on the Barhis of Kuṣha grass. (The Ruling Races of Prehistoric Times, Essay ii., p. 166; iv. p. 404.) The Drōn cake is like the Soma cask, with its infusion of barley, water, and the sap of the sacred Soma plant, that which contains the living and life-giving spirit of the living god, the Zi or life of the barley (zea). The Sesamum, yielding the sesame or holy oil of the Mandaites and Indian Telis, will not grow in Europe. Hence the deification of oil, which began with making the oil of sesame holy, was an Asiatic cult, and the Athenian goddess Athene was, as the goddess-mother of the olive-tree, an Asiatic goddess, whose worship was in Greece added to that of the flower-goddess Koronis, when the Eastern races came into Greece, bringing with them the ploughing bull. It was the sesame seeds, sacred to her, which were mixed with rice, barley, and beans, in the baptismal water, and by the Hindu Vaishya, sons of the Udumbara fig-tree (Ficus glomerata), to sanctify their children before they were shaved and consecrated to God by the sacrifice of their hair when they were seven years old. (The Ruling Races of Prehistoric Times, Essay iii., p. 279.)

[2] West, Shāyast lā Shāyast, iii. 32; S.B.E. vol. v. pp. 283, note 6, 284.

[3] This is the cross, the mark of the living God drawn, in Ezekiel's vision, on the foreheads of those 'who sigh and cry for all the abominations' of Jeru-

the cakes have been sprinkled and sanctified with the blood of life drawn from the pigeon, four of the deacons inside the Mishkna take it round to the north side of the altar, where they dig a hole, in which they bury its body. The chanting of the confession is then closed, and the Ganzivro takes the place of the 'Tarmido, and reads the ritual Massakhto of the renunciation of the dead, addressing the prayers to the North Star. This recitation lasts three hours till, as dawn approaches, the chief 'Tarmido signalises its completion by calling out, ' Ano ashorlakh, ano asborli na, Avather.' ' I mind me of Thee, mind thou of me, Avather.'

This utterance announces the beginning of the sacrifice of the wether, the father-totem of these pastoral and agri- cultural people who never kill ewes or eat their flesh, and the wether now to be offered is a substitute by the fire- worshippers of the unsexed father-god for the ewes offered to Hekate, the cross-road mother-goddess of the shepherd- sheep-race, and the child offered by the Haranite Sabæans. This sacrifice, a survival of an earlier creed, follows that of the mother-bird of the barley and oil-growers, the bird of Varuna, the god of conjugal union, which has replaced the primæval vulture, the bird of the North Star, and the animal victim is offered after the congregation of the united tribes of shepherds, agriculturists, and fire-worshippers have eaten of the bread and blood of life of the star-gods, purified by the holy fire. The wether now offered is a sacrifice to Avather Ramo, the judge of the world of darkness (*ramo*), the father Ram, the father of the sun-god Rã, the god of the East and West, and his companion deity, Ptahiel, the opening (*ptah*) god (*el*), the star of the morning in the eastern heaven. The victim is led into the Mishkna by one of the Shkandos as the North Star fades in the pale ashen grey of the coming dawn, and laid upon a bed of reeds, *with its head west and its tail east*, towards the rising and setting

salem by the man with the writer's inkhorn by his side, the mark that was to save those who bore it from destruction (Ez. ix. 4-9).

equinoctial sun,[1] while the Ganzivro stands behind it facing
the North Star. He first pours water brought to him by a
deacon over his hands and then over his feet, and one of the
Tarmidos stands by his left side and places his hand on his
shoulder saying, ' Ana Shaddakh.' ' I bear witness.' Then
the high-priest, bending towards the North Star, draws a
sharp knife from his left side, saying, ' In the name of Alaha
(the exalted; Heb. *Alāh*, to ascend), Ptahiel created thee,
Hibel Sevo (the shepherd (*sib*) god, the Hindu Shiva) permitted thee, and I slay thee.' He then cuts the sheep's
throat from ear to ear, allowing the blood to flow on the
reeds on which the animal is stretched, while the four
deacons go outside, wash their hands and their feet, and
then flay the sheep, cutting up the flesh into as many pieces
as there are worshippers. When these have been distributed
and eaten, the sacrifice to the bird-mother of the southern
sons of the North Star, and to the sheep-father of the fire-
worshippers, and shepherds of the East and West, the confederated sons of the upright right-angled cross, is ended,
and the assembly closes with a benediction pronounced by
the high-priest, after which the priests leave the Mishkna
in the same order as they entered it, and the congregation
disperses to their homes.

In this sacrifice, we see a commemoration of the union of
the Northern Finns, sons of the bird and the fire-drill, with
the Eastern and Southern shepherd and agricultural tribes,
who called themselves sons of the sheep-mother Idā, and
of the ram with the golden fleece, the heavens studded
with the stars of Varuna, the god of the rain (*var*),
and the dark night, to whom the rain was sacred. It was

---

[1] This position of the tribal totem with its head towards the West, or the
setting sun, is one taken from the Semite reckoning of time, in which the day
always begins at sunset; similarly in Egypt, the god Ptah, the Egyptian
equivalent of Ptahiel, the creator of the totem, according to the Mandaites,
is represented as a mummy, showing that he was the god of the dying or
setting sun, which opened the new day.—Lockyer, *Dawn of Astronomy*,
p. 209.

this confederacy which had made the olive tree their parent tree, and used as their sacred oil that pressed from the seeds of the *Sesamum Orientale*, which, as I have shown, the Indian Telis or oil-pressers say is the only oil which pure Telis can make.[1] These Telis, who are the growers of Northern crops, and therefore ethnologically distinct in their origin from the Southern growers of rice, claim to be the descendants of the eleven gods of generation of the twin races, to whom animal victims are offered at the Soma sacrifices, and these twin races are the fire and sun-worshipping growers of barley, and the star-worshipping shepherds, who also worshipped the moon as the mother of life who marked the months of gestation of their flocks. This reverence for the moon and the lunar periods is shown in the ritual of the star-worshippers in the number of their priests. These are the eight deacons, sacred to the holy fire and the eight-rayed star, the sign of seed, the sun of the East; the six Tarmidos, the representatives of the six gods of creation worshipped by the twin races, the four seasons of the year depicted in the four-armed right-angled cross, and the twin door-posts, their parent-gods, who guard the entrance into the Mishkna, the divine womb, or house of the birth of immortal or constantly reviving life. These fourteen have their counterparts in the days of the moon's phases of the lunar year, and besides these, there is the Ganzivro, or high-priest, who completes the fifteen, the seven days of the week added to the eight supreme gods of the eight-rayed star.

The evidence I have adduced to show the connection between the Euphratean star-worshippers and the Indian growers of oil-seeds and barley, is most strongly corroborated by the striking similarity between the Indian ritual of the Soma-worshipping sons of the fig-tree, and that of the worship of the polar star, which I have just described. For in the Soma sacrificial ground the Sadas, or house of the gods and priests, and the hut and baptismal bath of the sacra-

[1] *The Ruling Races of Prehistoric Times*, Essay ii., pp. 86, 87.

ficer, correspond with the Mishkna of the star-worshippers and its appurtenances, with variations which clearly show that the Soma ritual is a later descendant of the earlier and simpler form of worship used in Mesopotamia, and that the Indian ritual in its final form, as set forth in the Brāhmaṇas had been moulded by a race who had begun to look on religious ceremonies as instruments of individual moral regeneration, instead of thinking of them as forms of national worship of the parent-gods of the nation. Both in Mesopotamia and India, the holy tabernacle is an oblong shed, standing north and south, and is a temporary building erected for the sacrifice, made of mats woven of river reeds, for the first Soma sacrifice in Indian ritual is the Sautramani of the Ashvins, celebrated on the banks of the Sarasvati.[1]

The Hindu Sadas varies in length from eighteen to twenty-one, twenty-four, or twenty-seven cubits, and in breadth from six or ten cubits to one-half of the long side. It is, like the Mishkna, divided into two equal parts, not by the altar, but by the centre-line of the sacrificial ground running east and west, on which the Soma altar and the Uttara Vedi, called Āhavanya, or altar of libation (āhavana), are placed in the east, and the Vedi, or altar of knowledge (veda), to the west of the Sadas. Again, the Sadas has not, like the Mishkna, its door on the south side, but the east and west window openings of the Mishkna are changed into the doors of the Sadas, thus showing that the polar star, which was still worshipped in India by all Kushika married couples on their first night in their new home as the source and author of light, had been superseded in the annual sacrifices by the sun-god.

In the north compartment of the Sadas were the five dishnyas or extinguished sacrificial hearths of the earlier gods of the five seasons of the Hindu year, and in the centre was the Udumbara pole, the house-pole of the sons of the sacred fig-tree, the *Ficus glomerata*. This, when fixed in

[1] *The Ruling Races of Prehistoric Times*, Essay iii., p. 106.

taining, according to the Brāhmaṇas, 'the sap of all plants.'[1]
The space south of the central house-pole, the house-pole of
the Phrygian races, who, according to the custom of the
Bronze Age, dwelt in circular beehive huts with a pole in the
centre, was consecrated to Mitra-Varuṇa, the sun and moon
(mitra), and star-god (Varuṇa), the twin-gods whose priests
were the Praṣhastri, or teaching priests, the preservers of the
verbal Shasters, the national history, and guardians of the
national lore. They were the priests of the united nation
who, in the ritual of the star-worshippers, worshipped at the
south door of the Mishkna. But in the Hindu Soma ritual,
which was elaborated by the Northern sons of the sun-horse,
who had succeeded the sons of the sheep and cow, the men of
the upright right-angled cross, the bath and hut of the
sacrificer was not, as in the Mesopotamian ritual, placed on
the south side of the Sadas, but on the north, and in both
cases the hut was made with reed mats. In the Hindu ritual,
again, the Sadas was not, like the Mishkna, uncovered, but
it was roofed over with nine mats attached to the tie-beam
running north and south, and resting in the centre on the
Udumbārā pole, bifurcated at the top like the gable-ends of
the long houses of the sons of the horse race.[2] In these
arrangements we find clear evidence of the supersession of the
people of the round houses by the Northerners living in long
houses, and an explanation of the transfer of the baptismal
bath of the nation on the south of the temple to the regenerating bath of the individual sinner on the north. It was these
Northern people who completed the figure of the eight-rayed
star by adding the transverse rain-cross marking the position

of the sun at the solstices ✕ , and depicted on the

Hindu mother-altar, to the upright right-angled cross of the
sons of the cow and of the rivers running from north to

---

[1] Eggeling, Śat. Brāh. iii. 6, 1, 1-13; S.B.E. vol. xxvi. pp. 140-143;
p. 140, note 3. See also the plan of the Soma sacrificial ground at the end
of the volume.

[2] Ibid. iii. 6. 1, 22, 23; S.B.E. vol. xxvi. pp. 145, note 2, 146.

south, who grew first millets and oil-seeds, and afterwards
barley and wheat, and signalised the adoption of barley as the
typical seed of life by making the eight-rayed star the sign
of god and seed used by the barley-growers. Further proof
of the correctness of the deduction which proves that the
transverse cross was one marking the annual course of the
sun, is given in the figure of the eight-rayed star drawn on

an Assyrian bas-relief where the lines of the trans-

verse cross clearly denote sun-rays.[1] It was for the celebra-
tion of the solstitial festivals prescribed by the later ritual
of the sons of the barley-growers, that the additions to the
temple at Stonehenge which produced the plan to be traced
in the present ruins, were made ; and from this plan we can
see clearly that the race who began their year with the
summer solstice, and not with the autumnal equinox
beginning the Euphratean year, were the successors of the
people who in Mesopotamia worshipped the upright right-
angled cross. For in the line running from north-east to
south-west, marked by the Friar's Heel, the slaughter-
stone, and the opening in the circle of thirty stones through
which the beams of the rising sun fell upon the western
altar, we find the reproduction of the axial line of the
upright right-angled cross running from North to South.
This change in the direction of the axial line is accompanied by another which transferred to the West of
the gnomon the most holy place on which the victims were
slaughtered in the early ritual, and which formerly stood,
like the gnomon pillar of Hekate and the altar of the
Mishkna, in the centre of the consecrated ground. This
transfer shows us that there were, subsequent to the earliest
form of sacrificial ritual of the sons of the hammer and anvil,
when the victim's skull was broken by a blow on the forehead

[1] J. Menant, *Pierer Gravées de la Haute-Asie*: Paris 1886, vol. ii. p. 72.
D'Alviella, *Migration of Symbols*, Fig. 2, chap. i. p. 14.

from the sacred hammer or club of the fire-worshipping smiths, two stages in the earlier history of consecrated spots, the first when the gnomon or sacrificial stake was placed in the centre of the right-angled cross of St. George in the round hollow which was first the sacrificial pit of the Kabiri, and which afterwards became the circular baptismal water-

bath ——⊕—— . In this sacrifice the victim was, as in the

Hindu ritual of the Fathers, tied by the neck to the post,[1] and killed by piercing the jugular vein, so that the blood first spirted on to and vitalised the post before it descended into the earth to make it fruitful, and to supply the blood for the expiatory blood-baths used in Grecian and Phrygian ritual. The animals then slain were the moon-goat and the sun-sheep or ram. This stage in the evolution of the worship of the rain-cross was followed by that in which the sun and rain-god was not the ram of Varuṇa, but the sun-horse, and the rain-cross was not that denoting the four original points of the compass, but the diagonal cross formed by the daily paths of the sun, and by its rays when it rose and set on the solstitial days of midsummer and midwinter. In this innovation the axial line of the cross was changed from north to south to north-east to south-west to mark the line travelled by the rays of the sun rising at the summer solstice in their passage from the top of the Friar's Heel to the south-western altar of the holy circle. On this line the slaughter-stone, standing midway between the gnomon and the inner circle, replaced the gnomon Linga and Yoni of the sons of the sheep and cow.[2] At the south-western end of

---

[1] Eggeling, Sat. Brāh. iii. 8, 1, 15; S.B.E. vol. xxvi. pp. 189, 190; where the three ways of killing animal victims indicative of the three ritualistic ages are described.

[2] This line is still at Stonehenge, as it was in Hekalé's cross-roads, the axial line of a right-angled cross, for the line joining the S.E. and N.W. gnomon stones crosses it at right angles, passing through the slaughter-stone, the central point in which the original gnomon, the sacrificial stake, was placed.

the centre N.E. to S.W. line, the circle dedicated to the sun-
horse was erected with the altar at the south-west side.
backed by the two horse-shoes, the whole forming the
annexed figure, in which A denotes the gnomon pillar, B the

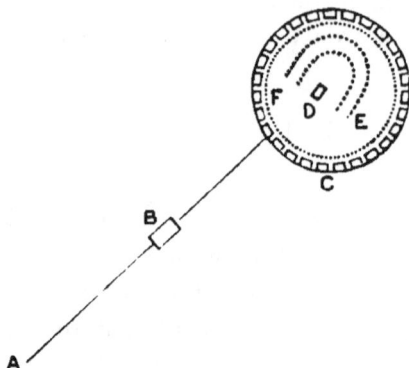

slaughter-stone, C and F the holy circles, D the second altar,
and E the horse-shoes. This sun-horse of Stonehenge, and
of the downs of Southern England, is shown by the Zend
mythology to be the direct descendant of the white sun-
horse brought from the north to the east. For as the white
sun-horse of Stonehenge began its yearly journey at the
summer solstice, so did also the white horse of the
Zend Tishtrya, the star Sirius, and in this latter myth, which
makes Sirius the rider on the white horse of the summer
solstice, and the herald of the Persian and Indian rains, it is
apparently probable that we can find a reliable date for the
maturity of the conception; for we learn from Stonehenge,
where it was the rising sun of the solstice which rode
the white-cloud horse—that the ritual of its worshippers was
based on the rising sun, and therefore the time when Sirius
was thought to become the conquering white horse, which
makes the black horse of the rain-cloud give up its rain,
was that when Sirius rose heliacally before or cosmically with
the sun at the summer solstice. This conjunction in the
heliacal rising of Sirius and the sun has been calculated by
Biot with reference to the Theban temples as occurring

about 3285 B.C.,[1] and the date would probably be about the same in the latitude of the Persian Gulf. But it was in all probability long before this date that the summer solstice, with which the Northern Indian rains began, was the season which opened the year both in India and the North; and in India the solstice had also been connected with Sirius before that date, during the age when Sirius was called the dog of Orion in Greece, and Saramā in the Vedic traditions. But it appears probable that the change which made Sirius the sun-horse instead of Saramā, the dog of the gods, dates from the time when Sirius rose heliacally at the summer solstice. Unfortunately for the use of Stonehenge as a chronological index, the direction of the rising sun at the solstices does not, as my astronomical friends tell me, vary like the direction of the rising stars, and the only guides to its age we can find are the round barrow tombs of the Bronze Age which abound in its neighbourhood, and in the fact proved by the race-course near it that the festivals solemnised at the temples were accompanied with games and horse-races, similar to those celebrated from a very early period in Greece, and also to those accompanying the Vāyapeya, a festival in India. These games were, as we learn from the Olympian calendar, in which the year began with the first new moon after the summer solstice, the national festivals of a people who, like the Persians, Hindus, and the builders of the solstitial form of the Stonehenge temple, began their year at midsummer. We also learn from the construction of its ground-plan that it must have been brought from the East, for it coincides with the plan of the Phœnician temples in Mashonaland and with that of Indian Soma sacrificial ground, in placing the holy shrine of the sun as father of fire, to the west.[2] This is in India called the Prāshīna Vaṃsa, meaning the eastern hall,[2] and from it we learn additional information to that given by the Mesopotamian Mishkna as to the history of ritualistic

---

[1] Norman Lockyer, *Dawn of Astronomy*, p. 197.
[2] Eggeling, *Sat. Brāh.* S.B.E. vol. xxvi. Plan of sacrificial ground at the end of the volume.

temple-building. For its interior arrangements, as depicted in the annexed diagram, copied from the plan given by Professor Eggeling in his translation of the Śatapatha Brāhmaṇa, clearly show it to be an adaptation made by the Mesopotamian worshippers of the upright right-angled cross, who took for the sacred line marking the path of the sun's rays through their temple, that which, like the sun at the equinoxes, went from due east to due west. The original model from which they started being the old round-house of the Finnish fire-worshippers and workers in metal, consecrated to the Hestia of the Greeks, the Vesta of the Romans, the goddess of the household hearth. The Prāshīna Vaṃsa, like the Mishkna,

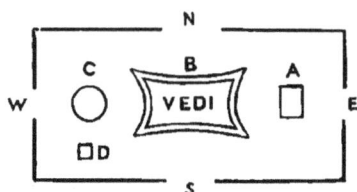

stands due north and south, and due east and west, and has, like it, the altar called in India the Vedi, the altar of knowledge (B),[1] facing the North Star, but it differs from it in having its longest sides running east and west; the change having arisen from the transfer of the plan of the temple built for the worship of the polar star, to one which, while it looked to the polar star through its northern door, was more especially consecrated to the worship of the equinoctial sun to which the Āhavanya altar (A), the altar of libation (āhavana) is especially consecrated. And this altar marks the substitution of a sacrifice of libations of milk, curds, and whey for the slaughter of the wether offered in the Mishkna. The Veda and Āhavanya altars occupied the front of the house marked by the eastern door, and behind these, to the west, is the Gurhapatya, or household (gurh)

---

[1] This was the altar originally made in India, in the form of a woman, with the gnomon, then used as the sacrificial stake, in the centre.—*The Ruling Races of Prehistoric Times*, Essay iii., pp. 163, 169, 170.

fire (C), watched by the Patni, or mistress of the house, the sacrificer's wife, from her seat (D). The whole building is, in short, an old Finnish round-house—with its circular house-fire changed into an oblong temple to the eastern sun by the worshippers of the polar star, who made their sacred figure the square or oblong formed by enclosing St. George's cross in the right-angled field used as their standard by the first tribes who measured land by measures of length. This was the four-cornered figure sacred to Varuṇa, the god of the starry heaven, to whom the ram was consecrated, the Chathru-gaosho of the Zendavesta, the land of Thraētāona, the Chatur Asrir Varuṇo, the four-sided figure of the Rig-veda, which is said, in the Rigveda, to have conquered and superseded the primæval sacred triangle.[1] This triangle had become the gnomon, and it was this which, when observed at the equinoxes, received the sun's rays and transmitted them westward to the Vedi altar and the household fire. When, in the elaboration of ritual, the Sadas, or priest's house, im-ported from Mesopotamia, was added to the original sun-temple, and when an especially sacred area to the east of the Sadas was dedicated to the worship of Soma, the sap of life, the original east and west house, though it was still called the east house, was placed to the west of the sacrificial ground.[2] We see from this analysis that the special con-

---

[1] Darmesteter, *Zendavesta Vendīdād Fargard*, i. 18 ; Introduction, iv. 8, 12 ; S.B.E. vol. iv. pp. lxiii. 9 ; Rigveda. i, 152, 2.

[2] The very strong conservative spirit which marks Hindu theology, and which is conspicuously shown in the gradually changing arrangement of the Soma sacrificial ground, also appear in the history of the Hindu Svastika. This, which, in its right-angled form derived from the cross of St. George, represents the annual course of the sun beginning at the autumnal equinox, is divided into two sacred forms answering to the division of the Hindu year, into the Devayāna and Pitriyāna periods of six months each. The Svastika proper, most holy Svastika, meaning that which is (*asti*), the sign of the Su or supreme soul, is that which has its arms bent to the right ⊡, and this is called by the Hindus the male form, sacred to the god of wisdom, Ganesh.

secration of the west, and of the sacrificial ground to the god of the eastern sun, arose from the use of the gnomon or obelisk, and from the custom born from its employment of orienting the sites for sun-worship so that the rays of the rising sun fell on the altar and thence on to the household-fire in the west.[1] From further observation of the gnomon, it was found by the sons of the sun-horse, who had brought

This is the Devayāna form of the barley-growing races. The other form,

which has its arms turned to the left , is the female form, sacred to

the goddess Kali, and it is that of the Pitriyāna. It is called Sau-wastika, or that which is sacred to the Saus, that is, the Shus or Khus, the sons of the mother-bird, the Indian trading-races. The two forms represent the annual course of the sun as conceived by the early cultivators, the Finno-Turano-Dravidian Māghadas and Kushikas. The male form representing the sun as starting from the East to the South at the time of the autumnal equinox, and the female form representing it as starting from the East to the North at the time of the vernal equinox. See, with reference to the meaning of the two Svastikas, (D'Alviella, *Migration of Symbols*, English edition, chap. ii. pp. 32, 40, 68 note 1 ; Birdwood, *Report on the Old Records of the India Office :* London, 1891, pp. x, xi). It was this double Svastika which was succeeded by the transverse cross of the sun-horse, showing in one symbol the course of the sun throughout the year.

[1] This gnomon, marking the daily and yearly path of the sun, appears in Hebrew historical mythology, as Jacob, the twin-brother and supplanter of Esau, the goat-god, the Babylonian god Uz, who watched the motions of the sun's disc and who was the god who measured time by the lunar months of gestation, which had always been carefully observed by the shepherd races, both were sons of Isaac, the blind-god, the ear of corn born of Abram, the sun-god, and Sara, the grain-mother. The goat-god was supplanted by the sun-god, who measured time by the gnomon, the Bēthēl, or House of God, set up by Jacob, and consecrated to the god of the revolving oil-mill of the heavens, the stars moving round the pole by the oil he poured on it (Gen. xxviii. 19.). This was the place of Luz, meaning the almond-tree (Gen. xlviii. 3), the parent-tree of the gardening races, answering to the Hindu mango-tree, and it was from the mango-stone, given to the two queens of the king of Māghada, who were daughters of the Kushika by Gotama, the son of the bull (*gut*), the moon (*chandra*) of the Kushikas, that the incarnate Hindu sun-god and king, Jāra-sandha, the union (*sandha*) by old age (*jārā*), was born, and the two halves borne by each queen were united by the old witch Jārā. He was slain by Bhima, the storm-god of the burning summer, ending

the solstitial year of the North to Asia, that the sun, at the
solstices, instead of rising due east, rose in the north-east
and south-east, and that, therefore, temples dedicated to the
solstitial sun must be oriented to these points. Hence, when
the Northern races, who had always, from time immemorial,
looked on midsummer as the season when the sun-god finally
overcame the winter frost-giants, became the dominant
power in South-western Asia, and when they found there
that midsummer was the season when the rains of Northern
India and the Persian Gulf began, they changed the beginning of their year from the winter solstice, when the young
sun-god was born, to the summer solstice, beginning the
Persian, Hindu, and Egyptian years. This change originated in the rainless and sunny lands of Arabia, the home of
the Sabæans, where the gnomon could be continually used as
a year-prophet without the fear of the intervention of clouds
which hid the sunlight, and it was in Arabia and Egypt,
where the obelisk, used in the Phœnician temples in Mashonaland to catch the rays of the sun of the summer solstice, was
especially sacred, that the first permanent temple enclosures,
oriented like that at Stonehenge, were erected, and it was the
trading Phœnicians, the crew of the Southern sun-ship Argo,
whose voyages I will describe presently, who diffused throughout the Western world the Asiatic developments of the
Northern worship of the sun-horse.[1] They were the earliest

in the rains of the rainy season at the summer solstice, and by Kṛishṇa, the
black antelope, the year-god of the discus, the ring of months. It was at the
place of the parent almond-tree that the sun-god of the gnomon-pillar first
appeared as the god marking the course of the year, the god who married
Leah, the wild moon-cow, and Rachel, the sun-ewe, daughters of Laban, the
moon-god of Haran, the place of the road (K'harran), where the years of
twelve solar and thirteen lunar months were first calculated by the gnomon,
the father of these twelve and thirteen children of the sun and moon, who was
told by God to go and dwell at Beth-el, the house of the sun-god of the
almond-tree (Luz).—Gen. xxxv. 1 ; xlviii. 3.

[1] The position of the apse sacred to the sun-horse at the eastern and
western ends of the temple vary, according as the worshippers were the sons
of the polar star, the star of the left-hand, who looked to the east as the
quarter sacred to the sun of the equinoctial year, or as they were a people

navigators of the Mediterannean, who made voyages to Cornwall to get tin, who first brought this Asiatic ritual to England, where it was first taken up by the Druid priests, who used this Eastern method of orientation in forming the plan of the renovated temple to the sun-horse which still stands at Stonehenge.

But this worship of the sun-horse of the North, which had made its way overland to Persia and India, and had given birth to the plan of the sun-temple, which was transferred from the East to Stonehenge, was also diffused through all European countries in the mythology of the holy wells. The most celebrated well of the sun-horse, in Greece, is that called Hippokrene, the well ($\kappa\rho\acute{\eta}\nu\eta$) of the horse, on Mount Helicon, which was said to have been brought to light by the footsteps of the flying sun-horse Pegasus, and this same myth reappears in the stories in the Isle of Man, which tell how the wells of St. Maughold and St. Patrick were brought to light by the horses of these saints. These wells, again, have the reputation of curing all diseases, and, therefore, the origin of the myth dates from the age when the ruling gods were the healers of the people, the divine physicians, like the Hindu Ashvins, the Zend Aryaman, the Greek Æsculapius, and Chiron, the Centaur, the sun-horse; and this was the time when, according to Greek mythic history, games and horse-races, such as those held on the Stonehenge race-course, were first celebrated in memory of the Fathers of the nation. These Manx wells, again, are supposed to be especially efficacious when

who dated their year from the summer solstice, and looked to the west where the sun's rays first fell on the ground as the hallowed spot, where its tracks were first seen, and the existence of the two customs, one originating in the temples of Syria, and the other in the oriented temples of Egypt and Greece, is shown in the design of the church built by St. Augustine at Canterbury. This, as Professor Willis has shown, had an apse at each end : the high altar of the presbytery, the priest's house of the Mandaites and the Sadas of the Hindus, all looking to the eastern sun, and the Lady Chapel devoted to the laity at the west. This was the quarter sacred in Egyptian, Greek, and in early Druid mythology to the Fathers.—R. Hughes on Ecclesiastical Buildings, *Social England*, vol. i. ' Britain under English and Danes.' p. 196.

visited on the first Sunday in harvest, and on St. John's Eve at the summer solstice,[1] these being the times when the Hindu rain-festivals were held in June, July, and August, and when the Stonehenge year began.

In the ritual of this new worship of the sun-horse, who brought to light the life-giving and healing water, the blood of god hidden in the earth, no animal sacrifices were allowed to be offered within the circular enclosure especially dedicated to him, in temples like that at Stonehenge, and in the Indian Soma sacrificial ground ; and the slaughter of victims was only permitted on the slaughter-stone outside, in temples of the Stonehenge pattern, and at the sacrificial stakes erected to the east of the Indian sacred enclosure.[2] It was the circular enclosure, with its horse-shoes protecting the altar, which became the western apse of temples oriented to the rising sun, in those countries where the architecture founded on the veneration for the shoes of the sun-horse, and the round-house of the fire-worshippers had not given place to the square architecture copied from the Hindu temple enclosure.[3] It was also in those temples of the sun-horse that the Soma sacrament of the cup originated. The mythic author of the reformation which sanctified the Hippokrene or well

---

[1] Moore, 'Water and Well Worship in the Isle of Man,' *Folklore*, vol. v. Part 3, September 1894, pp. 221, 224, 225.

[2] Dr. Stukeley, who, in 1723 A.D., dug up the surface of the ground in the western circle of Stonehenge to see if he could find there the remains of animal victims, only found one ox's tooth there ; and Mr. Hayward, the owner of Stonehenge, who dug about it in 1702, found the heads of oxen and other beasts, but it does not appear in the published guide-book whether these were found within the circle or by the slaughter-stone. Mr. Webb, who dug in 1637 A.D., found what he thought was the cover of a thuribulum, and it certainly is most probable, judging from the Eastern ritual of the sons of the horse in India, that it was incense and libations which were offered on the altar in the inner circle. Judd, *Stonehenge*, Part II. p. 35.

[3] The western position of the apse, or holy circle, at Stonehenge, and of the originally circular house of the Hindu Vedi altar—which was, as I have shown, caused by the custom of orienting the temples to the rising sun—has been reversed in Europe, where the horse-shoe apse is always now placed at the east end.

of the horse as a healer of diseases more to be relied on than the oil of the earlier surgeons, and the blood of sacrificed victims was, in Greek mythology, Bellerophontes, the slayer of the Bellero. This latter name, as Max Müller has shown, is the Greek form of the Sanskrit Varvara, meaning woolly, rough, shaggy, from the root *vri*, ' to enclose,' and from the same root comes the Var of Varuṇa, meaning the cloud enclosing the rain ; and also Urā, the sheep, and Ura-bhra, the ram,[1] the animal producing coverings, the ram of the woolly fleece of the clouds of Varuṇa, the ram of the Golden Fleece. It was the rider on the sun-horse who slew the ram, who was the rain-god who produced the springs rising up in the holy cup, the healing well. This cup of healing, the holy well of the race who made the divine physician their father-god, became in the ritualistic services the Soma cup of Hindu mythology, and the Kukeōn (κυκεών) of Demēter in the Eleusinian mysteries. These sacramental cups are shown by their ingredients to descend from the ritual of the barley-growing races, the sons of the seed of life, the eight-rayed star, who made the deer-god their divine guide, and worshipped the moon and sun-cow, as well as the sun-horse. For in the Vājapeya and Sautramani sacrifices of the Ashvins or heavenly horsemen the Soma cup was made of water mixed with Kusha-grass, the sacred grass of the worshippers of the sun-antelope, young ears of barley, and roasted barley.[2] Also in the earliest official Soma sacrifice of the Brāhmaṇas and Rigveda, the Tryāshira, or Soma of three mixings, called Gavāshir, Dadhyāshir, and Yavāshir, is made (1) of water mixed with milk, the spring mixing ; (2) with milk clotted by heat, that of summer ; and (3) the autumn mixing of water and barley (*yava*).[3] The κυκεών, the cup given in the *Iliad*

---

[1] Max Müller on Comparative Mythology, *Chips from a German Workshop*, vol. ii. pp. 172-188.

[2] Hillebrandt, *Vedische Mythologie*, pp. 253-254 ; *Şat. Brāh.* xii. 8, 2, 3; *The Ruling Races of Prehistoric Times*, Essay iii. pp. 206, 207.

[3] Hillebrandt, *Vedische Mythologie*, p. 209 ; *The Ruling Races of Prehistoric Times*, Essay iii., p. 242.

to his guests by Nestor, one of the heroes who went in the
Argo to bring back the Golden Fleece, was, as we are told in
the description in the *Odyssey* of the κυκεών of Kirke (Circe)
the hawk (κίρκος) mother-bird, made of barley meal, grated
cheese, Pramnian wine, with honey added to it.[1] To these
ingredients the hawk-mother, according to the *Odyssey*, also
added magical drugs when the contents of the cup were given
to the transformed animals she kept around her as the
father-totems of an earlier age. In the cup of Demētēr the
barley-mother, which became the constellation Kratēr, the cup,
the wine which, in the age of the twin horsemen, was the
intoxicating honey mead (*madhu*) of the Ashvins, of which
they poured out one hundred casks in the Rigveda,[2] became
water, and her Eleusinian cup was made of water, barley, and
meal.[3] Thus we see that the ritual of libations of milk,
water, barley, curds, cheese, honey, mead, and wine, drunk by
the sacrificers, and poured on the western altar, which was
never profaned by blood, was one which distinguished the age
of the sons of the seed, the eight-rayed star, from that of
their predecessors, the sons of the ram, and the mother-bird,
when the sacrificers drank not the fruits of the earth, and the
cow, but intoxicating drinks, and the blood of the totem
victims which they also sprinkled on the earth of the temple
enclosure, taking it from the sacrificial pit in which the
victim was slain, and the blood collected. This reformed
age of the worship of the cow, the rain, and well-water ended
in India with the rule of the water-drinking sons of the
horse, the Sabæan Arabs, who made the upper classes of the
country what they still are, abstainers from strong drink. It
was also that which saw the first beginnings of extended
maritime enterprise, which, starting from the western coasts
of India, laid the foundations of the commerce of the
Euphratean ports, of that of Egypt, and of the Phœnician
colonies on the Palestinian coasts, and ushered in the age

---

[1] Homer, *Iliad*, xi. 624, 641; *Odyssey*, x. 234.
[2] *Rigveda*, i. 117, 6.   [3] Homer, *Cer.* 208.

when trading cities were built, and when temples ceased to
be consecrated open spots, and became covered buildings,
approached by long corridors of columns marking the path
of the rays of the rising or setting sun, or of the star to which
they were oriented.

It is the history of the beginnings in Europe of this age,
when the sun-horse became the mother-star-ship, that we read
in the myth of Jason. This mother-star-ship, the Argo, was
the southern constellation called in the Zendavesta Ṣata
vaēsa, or the hundred (ṣata) creators (vaēsa), showing that
the astronomical mythologists who christened it, were the
descendants in religious belief of the hundred sons of
Gandhārī, the bird-mother, who wets (dhāra) the land, the
Brunhilda, or goddess of the springs (brünnen), of the history
of Sigurd, and the Hekate of the Greeks. And it was the pupil
and priestess of Hekate, Medea, the guardian (μῆδ), who was
the counsellor, helper, and bride of Jason, the healer (ĭaς), the
leader of the Argonauts. The year, as measured by the
worshippers of the sun-ship, was that which in Zend
chronometry started from the appearance of Tishtrya, the
star Sirius, and this year-star of the summer solstice, which
had, in the early astronomy of the year of Orion, been the
dog of Orion and Saramā, the bitch of the gods, became, in
the pictorial astronomy of the sons of the horse, the white
horse of Tishtrya, the symbol under which it is still depicted
in astronomical pictorial maps.[1]

The course through the stars of the ship of the sun-horse,
led by the offspring of Gudrūn, the witch-bride of Sigurd,
and measured by the Babylonian astronomical mariners, the
Phœnician traders of Turos on the Persian Gulf, was,
according to the oldest forms of the Argo legend, from Æa
on the Phasis, in the extreme east of the Black Sea,[2] where
the Golden Fleece was won; that is, from the land of the

---

[1] See diagram, Essay iv. of *The Ruling Races of Prehistoric Times*, p. 333.

[2] Phasis, the modern Rhion. The town of Poti at its mouth is the
ancient Æa.

Chalybes, who ruled this country, and substituted iron and steel (χάλυψ) for bronze, to Oceanus, the Indian Ocean on the south, and this must have taken it down the Euphrates, whence it went to Libya in the west. Thence it was carried overland westward to the mythical lake Tritonis for twelve days, the twelve days' winter rest of the sun in the house of Agohya in the Rigveda, and the twelve days' winter orgy of the North, ending with the drama of the death of the deer-hunter, Orion. From Tritonis it was led by Triton, the Trita Āptya of the Rigveda and Thraëtaona of the Zendavesta, the third (*trita*) god of the waters, the rain-god of the spring season, to whom Tritonis was sacred, to the Mediterranean, whence it returned eastwards to Iolchos.[3] But it is in the story used by Sophocles, which made the Argo not go sunwards from north to south and back to north by west and east, but took it northwards to the North Sea, that we find the history of the myth set forth most clearly.[2]

The leader of the expedition, Jason ('Ἰάσων), the healer ('Ἰάς) was the mythological counterpart of the divine physician Æsculapius, the son of Koronis, the flower-garland, forming the year-circle or coronet, and Ischus, the Greek equivalent of the Sanskrit Ishā, the pole of the oil-press, sacred to Athene, goddess of the olive-tree, the mother tree of the Semite race; for both Æsculapius, the divining intellect, and Jason, the practical worker, were brought up by Chiron, the Centaur of the healing hand (χείρ), and both sailed in the ship Argo. Jason belonged to the race of the Minyæ of Orchomenos, the race of measurers (*men* or *min*), the great irrigating race, who buried their dead in the round bee-hive tombs, representing the Phrygian bee-hive huts, which still remain near Orchomenos, and which are precisely similar to the round barrow graves of the Bronze Age, so that it is to

---

[1] *Encyclopædia Brit.*, Ninth Edition, Art. 'Argonauts,' vol. ii. pp. 496-497.

[2] The mythic incidents of the Jason myth are, in the following account, taken almost entirely from the very accurate transcript of the Grecian story in William Morris's poem, called *The Life and Death of Jason*.

this age that the Jason myth and Stonehenge ritual belong.
These irrigating Minyæ and incipient physicians were the
first people who used weights. These originally were used in
bartering goods ; next, in determining the quantity of seed to
be sown in a given area of land ;[1] and lastly, in the com-
pounding and giving of drugs. They were also the people
who measured time, not only by days, nights, weeks, months,
seasons, and years, but who also divided the day and night,
first into watches measured by the water-clock, and afterwards into the more minute divisions of hours, minutes, and
seconds, forming the perfect time-circle of the Babylonians,
and who called the moon Mene, the measurer. They first
came to Greece from Phrygia, as the followers of Kadmus,
the Eastern (Heb. *kedem*) arranger (καδ), whose conquest of
the country is told in the story of his killing the great parent-
snake of the Achæi of Bœotia, the sons of the serpent (ἔχις),
which guarded a spring, the spring of the sun-horse supplying
water to the village, encircled and protected by the snake or
ring of cultivated land surrounding the mother-grove of the
matriarchal races. This cultivated land he ploughed with
the bulls of the sons of the moon-cow, the Celtic shorthorn
(*Bos frontosus*), and the domestic ox (*Bos taurus*), brought
by the Neolithic Basque cultivators from Asia Minor
into Europe,[2] instead of tilling it with the mattock,
and thus made the ploughing-bull ruler of the land
instead of the snake. By the command of Athene,
the flower (ἄνθος) goddess of the olive-tree, counterpart of Koronis, the goddess who makes the flower-garland
marking the stages of the revolving year, Kadmus sowed
the land which he ploughed with the teeth of the serpent,

---

[1] There can, I think, be no doubt that weights were generally used before
measures of length in determining areas of land. They certainly preceded
them in India, for at the present day all the least advanced tribes who
measure land at all measure it by the weight of the seed sown in it. Measures
of length were invented by the carpenters and builders, followed by the
weavers and potters.

[2] Boyd-Dawkins, *Early Man in Britain*, chap. viii. pp. 297-299.

the seeds of the cultivated land, and from them there sprang up armed men, called the Spartoi, or the sown (σπείρω), the race of the sons of the seeds, the eight-rayed star. They fought together, only five surviving to be the sons of the ploughing-ox (βοῦς), the Bœotians, who reproduced in Greece the five races of the Turano-Semite national mythology, as set forth in the early Zend and Vedic doctrine of the national descent from the five parent-gods, the five seasons of the Hindu and Zend Kushika year, formed from the union of the three seasons of the Northern year with the twin seasons of the rains and autumn. These twin seasons reproduced, in the mythology of the year, the twin-gods, the parents of the Semite, Zend, Hindu, and Dorian races, and the last of these in Greece became merged in the ruling race of Sparta, the sown land, who traced their descent to the twins Kastor and Polydeukes.

It was from this race that Jason was descended, for he was the son of Æson, and the grandson of Kretheus, the head (κράς), the god Njord of the Edda, who came from the South in exchange for the sun-horse Hœnir, and his wife, Tyro or Turo, whose name shows her to be the mother-goddess of the sons of the Tur, the gnomon or revolving pole. The correctness of this deduction is proved by the name of her other son Pelias, born before she wedded Kretheus, whose father was a god. For Pelias means the son of the potter's clay (πηλός), the clay fashioned by the revolving potter's wheel, the offspring of the creating and fashioning god of the artisan race, the Egyptian god Chnum, meaning the architect, represented as the four-headed ram,[1] who was the pole revolving with its circling stars, like the potter's wheel, the god whose worship was widely diffused among the manufacturing Minyæ. Pelias is the unformed clay revolving with the potter's wheel, while it is being shaped by the artist into the form he desires to create. Æson, the father of Jason, is the counterpart of

[1] H. Brugsch, *Religion und Mythologie der Alten Ægypter*, pp. 292, 293.

Hesus, the god of the woodland Druids,[1] worshipped by them, together with the creating spirit Hu, the Zend form of Khur, the mother-bird, for Æson and Hesus are both formed from the root of the Greek αἶσα, destiny. Thus fate or destiny, the immutable divine law of the sequence of natural phenomena, was the father of the healing sun-god Jason, the god of the wise race of the physicians, the discoverers and interpreters of the secrets of nature, who succeeded the god of the potters and weavers as the parent-god of the twin races. Æson, when he was ousted from the rule of Iolchos in Thessaly, the town at the foot of Mount Pelion, the mountain of the potter's clay, by his half-brother Pelias, sent Jason, during the rule of the potters, to Chiron, the Centaur, who belonged to the race of the Northern gods, and that of the sun-horse, for it was he who gave to Pelias, the father of the young Achilles, the sun-god of an age succeeding that of Jason, the ashen spear cut from Mount Pelion, the Greek form of the parent ash-tree of the North, the Yggdrasil of the Edda. The original ash-tree was the tree of the mother-grove of the Northern woodland race of the limestone country, which grew without effort of its own, according to the divine ordinance, while the spear taken from it and used by the Northern warriors in their wars of conquest, was made by the wit and skill of the sons of the North, who inherited the imperious will of the god they worshipped, the god who made the miraculously-born fire. Like him, they refused to wait for the fulfilment of the decrees of blind fate, and found in their own strong hands and fertile brains the means of redressing and making less unendurable what seemed to them to be the injustices of

[1] The woodlanders, the Volsungs of the Sigurd story, was one of the tribes of the Dorian confederacy called the Hylleis, or the wood (ὕλη) people, and Dryopes, or sons of the Druid tree (δρυ). This tree was the sacred oak of Dodona, from which Jason made the prow of the Argo, the tree of the sons of the dove, the sacred pigeons of the shrine of the sons of the fish-god which superseded the ash-tree of the Centaurs, the sacred tree of the sons of Odin's ravens.

nature. Their belief that gods and men were masters of fate was expressed in their mythological theology, ascribing natural processes and their causes to the will of the gods, and in the doctrine of the revolving pole. This, the mighty ashen parent-spear, which in the Edda is the Ash Yggdrasil with its root in heaven, under the mother-mountain of the sky, in the rain-fountain of the Urdar or gnomon stones (urdr), was, in Greek mythology, the pole reaching to the polar-star, which does not turn itself, but is turned by the twins Day and Night, the parent-gods of the Dorian sons of the spear (δόρυ). This was the weapon which could not be baffled, that which Karṇa, the horned (keren) sun, the united moon and sun god of Hindu theology, took from Indra, the rain-god, in exchange for his invulnerable coat of golden mail, which was also the defence of Sigurd, the rider of the sun-horse of the North. This pole, turned by the will of the heavenly twins, the divine physicians, carried round with it the stars represented by the mother-star-ship Argo, and was the fire-drill of the earth, generating the life-giving heat of summer, and the fertilising rains which followed the heat.

The tokens given to Chiron, the sun-horse, together with the child Jason, were an ivory horn, the emblem of the crescent moon, the ram's horn of the Semite worshippers of Varuṇa, the god of conjugal union; and a red ring, the sign of the rising and setting sun, the maker of the rings of time, the days, months, and years, the red rings of the treasure of Andvari, placed by Sigurd on the back of the grey sun-horse Grāni.

When Jason had grown up, and when his mind was imbued with the golden lessons of the wisdom of past time, the heavenly lore taught by the red rings borne by the sun-horse, the rings hung on the wall of Völundr's (Wieland's) cave, he went, by the command of Hera, the mistress (ἥρα), the ruling moon-goddess of the year of the ten lunar months of gestation, and the protectress of the new-born sun of the sun-ship, to Iolchos, to claim his father's throne, and carried with him the horn and the ring, the tokens of his birth. On his way he was taken across the flooded Anauros, the windless (ἄναυρος) river leading from conception to life, by Hera, the goddess of births, in the guise of an old woman. When passing through the river, he dropped into it one of his sandals, and became the one-sandalled (μονοκρήπις) sun-god,[1] the descendant of the one-footed fire-drill, the beam of the oil-press of the wheel-year, and not the two-footed moon with its double crescents, the shoes of the sun-horse. He appeared before Pelias clothed, by Hera's command, in Magnesian garments, the garb of the sons of the great witch-mother Māga, brought by them when they emigrated from Magnesia to Thessaly, and left the city of Lydia on the river Hermes consecrated to Niobe, the snow-mother-goddess of Mount Sipylus, which rose above the city of the Magnetes, called by Plato the mother of laws.[2] This land, from which the Magnetes came to Greece, was that ruled by the sea-faring Ionians, the descendants of the matriarchal Amazonian races, who first started on their national career as voyagers on the seas of time in the ships they built on the forest-clad mountain coasts of Western India, whence they made their way to the Persian Gulf, and afterwards up the Euphrates to Asia Minor, substituting in their landward journey barley, made from an indigenous grass of the Mesopotamian plains, for rice, the seed-grain of India.

At the banquet given to celebrate Jason's succession, and the restoration of his father Æson to power after Pelias had acknowledged his rights, Jason was invited to undertake the adventure of winning the Golden Fleece by the tale told by Pelias of its origin. The Golden Fleece was the fleece of the winged ram, the successor of the mother-bird, sacred to Varuna, the Greek Ouranos. Ouranos, or Varuna, was the father-god of the races who looked first on the rain-cloud (*var*), and afterwards on the heavens, studded with the star-

[1] Pindar, *Pyth.* iv. 75 (133), the ode telling the story of Jason.
[2] Jowett, *Plato*, Second Edition, 'The Laws,' vol. v. p. 418.

measurers of time, as the source whence the germ of life
descended on the earth, incorporated itself in the seed, and
made it grow. Varuṇa, the god of conjugal union in Hindu
mythology, is in that of the Zends the god Varena, the god
of the squared field divided by the right-angled cross, the
father-god of the five parent-gods of the twin-races, the five
Kadmæan ancestors of the Bœotians, and the Hindus of the
Rigveda. The twin father-gods of these sons of the cow and
the ploughing-ox are called, in the Rigveda and in Zend
mythology, the Nāsatya, or those who do not deceive, the
twins Day and Night.[1] In the Zend mythology of the
Varenya Devas, the twin parents of their offspring, the sons
of the Nāunghaïthya or Nāsatya, are Tauru and Zairi names
which reproduce the Hindu and Akkadian Tur, the revolving
pole-father of the Tur-vasu, the people whose god (*vasu*) is
the pole (*tur*), and that of the Hindu father-god of the
Yadavas, twin brethren of the Tur-vasu, the god Hari or
Hairi, the yellow storm-god parent of the yellow races. It
was these people who had, in their earlier mythology, been
sons of the Dravidian sheep-mother Eḍā or Iḍā, the black
ewe lamb offered to Hekate, who became, owing to the union
between the patriarchal sons of the North, the conjugal
monogamists, and the matriarchal races of the South, the sons
of the father-ram, whose golden fleece was the dark sky
studded with the stars of heaven.

The golden-fleeced ram, the totem-god of the sons of the
winged bird-mother, the hawk-goddess of the West, called
by the Egyptians Hathor, the mother of Horus, had flown
away with Phryxus and Helle, the children of Athamas, king
of Thebes, and Nephele, the cloud-mother, the ram's mistress, thus saving them when they were about to be sacrificed
to the mother-earth by Ino, the daughter of Kadmus, the
rival wife of Athamas, the second of the two wives allotted,
in ancient genealogical mythology, to the national father-
gods. These wives were the mothers of the sons of the ram,

[1] *The Ruling Races of Prehistoric Times*, Essay v., pp. 429, 430.

the sheep race, the prophet-sons of Rachel, the ewe, the Hebrew mother, answering to Iḍā, the sheep-mother, and the daughter of the bull-father Kadmus of the Eastern (*kedem*) race, answering to the Hebrew Leah, the wild cow, the Hindu cow-mother Gauri, the race who sacrificed children to the mother-earth.

Phryxus and Helle mean the ripened or roasted (φρύσσω, φρυξ) grain, the ripe or roasted barley of the Hindu offerings to the Fathers, the sacrificial offering of the barley-growers, and the sun-maiden, whose name, like that of Hellene, the dawn, and her Sanskrit counterpart of Saramā, the sacred bitch of the gods, is connected with that of Sar, Sare or Zare, the osier (*sarats*) basket-mother of the sons of the rivers, the Iberian Basques of Asia Minor, the sherd or shard of the sons of the potters, the race of Pelias, Shar, the cloud-goddess of the Armenians, I-shara, or the house (*I*) of Shar, the mother-goddess of the Akkadian Serakh, god of corn,[1] who became in Northern mythology the Nephele, or cloud-mother of Helle, the German Nebel, or mist-mother of the western sons of Niblung, and among the Dorian Spartans, the sons of the seed sown by Kadmus, Helene, the immortal sister of the immortal Polydeukes, the great (πολυ) wetter (δευκης), the rain-god, one of the parent-twins of the Dorian race. She, as the dawning and glimmering light heralding the rising sun, and the sun emerging from the rain-clouds, formed, with the rain-god, the dual offspring of the cloud-mother Sar of the corn-growers, and she was worshipped in the Dorian island of Rhodes as a goddess, shown by the epithet δενδρῖτις,[2] to belong to the race of the sons of the tree. She was the dawn-goddess of the barley-growing sons of the East, who in Sparta reckoned time by the Syrian year, beginning with the sun travelling from the due east to the west at the autumnal equinox. This juxtaposition of the two children of the cloud-mother, who nourishes the grain, and one of

[1] *The Ruling Races of Prehistoric Times*, Preface, pp. xx. xxi.
[2] *Encyclopædia Britannica*, Ninth Edition, vol. xi., Art. 'Helena,' p. 629.

whom is the ripe barley seed, contains a mythical statement of the Hindu patriarchal creeds of the race of the barley-growers, which, as stated in many places in the Mahābhārata, asserts that ' the mother is but the sheath or husk in which the father begets the son,' who is indeed the father himself reborn, or, as it is put literally in the poem, 'the father is himself the son.'[1] The pair born from the rain generated from the cloud and the lightning flash were the twin-parents of the sons of the fire-god Phur, the father of the Phrygians,[2] and of the seed-grain of the barley-growing races, the off-spring of the husk-mother Sar, who became the cloud-mother called Nephele in the Northern speech of the Grecian mytho-logists, who had adopted her as a goddess-mother taken from the Semite Armenians, who looked on her as the cloud-mother Shar. It was her daughter Helle, the sister of the barley-seed, Phryxus, who, in her transit through the air on the back of the flying ram, fell off in the Hellespont between Greece and Asia, as the husk falls off the ripened seed grain when it is gathered and winnowed in the wind.

This was the strait traversed by Europa, the sister of Kadmus, when she came on a bull from Asia to Europe, bringing with her the knowledge accumulated by the Eastern sons of Assu or Asia, the mountain (as) land of the rising sun ; knowledge which was commemorated in the names of her sons Minos, the measurer (min), Rhadamanthus, the judge or diviner (mantha), who judges by the rod (rhodon), and Sarpedon, the cleanser, from sair, sar, to sweep,[3] the collector of facts. Their teachings were to be imparted to the people of Europe, the Western land, Ereb,[4]

---

[1] Mahābhārata Ādi (Sambhava) Parva, xcv. p. 284. It was this conception which made the goddess Sar, the husk-mother of the seed-grain, the withered husk of the historical myth of Abram, the sun-god, and Sara.

[2] The Ruling Races of Prehistoric Times, Essay i., p. 37.

[3] Ibid.

[4] The name Ereb (meaning the West), the Hebrew ereb, or erev, the evening, is preserved in the Arab names of the land conquered by them in North-west Africa and Spain, called El Gharb, Maghreb, Algarves.

by the bull race, the sons of the bull Taurus, the revolving pole, Tur. It was this ploughing-bull of Kadmus the arranger, the god of the East (*kedem*), which first civilised the land of the West, Ereb, and the evening sun. And it was as the tamer of the bull that, as I have already shown [1] when writing of this myth, that he was made king of Bœotia, the land of the ploughing-ox (βοῦς), by Athene, the goddess of the Eastern olive-tree, and became parent of the sons of the seed-grain, whose husk was Helle, the sister of Phryxus, the ripe seed.

Leaving the useless husk behind, Phryxus, the seed barley, was borne by the ram to Æa, on the Phasis, the city of the metal-workers, who first used copper, and afterwards of the iron, which gave to them the name of Chalybes, or men of steel (χάλυψ). This was ruled by Æetes, the eagle, or wind-bird (ἀετος, ἄημι, to blow), the son of Helios, the sun-god, who, as the bird, was the star-god of the pole-star Vega in the constellation of the Vulture. It was in the Eastern city of the winds, which in ancient mythology drove the stars round the pole, and in the land of the race descended from the first growers of barley, that Phryxus died, and was buried as the seed of the future crop, and that the ram was sacrificed to the sky-father, to whom the land was consecrated by his blood, and it was to bring back the seed of a new time that Jason was to start on his voyage for the recovery of the fleece of the sacrificed ram.

Among the crew who manned the Argo were the Dorian twin-fathers, Kastor and Polydeukes, and the winged twin-brethren Zetes and Kalais, sons of Boreas, the north-north-east wind, the wind coming from the east of north, whence the sun of the summer solstice, the sun of the sun-horse, rose, and of the daughter of Erechtheus the thunder-god.[2] They, as the rain-bearers of the North, replaced, as we shall see when we examine the myth of Phineus, the sea-eagle, the vulture wind-birds of the mythology of the South, the home

---

[1] See p. 182.

[2] The Tearer or Breaker (from ἐρέχθω, ἀράσσω, *to break or tear*).

of the constellation Argo, only visible to those who dwell south of the latitude of Greece. The north-east and north-west winds were the two sons of the north wind, whose track marked the path of the sun-horse at the two solstices, when he rose in the north-east at the summer, and the south-east at the winter solstice, setting in the north-west and south-west. It was these winds of the North which, when the sons of the North united with the mariners who sailed the Southern star-ship Argo, formed, like the sun of the solstices, the rain and sun-cross which, when placed on the right-angled cross of the father points of the compass, sacred to the fire-god and the pole-star, formed the eight-rayed star the sign of god and seed of the barley-growing sons of the bull and the wild cow, the Gaurian race of Girsu, the god of creating fire, whose sacred number was eight, and also denoted the eight points of the compass. This became the sign of the sons of the seed-grain, when they had become worshippers of the solstitial sun as well as that of the equinoxes, had changed the number of the seasons from three to four, and when, as mariners and observers of the weather, they found it necessary to use as their guide a better wind-chart than that of the four points of the compass. They then changed the six-rayed star of the Hittites and Cypriotes, marking, when made to point east and west, the solstitial

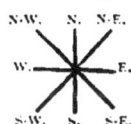

and equinoctial points ,[1] into the eight-rayed star.

These worshippers of the eight-rayed star were the sons of Kadmus, the arranger, when his wife was called Harmonia,[2]

---

[1] But the original form of the six-rayed star, as I have shown above, pp. 53, 54 , derived from the Egyptian five-rayed star , when the South pole was fixed by the astronomers of the Indian Ocean in the island of Lunka, Ceylon, sacred to Agastya, the star Canopus of the constellation Argo. It was in this form, as in that of the star of Horus, the sacred star of the worshippers of the solstitial sun.

[2] She was one of the triad gods of the Kabiri—Dardanus, Jasion, Harmonia ;

a name producing her functions. They were the united sons of the father sun-god of the North, Rā, and the grain-husk and cloud-goddess Sar, the parents of the seed-born, from the union of the sons of the Northern father with those of the Southern mother, who had, from the Southern village and province, formed the stable government of the con-federated kingdom united under the sway of the ruler dwelling in the central province.

The first of the volunteers who came to Iolchos to take part in the expedition was Argus, the builder of the ship Argo, the mythic representative of the Southern constellation of the mother-ship. He was in Greek mythology the watcher of the stars, the star Canopus of the constellation Argo, the Indian star-father-god Agastya who guarded Io, the queen of the dark night, till he was, after the age of Jason and his voyage, supplanted by Hermes, who slew Argus with the Harpe, the crescent-shaped sword of the crescent moon. Argus when building the ship Argo for its northern voyage, told Jason to take for its prow the second rafter in the royal hall of Pelias, the oak-beam which had come from Dodona, the temple of the sacred doves, and had superseded the ash of Mount Pelion, the sacred tree of the Centaur sun-horse. Thus the ship was consecrated to the oak, the mother-tree of the sun-worshipping Druids of Asia Minor, who united themselves to the Southern mariners who came to Greece in the Argo.

This age of the installation of Jason, the physician sun-god, as ruler of the star-ship Argo, the mother-stars of the Euphratean maritime traders and astronomers who traced the annual path of the sun round the eight points of the compass, the eight-rayed star, is marked by the deposition of Pelias, and the close of the rule of the potters. This change in the reckoning of time from its measurement by the changes of seasons, and the revolution of the stars round the pole is shown by the incidents of the mythic story to have been made

and her husband Kadmus was the fourth Kabir, Kadmilos, who added the year of four seasons to that of three.

under the influence of the Southern traders who worshipped the essentially Southern constellation Argo. The year which they introduced after their union with the sons of the sea-horse, who had already divided the year into four seasons, was the wheel-year of 360 days, the year of Stonehenge. The fundamental idea of the year-wheel is shown by the myth of Ixion, who was condemned to turn a wheel in the heavens, to have originated from the oil-press used by the immigrants from the East, who, as the sons of Kadmus, brought the plough and its yoke of oxen to Europe. They arrived in the Neolithic Age, when the oxen of Asia Minor came into Europe and brought with them the worship of the goddess of wisdom, the flower-goddess Athene, the virgin mother-goddess of the sons of the olive-tree, as distinguished from Koronis, the twin sister of Ixion, who was the mother of Æsculapius, and from Freya, the seed (*frio*) goddess of the North, the hawk who became the hawk-mother of the West, the successor of the vulture-bird-mother, the star Vega, the polar star from 10,000 to 8000 B.C. The virgin-mother of plant life was the star-goddess worshipped by the Minyan barley growers as the star Virgo, the star of the Egyptian goddess Min, depicted as holding an ear of corn. This was the second star of the Akkadian Lumasi, or seven creating stars.[1] This Minyan parent-star was the mother-star of the land of Khem, the native name of Egypt, and it was her worship which was united with that of Argo when the crews of the Southern trading ships brought to Greece the cult of Canopus, the star of the south pole, the chief star in Argo, and this is shown to have come from Egypt by the dedication to this star of the earliest Egyptian port at the mouths of the Nile, which was named after it.

We have thus, in determining this chronology of the age, two factors to deal with, the first representing the theology and civilisation of the barley growers, worshippers of Min, the goddess of Asia Minor, who measured the year by four

---

[1] *The Ruling Races of Prehistoric Times*, Essay iv., pp. 359-362.

seasons, and the second, the later influence of the Southern traders, who used the wheel-year.

In the first inquiry we have as guides the traces of the rise to power of the predecessors of the Dorians, the Æolic races, the sons of the wind-god Æolus, formed from the union of the Phrygian agriculturists and shepherds, the growers of dry crops, with the Kabiri, the artisans and builders of Mysia and its islands, consecrated to the dwarf fire-god Hephaistos, who is Thor in the Edda, the god drawn by the Phrygian goats; Il-marinen, the bear-god, in the Kalevala, the god of the stars of the Great Bear; Wieland in the Norse tradition, who became the god Ptah of the Egyptians, and Japhet of the Semites. He is in all these mythologies the bearer of the

hammer † the earliest form of the upright right-angled

cross. It was with this father-god that the worship of the virgin goddess of seed, Athene, the goddess of the olive-tree, was most constantly associated after the mother of fire, the fire-socket, became the mother of barley, the seed of life. The mother-goddess of the olive-tree was the mother-goddess of the cultivators of Asia Minor, who first grew the sacred sesame (*Sesamum Orientale*), used in the ritual of the Mandaites who adore the polar star as Shemol, the star of the left hand, and who look to the equinoctial sun of autumn, rising in the due east, as that which ushers in their year. It was they who took from the revolution of the stars round the pole, and from the twirling of the fire-drill, the first idea of the oil-press which was used in Asia Minor, as it has been used from time immemorial in India to extract the oil of the sesame seeds. It was these people, the Iberian sons of the rivers, who were also the inventors of the potter's wheel, which is shown by the myth of Pelias, formed from the potter's clay, who was the son of Turo, the revolving pole, and a god, to have been adopted at a very early age as a symbol of creation. As the first anthropomorphic creator, the master-smith, the turner of the fire-drill and the bearer of the hammer, was a god of the fire-worshipping Kabiri, the building race who ruled the Æolic confederacy of Mysia, it is also probable that the god who made man from clay on the potter's wheel belonged to the same mythology. This supposition is confirmed by the researches of Dr. Schliemann, who found in the lowest city on the site of Troy, where almost all the implements unearthed, except a few bronze knives, a ring, and some bronze hairpins, were of stone, a jug turned on a potter's wheel,[1] thus proving that it was used in Mysia at the very beginning of the Bronze Age. From thence it went to Egypt with the worship of the Phrygian goddess Min, and in Egyptian mythologic art, Chnum, the architect, the god with the four ram's heads sacred to Varuṇa, and Ptah, the dwarf mason-god, who also carries a hammer, are depicted as creating by the aid of the potter's wheel. It was the worship of this god which existed side by side with that of the phallic god of the shepherd races, the god of generation, whose influence, as I have shown in the Essay on the astronomy of the Veda, is so especially visible in the Indian chronometry, and the conception of god the creator as the potter, is one that coincides with the poetical creed of the fire-worshippers, which abhorred impurity, and thought sterilisation to be a religious duty incumbent on those who have children living to perpetuate the family.[2] The cult of the creating potter was probably taken southward by sea, for Asia Minor must always have been a country where maritime commerce flourished from the first dawn of civilisation ; for its earliest matriarchal population were all sea-faring folk, and the country, like Western India, has forests clothing the hills which come down to the coasts, both in Northern Mysia and Southern Cilicia. It was the matriarchal dwellers on the

[1] Schuchhardt's Schliemann's *Excavations*, p. 37, fig. 16, p. 40.
[2] This custom is called by Herodotus ' The female disease,' and he ascribes it to the Scythians, Herod. i. 105. It is still common among the Lippovan Tartars and among certain Russian dissenters. It seems to be of Slavonic origin.

coasts of Palestine, where Ashkelon was the city especially consecrated to the mother-goddess of the earth, who established the commerce in timber, which brought the cedars of Lebanon to Girsu,[1] at a date before the building of the Pyramid of Gizeh, about 4000 B.C., and it was the traders of the Eastern coasts of the Mediterranean, and the cultivators of the fertile lands of the river valleys of Asia Minor and Syria, who took the potter's wheel, together with the oil-press, in all their migrations. Hence we find identically the same form of potter's wheel used in India, Assyria, Egypt, and Greece. Egyptian pottery made with the wheel is said to date from about 4000 B.C.,[2] but the invention of the wheel must have been much earlier than the time when it came into common use in Egypt, and it must have been long before that date that the son of the potter's clay, born of the revolving mother Turo, first invented and popularised the potter's wheel in Mysia and Phrygia, whence it was to be brought to Egypt and made in that conservative country the symbol of creating power. It was long after the building Kabiri, who used the potter's wheel, reached the Euphratean Delta that they instituted the commerce from Eridu, which brought from Sinai the diorite of which the Girsu statues are made, and this commerce is shown by the tin and copper which they imported, together with gold, timber, and stones, to have belonged to the Bronze Age,[3] while the identity of style and the inferiority of execution shown by comparing the Girsu statues of Gud-ia with that of the Egyptian king Kephren, which was made about 4000 B.C.,[4] shows that the inferior Euphratean statues must be much the older of the two.

Thus, from a review of the evidence furnished by the theology and archæology of Asia Minor, Egypt, and Assyria, it seems probable that the Kadmæan and potter's age preceding that of Jason must date from before 4000 B.C. This age,

---

[1] *The Ruling Races of Prehistoric Times*, Essay iii., p. 282.

[2] *Encyclopædia Britannica*, Ninth Edition, Art. 'Pottery and Porcelain,' vol. xix. pp. 601 ff.

[3] *The Ruling Races of Prehistoric Times*, Essay iii., pp. 282, 283.

[4] Sayce, *Hibbert Lectures* for 1887, Lect. i. pp. 32, 33.

which corresponded with that of the rule of Dan, the judge, in Syria, the Euphratean lands, and India, was succeeded first by that of the sons of the Northern sun-horse, the Assyrian, Zend, and Indian Ashura, or sons of Ashur, and next by that which in Grecian history opened with the coming of the Southern traders, worshippers of the mother-constellation Argo, who perpetuated her memory in the name Argos, which they gave to Southern Greece, and which meant originally the land of the snake, the Doric Argas (ἄργας). It was when the sun-god of the barley-growing races of the upright right-angled cross, symbolising the earth, became Apollo, the dolphin, the fish-sun, which was first the mother-ship Argo, that he took the epithet of Argeiphontes, meaning the slayer of the snake. The crew of this mother-ship were the Dorian race, who, as I have shown, introduced Indian Turano-Dravidian customs into Crete and Southern Europe,[1] ceased to map the line of march of the national history by the sign marks of the plastic arts of anthropomorphic religion, and who measured the year by the circuit of the sun round the wheel formed by the eight-rayed star, depicting the tracks of Boreas and his sons, and those of the monsoon winds which brought up the rains of the Indian Ocean. This was the year of Argus, which was superseded when the reckoning of time was changed by the introduction of the lunar-solar year of thirteen lunar months, the year beginning with the lunar crescent, the Harpe, with which Hermes killed Argus, the builder of the Argo, when Hera changed him into an Indian peacock, displaying on his tail the stars through which the annual path of the moon and sun was tracked by the Euphratean astronomers, who made the moon the nurse of the sun during the three months from November to February, and traced the path of the sun as an independent god for ten lunar months, from February to November.[2] This was the year of the sons of Leah, the wild

[1] *The Ruling Races of Prehistoric Times*, Essay iii., pp. 296-298.
[2] *Ibid.*, Essay iv., pp. 376-387.

cow, and of Rachel, the ewe, of the thirteen children of Jacob, the husband of the daughters of the moon-god of Haran, the city of the road (*Kharran*), which ruled the international land traffic between the Euphratean lands of the East and Asia Minor, and Europe in the West,[1] and also the year of the fish-god, the Akkadian Ia and Sal-manu, the Greek Perseus, which I shall describe later on, when I have finished the narrative of the Jason myth. We must now, after this long digression, in which I have tried to work out the chronology of this mythic history, return to the voyage of the ship Argo. On her outward voyage from Iolchos she passed by Lemnos, the mother-island of the matriarchal maritime races, whose unmatrimonial customs are alluded to in the story telling how, on passing the island, the Argo took on board the last survivor of the husbands of its women, the rest having been slain by their wives. Thence, after stopping on the Mysian coast, they sailed into the Sea of Marmora, the Greek Propontis, reaching the city Cyzicus, whence, by the advice of their steersman Tiphys, they went to the Thracian town of Salmydessa ruled by Phineus. It was he who, according to Tiphys, could teach them how to pass unharmed through the Strait of the Symplegades, meaning the clashing rocks, into the Black Sea, the Euxine, or Sea of Pontus.

It is in the story of Phineus and the Argonauts that we find one of the clearest clews to the understanding of the history told us in Jason's voyage. Phineus, the blind old king of Salmydessa, is the Greek counterpart of the blind Hindu king Dhritrāshtra, the gnomon pole, the husband of Gandhārī, the mother-bird, for his name means the son of the sea-eagle (φίνις or φήνη), the pole-star, descended from the vulture, the star Vega. He was the victim of the persecutions of the Harpies, the three vulture birds, who were called ῞Αρπυιαι, the tearers, the three seasons of the year of the worshippers of the mother-bird. They were the mother-goddesses of the Lapithæ, the sons of the destroying storm

---

(λαπ), the predecessors and enemies of the Centaurs, the
worshippers of the sun-horse. The Lapithæ were the races
who, before the myth of the sun-horse, and the belief in the
permanence of natural order was evolved, looked on the
capricious storm-god as the ruler of heaven and earth. In
the Jason myth Phineus, the ruling sea-eagle, whenever he
sat down to eat, was buffeted and pecked at by the Harpies,
the storms which ushered in the changes of seasons. It
was the two sons of Boreas, the North-east and North-
west wind, starting from the points where the sun of the
summer solstice rose and set, who drove these unclean birds
out of the realm of Phineus, and it was the discoverers of the
path of the solstitial sun round the eight-spoked wheel, and
of the sun-cross uniting diagonally the rising and setting sun
of the summer and winter solstices, who established, together
with a year of four seasons and a solar astronomy more
accurate than that of the worshippers of the pole-star, a
more scientific meteorology based on the directions from which
the winds blew at different seasons of the year. These
lessons had been learned without difficulty in the region of
the monsoons of the Indian Ocean, whence the Argo started
on her first voyage, but they were found to be more complicated in the Northern seas. We learn also in this myth how
the anthropomorphic and individualising instincts of North-
western myth-makers changed the mother-birds and winged
beasts of Eastern, Finnic and Southern mythology into the
winged angels of the new Northern faith. It was these wind
angels of the North-east and North-west winds who drove
the Harpies to the Strophades or Turning (στρέφω) Islands,[1]
turning with the pole, which was from henceforth to be the
prison of the winds, guarded by the pole-star of the North,

[1] The connection between the Strophades and the revolving circumpolar
stars, which measured time before the year of the fish-god, the year of the
moon-nurse and sun-father, is clearly marked in the line of Sophocles, telling
of the Ἄρκτου στροφάδες κέλευθοι, the circling paths of the Great Bear, Soph.
Tro. 131.

the Greek Boreas, in the North-north-east, which, like the true pole-star, varied a little from the meridian. Under this new rule the heavens were governed by the circumpolar stars traversing in undeviating order the track marked out for them, not by the capricious winds of the earlier mythology but by the all-wise creator, the master-smith of Northern belief. The blind king who was freed, the prophet-bird of the blind gnomon or prophetic stone, the rock of the sea-eagle, the Bethel or House of God of Jacob,[1] gave to the crew of the Argo as their guide through the Symplegades the grey dove, the sacred bird of Varuna and Dodona, whence the prow of the Argo came. This bird, the Yonah or prophet-dove of the Jews, was the sacred bird of Nineveh, the fish-town of Ashteroth, the moon-goddess, and of the Greek Aphrodite, the daughter of the sea-foam, and it was this prophet-dove who, after the raven, the bird of Odin, the father-god of the sons of the sun-horse, had failed in its mission, bore in its beak to Noah the olive leaf, the leaf of the mother-tree of the Semite and Minyan race;[2] and Noah was the Hebrew counterpart of the Akkadian Dumu-zi, the son (*dumu*) of life (*zi*), the god who made his yearly voyage in the sun-ship as the father of the sons of righteous-ness, born under the mother-tree of Eridu, of the virgin-mother Istar, in the temple 'into which no man hath entered,'[3] the heavenly temple of the star Virgo. This voyage of Noah, or Dumu-zi, in which the sun-ship was led by the dove, was that of the year when the mother-mountain, the home of the sons of the oil-tree, rose out of the life-giving and consecrating waters, and the new birth on earth of this regenerated race was marked in Greek mythic history by the bestowal of the name of the Peleiades or dove (πέλεια) constellation on the Pleiades, the stars led by the Hindu mother-star Amba, and called by them the Krittakas or

[1] Genesis xxviii. 18, 19.   [2] Genesis viii. 7-11.
[3] *The Ruling Races of Prehistoric Times*, Essay iii., p. 150 ; Sayce, *Hibbert Lectures for* 1887, Lect. iv. p. 238.

spinning (*krito*) stars, the wives of the Northern constellation of the seven bears.[1] It was the union of these spinning mother-stars of the Hindu matriarchal races, the companion stars of Rohini, the red antelope and cow-mother of the Hindu Gautuma, the star Aldebaran, worshipped by the Minyo-Sabæan Arabs of Southern Arabia, as the Turāyya or stars of the Tur, with the Northern bear or antelope (*rishya*) stars, which formed the web of time woven by Penelope, the wife of Odusseus, the wandering sun-god.[2] Under the name of the dove-stars, the Pleiades, on the transfer of belief from the matriarchal to the patriarchal creeds, remained the mother-stars of the new faith, but they were consecrated to the bird of conjugal union instead of to the mango (*am*), the tree-mother of legalised tribal concubinage. It was this bird which was to steer the ship of nuptial faith through the stormy straits of the Symplegades, and to be the herald of its successful passage, for it was when the dove had flown unscathed through the rock-bound cleft that the ship was to follow in its wake. After the crew of the Argo had passed through the Symplegades, under the guidance of the dove, they reached the realms of King Lycus, the wolf fire-god (λύκος), and it was there that they lost Tiphys (τίφυς), their pilot, who was bitten by a snake and died. This symbolic death tells us how the Argonauts ceased to worship the mother-pool Tiphos (τῖφος), the healing-well springing up under the hoofs of the sun-horse, which formed the sources of the parent-rivers of the sons of the barley seed, the sacred pool of the

---

[1] *The Ruling Races of Prehistoric Times*, Essay v., p. 426 ; Eggeling, *Sat. Brāh.* ii. 1, 2, 4 ; S.B.E. vol. xii. pp. 282, 283.

[2] The worship of the Pleiades as the weaving stars seems to have arisen among the weavers, who were united with the potters as the ruling races when the artisans were the organisers of trade. It was the weavers and potters who were the sons of Shelah, the son of Arpachsad, or Arpa-Kasad, meaning the land of the conquerors (*kasidi*), the name given in the genealogies of Genesis to the land of Armenia, and Shelah, their father, is also called a son of Judah before his marriage to Tamar, the date palm-tree, when he was the fire-god, the god of the sons of Caleb the dog. See *The Ruling Races of Prehistoric Times*, Essay iii., pp. 179-189; 1 Chron. iv. 21-23 ; Genesis x. 21-24.

lotus (*Nymphea nelumbo*), the Indian plant of life of Hindu, Zend, and Egyptian mythology,[1] the water-mother of the Southern crew of the ship Argo ; and how they took for their guide the counterpart of the master smith of Northern belief, Erginus, the active worker (ἔργον), the son of the sea-god Neptune, who led his followers to fresh discoveries. This new steersman was the son and representative of the god called Neptune by the Latins, Napāt Apām, the son (*napāt*) of the waters (*apām*) in the Zend and Hindu mythologies, to whom the Indo-Greek immigrants from the the South gave the name of Poseidon, the god formed in the image (εἶδος) of Pūsh, the Hindu barley-eating god, who makes plants to grow (*pus*) the Lithuanian Purk-an,[2] the black bull Pashang of the Zends, the animal which was offered to Poseidon as his totem animal at Ephesus.

Thus this change of steersmen tells us of the substitution of the father-bull, the ruler of land and sea, and the father of Thetis, the mother-moon, the mother and nurse of the sun-god Achilles, for the mother-pool, which gave life to the cultivated land and its offspring, while the choice of the son or alter ego of Poseidon as the guide of the heavenly ship, marks also the age in which the sea-god is represented in Greek mythology, as contending with Athene, his predecessor and the mother-goddess of the earlier cultivating race of Semites, growers of the oil-plant and tree, and the parent-barley, who were ousted from power by the new race of maritime merchant traders, sons of the twin-gods. Erginus brought the Argo to Æa, where the Fleece was to be won, and whence the sea-going sons of the circumpolar stars and of the barley seed were to start on their voyage as the rulers of the year measured by the voyage of the sun-ship round the eight-rayed wheel-star, ruling the land and sea, instead

---

[1] The *Nymphæa nelumbo*, the rose-coloured lotus of India, though depicted in Egyptian art as the Plant of Life, does not grow wild in Egypt (D'Alviella's *Migration of Symbols*, p. 31).

[2] *The Ruling Races of Prehistoric Times*, Essay v., pp. 437-440.

of by the land-journey of the sun-horse to the solstitial and
equinoctical points, which succeeded the year marked by the
migrations of the bird-measurers and prophets, who told of
the solar movements which brought spring and winter to the
North and the rains of the summer solstice to India and the
Persian Gulf.

It was in Æa that Jason obtained the aid of Medea, the
priestess-guardian of the Golden Fleece, the daughter of
Æetes, the wind-eagle son of Helios, the sun, and it was by
her help that he performed the tasks allotted by her father's
order to whoever would win the Golden Fleece, dethrone
Varuna, the ram-ruler of the heavens, and replace him by the
protégé of the moon-cow, the Southern feminine and not the
Northern masculine moon, the descendant of the tamer of the
ploughing-bull. For the performance of these tasks he was
consecrated by the water of life brought to him by Medea
from the sacred pool, and made from the plants yielding the
Soma or sap of immortality. This she had been taught to
make by Hekate, the three-formed goddess of the ever-
reviving year of three seasons, the bird-mother of the
hundred (hekaton) sons of the bird and tortoise race (Khur),
the Kurds of the Iberian land of Mount Ararat, the goddess
whose sacred symbol was the perpendicular right-angled
cross. When Jason was proclaimed as the sun-god by being
anointed with this impenetrable and incombustible protector,
the golden armour of the god who lived unharmed amid the
burning flames of the blazing sun, he was able to do all that
was required to gain his annual victory over his foes. His
first task was similar to that of Kadmus, except that
Kadmus, the master-smith and ploughman, who made the
generating fire-drill of the earth, the plough,[1] turn in per-
forming its annual task, found the bulls which breathed the

---

[1] The plough, Germ. *phlug*, is formed from *Phru-ga*, the manifestation (*ga*)
of the fire-god Phru or Phur, just as the Greek Phlegyas is formed from
a Phrygian form *Phre-guas* (*The Ruling Races of Prehistoric Times*, Essay i.,
p. 39).

fire generated by the heavenly fire-drill after its revolutions ready to move at his will. These had to be trained to know their new master, Jason, the sun-god, to be yoked to his plough and forced to plough the fields of heaven, not in obedience to the caprice of the thunder-god of the sons of the vulture-bird, but in the regular order necessary to produce the growth and maturity of the produce of the serpents' teeth, the seeds grown from the ring of cultivated land, the snake encircling the mother-grove of the primæval village. When the men born from the seed grew up Jason threw among them the crystal ball of time, the prophetic crystal given him by Medea. This rolling ball of the votaries of the eight-rayed star is the successor of the prophetic gnomon, and the ball given in so many of the Russian fairy tales to Ivan, the youngest and most successful of the three brother-seasons of the fairy year. It is this ball, the gift of the counselling witch-mother, Baba Yaga, which is to roll before him to show him the way to his destined bride, the sun-princess, the mother of the future year. In Jason's story, as in that of Kadmus, the armed men born from the seed all killed each other in mutual combat, only five surviving in the Kadmus and none in the Jason story. This part of the story partly tells how each conquest made by an alien race, only becomes stable and effectual in producing co-operation between the old and new comers after years of disturbance and distrust, ending with the time when a new generation born from the union of originally alien parents grows up, while in its historical aspects it tells us how the advances made in solar astronomy and practical knowledge by Jason, and his obedient mariners were a continuation of the work begun by Kadmus and the five Bœotian fathers left alive, from whom Jason and his companion Minyæ were descended.

When Jason, the sun-god, leader of the united people who had forgotten the animosities of their parents, remained alone on the fields ploughed by the heavenly bulls he had tamed,

he was the conquering leader of the new nation, and therefore entitled to receive the control of the seven keys, the
seven days of the birth-week ushering in the new year of the
sun-ship. These he received from Medea, who went with
him as his guide when he unlocked the seven successive gates
of the temple of the far-darter, Apollo, the thunder-god,
and, in the prominent place assigned to Medea, we find the
beginning of the national ritualistic organisation which,
among the Finns, who looked on life as born from the egg of
the mother-bird, made the mother the guardian and priestess
of the Jouha or household-fire. It was from this starting-
point that the inspired priestess became the prophetess, ruling the Jews, and the consecrated announcer of the oracles
of god in the shrine of Delphi, consecrated to the mother-
dolphin, the fish-goddess-mother, a variant form of the
star-ship Argo, and this consecration of the house-mother
branched off in another direction into that of the Vestal
virgins, priestesses of the goddess Vesta, the Greek Hestia, the
goddess of the hearth (ἑστία), the virgin guardians, being a
reproduction of the virgin tree-mother of the sons of the
barley-seed.

Medea, the guide and helper of Jason, was the high-priestess
who had been guardian of the worship of the ram, the sun-
god of those who looked on the rain-thunder and Northern
star-god Varuṇa as the ruler of heaven, and she now became
the co-partner of Jason, the healing-god of the sun-ship. It
was she who placed her foot on the guardian dragon, the
circumpolar stars of the constellation Draco, and charmed it
to sleep while Jason seized the Fleece. This story is precisely parallel with the astronomical pictorial myth depicting Hermes or Hercules, the fire-god, the constellation
Hercules, as climbing up the mother-mountain to the
summit of the heavens and placing, as he climbs, his left
foot on the head of the dragon constellation Draco, while he
reaches forward to touch the constellation of the Vulture or
the Tortoise, and to give it the seven strings, the seven locks

of the temple of the thunder-god, which was to make it the constellation Lyra of the astronomical observers of the annual course of the sun-fish. It was after the victory of the sun-god as the god who, first as the sun-horse, heralded the changes of the year by his journeys through the heavens from the north-east at the summer, to the south-east at the winter solstice, and thence back again to the north-east, was gained, that Jason, the pupil and heir of the conquering god, set forth on his year's voyage, not as the rider on the sun-horse, but as the healing physician, the captain of the sun-ship Argo.

It was when the sun-ship started on its voyage from the river Phasis that Absyrtus was slain. He was the brother of Medea, whose name means he was washed down (συρτός) by the rivers, and became the quicksand (σύρτις), or unstable foundation of the believers in magic and witchcraft as controlling natural laws, and of the time-reckoners of the sons of the rivers, who measured time by the migrations of the birds, the leaf and flower clock of Koronis, the movements of the circumpolar stars, and the journeys of the sun-horse from east to west and south to north to find the equinoctial and solstitial points. He, when trying to stop the Argo, was slain, according to one account by Medea herself, the converted witch, who also scattered his limbs over the waters of the Phasis, and to another, by the spear of Jason, the unerring lance of the sun-god, and by the arrows of Arkas the Arcadian, the constellation of the Great Bear, the stars of the Arcadian goddess Arktos, who became Artemis, the true (arta) goddess of Phrygia.

It was the voyage thus begun which, in the original Greek story of Argo, led the sun-ship from its Northern station at the summer solstice to the Southern ocean and Libya, and thence home again by the west to its setting place in the north-west at Iolchos. But in the later myth, as altered by the earlier geographers, whose memory reverted to the Northern origin of the sun-horse, who became the sun-ship, the course of the Argo seeking the North-western lands of the original sun-horse is represented as going against the deasil, or sun-path, from the east to the north,[1] and thence home to Iolchos in the west by the pillars of Hercules, the Straits of Gibraltar, Libya, and the Italian coasts. According to this story, when Jason heard that Æetes, having lost all trace of the Argo in the fog which arose when she left the Phasis, had stopped their way home through the Hellespont with ten ships, he, by divine command, caused the Argo to be steered to a river running from the north, up which the voyagers made their way. This river led the Minyæ to a tunnel through the Carpathian Mountains, and thence they bore the Argo on a sledge to a river flowing northwards. They wintered on its banks, and in the spring they sailed down it to the North Sea. From thence, by the help of a north wind brought by the angel sons of Boreas, they reached the Pillars of Hercules, and passed into the Mediterranean Sea. There Medea visited Kirkē (Circe), the hawk (κίρκος, κίρκη), the mother-goddess of the West and sister of Æetes, the eagle or vulture-bird, the parent of the East. The Western mother-goddess was the mother-bird of the mining races, the Egyptian hawk goddess Hat-hor, mother of the hawk-headed Horus, and of the sons of the wild boar and sow, into whose shapes she changed her lovers, the dead totem-fathers of past ages. It was Circe, the goddess of the Western sun-worshippers, and the possessor of the κυκεών, the cup of healing, who absolved Medea and Jason from the guilt of the death of Absyrtus, and the subversion of the ruder creed and scientific methods of

---

[1] This northern voyage of the Argo may also be a reminiscence of that told in the traditions of the Cymri, who say that they came to England from Defrobani or Greece, where Constantinople now is, by the way of the Hazy Sea, the German Ocean. Thus the Northern form of the Jason myth tells of the path by which the discoverers of the wealth of Britain made their way to the north-east ports of the German Ocean whence they sent forth colonies, which, like the later Saxons, occupied and ruled the country, and made the ritual of Stonehenge to coincide with that required by the wheel-year of Jason.

the sons of the rivers, who worshipped the polar star, and it was she who made the healing sun-god supreme god of the enlightened West, the god of the magic cup containing the immortal essence of life. This visit to the West tells us not only of the victory of Western over Eastern creeds, and of the opening up the stores of mineral wealth brought, in the Bronze Age, from the West to the East by the traders who superintended the working of the copper and tin mines in Spain and Cornwall, but also of the recognition of the apparently complete solution of the problem of the annual course of the sun which, by adding the solstitial to the equinoctial points, changed the computation of the year from that of Syria, Asia Minor, and Macedonia, which began the year of the votaries of the upright right-angled cross, beginning with the autumnal equinox, to that of the people who constituted annual games in honour of their barley-growing fathers, and made their year, like that of Egypt, Athens, and the founders of the Olympian Games, to begin with the summer solstice.

It is after the return to Iolchos that the myth of Jason and the Argo becomes one telling how the belief in the sungod circling the eight points of the compass in his yearly course succeeded that which looked on the Creator, not as the hidden power which made the seed to grow, but as the god who, as the creating potter, brought about the changes in the growth of all things endowed with life by moulding the clay of which they were made in the lunar months of gestation and the seasonal changes. It was this latter belief which is expressed in Egyptian theology by the six creating apes called Keftenu, or the Phœnician, and Uetenu, the apes of the green (*uct*) land, or India, which preceded the eight creating gods, first depicted as the apes, and afterwards as the eight mysterious spirits, headed by the Nun, or soul of life in water, who appear on the temple walls as frogs and snakes.[1] The Jason myth tells us how the faith of the twin races in the anthropomorphic god was destroyed by Medea,

[1] H. Brugsch, *Religion und Mythologie der Alten Ægypter*, pp. 153-160.

the sun-sorceress. She, disguising herself as an old woman, persuaded Alkestis, Eradne, and Amphinome, the daughters of Pelias, that their father, the son of the potter's clay (πηλός), would, if slain by them, be made young again by her magic art, and in proof of the truth of her assertion, produced a lamb from the caldron into which the daughters of Pelias had cast the limbs of a ram sacrificed by Eradne, the earth (ἔρα) goddess of the winter and mother of the coming spring. Having thus induced them to kill their father, Medea, when he was dead, instead of reviving him, summoned, by a signal torch, Jason and the crew of the Argo, who were hidden near the palace, to succeed to the throne.

This story tells us that the year of those who looked on the Creator as the divine potter was one of three seasons, which, in their revolving course, killed the father who had begotten them, the revolving pole, or the wheel of the potter. He, when he had finished his work, was succeeded by the sun-god, who, in the year of Jason and Medea, beginning with the summer solstice, marked its course round the eight points of the compass, the completed figure of the eight-rayed barley-seed. This new age, which united the monotheist creed of the worshippers of the polar star, whose year was divided into three seasons, with the dualistic faith in the sun-god, the revolving pole, and the year of four seasons of those who had changed the original dualistic parent-gods, the twins Day and Night, into the stars Gemini, the twins Kastor, the Akkadian Tur-us, or god of the pole, and Poludeukes, the much raining, the rain-god, who were on board the Argo, and the two children of Medea.

It is the story of the close of the epoch of the wheel-year, and of the substitution for it of one measured by the stages of the path traced by the moon and sun through the stars, which is told in the desertion of Medea by Jason, the dying sun, and of his love for Glaukē, the blue-eyed daughter of Creon, king of Corinth, the rising sun of the new age, who was consumed on her wedding day by the burning of the garment made for her by Medea, the brilliant hues of sunrise and sunset. It was in these that the new sun, the sun-maiden, married, in the Rigveda, to the male moon-god of Northern mythology was, at her rising, borne in their chariot to her new home by her twin-brethren, the Ashvins, the sun-horsemen, who were the stars Gemini,[1] and the brethren of Helene, the Greek goddess of the dawn. It tells of a mythology which taught how Hera, the mother-moon, the queen of the sky, and patron goddess of Jason, had to resign her dignity to the male-moon and sun-god Herakles, the son of Alkmēnē the moon-bow (*alk*, ark), the crescent-moon, whom she persecuted throughout his career. Herakles was the transformed fire-god of the sacrificial fire wedded to Omphale, the navel, or the fire in the centre of the altar, who came from Crete to Delphi in the dolphin sun-ship with the Dorian priests. It was they who depicted him in the

Delphic Trisūla placed over the gate of the temple as the Semite Ashēra, or creating-pole, the mast of the moon-boat. It was these reformers who abhorred the aberrations of the anthropomorphic creed of the wor-shippers of the creating-potter and the phallic god, and who were thus led to declare it to be impious to represent the Creator in human guise. The mystic creed arising from this doctrine, which succeeded that of the experimental physicians and the Western students of undisguised nature, represented by Jason and the crew of the Argo, came, like the more materialistic doctrines of the Herakles myth, from the East.[2] For Herakles was born at Tiryns, and was the

---

[1] Rigveda, x. 85, 8, 9.

[2] It was in Eastern India that the ritual of the altar representing the mother-earth, with the sacred fire in the centre, originated, and it was in its passage to Greece that the myth of Heracles, which originally depicted the twelve months of the wheel-year, was moulded into that which makes him the righteous sun-god by Kusho-Semitic brains. This is shown by the epic of his Akkadian prototype Gilgames, which, in the close of the sixth book, tells how

Grecian form of the Akkadian sun-hero Gilgames, and the descendant of Perseus.

Perseus was the son of Danaë, of the race of the Dorian Danaoi, sons of Danu, the judge, who was, as I have shown in the earlier part of this Essay, the father-god of the worshippers of the pole-star. She was the daughter of Akrisius, the mountain-top (ἄκρις), the Akropolis of the mother-city, and the mother-mountain of the sons of Ararat. The father of Perseus was Zeus, the golden rain of the sunlit shower, and thus Danaë was the mother-mountain of the land of Argos, made pregnant by the rain-cloud which encircled its top. His reputed father-land, Argos, shows by its name that the myth belongs to the series of the myths of the fish-father-god, who succeeded the fish-mother-ship, the dolphin, for the cognisance of the land of Argos was a fish.

It was as the son of the fish-god of a seafaring race, who made the ocean the mother of life, that the infant Perseus and his mother were placed by Akrisius in a box—the womb of darkness whence the god of light was born—and was thrown into the sea instead of into the river, to which the earlier mythologies, framed before the earth became the Kushite-tortoise, with the mother-mountain in its centre resting on the primæval ocean, intrusted the father-gods of the sons of the rivers. Danaë and her son were rescued by a fisherman Dictys—meaning the net. He is the mythological counter-part of the fisherman in the story of Sakuntalā, the bird (sakuna) mother of the Hindu Bhāratas or Bhārs, whose totem is the star-peacock, into which Perseus' father Argos, who became the star-watcher Argus, was transformed by Hera. It was a fisherman who found, in a fish he had caught

he refused the advances and despised the allurements of the mother-goddess Is-tar, who had began her career by being the mother-goddess of the matriarchal worshippers of the mother-earth. See the passage, which contains a mythic history of the evolutions of national beliefs, translated by Dr. Sayce in the *Hibbert Lectures for* 1887, Lect. iv. pp. 246-248.

in the river, the year-ring given to Sakuntalā by Dushmantā, the discoverer or unraveller (*mantā*) of hard sayings (*dush*), when her son Bhārata was begotten, and afterwards lost by her in the river. The finding of the ring made Dushmantā recognise Bhārata as his son, and the story is reproduced in a variant form in the myth of Perseus, whose name denotes a kind of fish,[1] and who was the fish-sun-god.

Perseus was the fish-god whose worship as the river-eel, the river Sal-fish fish, and the tank-fish Rohu of Indian mythology, originated, as I have shown, in India,[2] and who was borne in the ship of the mother-fish, the constellation Argo, to Eridu, as Ia, the fish-god : he thus became the parent-god of the Persian kings, who, according to Herodotus, traced their descent from Perseus.[3] The name is undoubtedly connected with the Vedic Parshu, meaning the ribs, or the sickle-shaped sword,[4] the Harpe which Perseus got from Hermes, the fire-god, and with which he slew the sea-monster who threatened to devour Andromeda ; and the use of the Harpe, when compared with the story of Hermes and Argus, shows that the myth of Perseus belongs to the age when men ceased to measure time by the apparent move-ments of the stars, or by mythic conceptions of the sequence of time, and the seasons connected with certain stars or groups of stars, such as Draco, the Pleiades, Orion, Sirius, Leo, the Crow, the Cup, and Hydra, and to use as their guides the stars shown by observation to lie so near the track of the moon and sun through the heavens as to mark the stages of their monthly and annual journeys.

These stars were the eyes in the peacock's tail of Argus, the star-watcher, father of Perseus. Perseus began his career as a sun-hero by taking from the Graiæ, the grey

---

[1] Ælian, *N. A.*, 3, 28 ; Liddell and Scott, *Greek Lexicon*, s.v. 'Perseus.'
[2] *The Ruling Races of Prehistoric Times*, Preface, p. xli.; Essay i., p. 22, ii., p. 285.
[3] Herod. vii. 61.
[4] Grassmann, *Wörterbuch zum Rigveda*, s.v. 'Parshu.'

old women, the clouds, otherwise called the Gorgons, or
storm-goddesses, the Harpies of the myth of Jason, the
three seasons of the earlier year, their one eye, representing, in Egyptian pictorial mythology, the sun of Osiris,
the barley-god. He exchanged it with the nymphs for
the winged sandals worn by Hermes, the wallet (κίβισις),
the seed receptacle of the sun-god of the sons of the
barley-seed, and the sun-god's cap of invisibility. With
these he attacked and slew the Gorgons as they slept, and
gave the head of Medusa, their leader, to Athene, the
goddess of the Semite parent olive-tree, who made Kadmus,
king of Bœotia. It was on his return from this feat that he
freed Andromeda, the daughter of Kepheus and Kassiopæia.
They were the king and queen of Æthiopia, the land of
Abyssinia, colonised from Southern Arabia, the land of the
Minyo-Sabæans, the irrigating race who formed the crew of
the Argo, as the sun-ship of the dwellers on the shores of the
Indian Ocean. Andromeda was chained to a rock as a
sacrifice to the devouring sea-dragon which ravaged the
country, the constellation Draco of the circumpolar stars
lying in the inner circle, outside which are the four constellations called after Andromeda, her parents, and Perseus.
This dragon—the Hindu constellation of fourteen stars,
representing the days of the lunar phases—he slew with the
Harpe, or crescent-moon.

Andromeda was engaged to be married to Phineus, whom
we have seen, in the Jason myth, to be the sea-eagle (φίνις),
and the victim of the persecutions of the Harpies, the
vulture mother-birds, the Graiæ and Gorgons of the Perseus
legend. But Perseus, unlike his mythological predecessor
Jason, received no aid from Phineus, and obtained, as the
moon-champion, the rule of the heavens, in spite of the prior
claims of the parent-bird of an earlier faith. Perseus thus
became the fish-sun of the race who worshipped the crescent-
moon as the nurse of the young sun-god. Thus their
theology differed from that of the twin-races, the Hittites,

or Khati, meaning the joined (*khat*) race, who made the six-rayed star, the solstitial sun and rain-cross, intersected by the equinoctial east and west line, ✳, and the crescent-moon, the symbol of the father-god and mother-goddess of heaven, the mother-moon ruling the months of gestation, their consecrated year, called by the Romans the annus, or ring. This star, drawn with the perpendicular instead of the horizontal line, the star sacred to the worshippers of the pole-star ✳ is still, as I have before observed, borne on the Turkish banners. It is this last star which is a direct descendant of the Egyptian five-rayed star ✳ of Horus, representing the solstitial sun and rain-cross and the gnomon-pole, and it is this star which is probably that of the six creating apes I have mentioned above, p. 208.

This myth of Kepheus, Kassiopæia, Perseus, and Andromeda is fully set forth in the four constellations called by their names, which follow in the order I have stated, Hercules, Lyra, and Cygnus, as circumpolar constellations, lying outside Draco. From a comparison of their position in the star-chart with the three grouped constellations they follow, and the star-myths of Hercules, the Lyre, and Cygnus, and the Perseus myth, it is clear that they both relate to a period in astronomical history when the sun-god, called Hercules, Kepheus, or Perseus, was represented as going round the pole, which is personified in the polar constellation called successively the Vulture, Tortoise, and Lyra. It is to the group of constellations supplying the pole-star that Cepheus belongs, for the stars γ and a Cepheus were polar stars from about 21,000 B.C. to 19,000 B.C.; and the polar star, after it had ceased to be one of the stars in Cepheus, became successively a star in Cygnus about 15,000 B.C., Vega 10,000 B.C., two stars in Hercules about 7000 B.C., one in Draco 5000 B.C., a Draco 3000 B.C., from whence it passed into the constellation of Ursa Minor, in which it now is.[1] The passage in Herodotus which tells us that Perseus was the totem-parent of the Persian kings, shows that the myth belonged to the birth-stories of a people who cultivated gardens and fruit-trees, for he goes on to say that the Greeks called the Persians Kephenes, a name which means the sons of the drone-bee (κηφήν), that is, the race who made their mother-prophetess Deborah, the speaking-bee; and who, like the Northern sons of Odin, and the Ashvin worshippers of the Rigveda, attributed prophetic inspiration to the virtues of the mead distilled from the honey made by the garden or orchard (κῆπος) bee. This was the race who became the Minyans in the Minyo-Sabæan confederacy of Southern Arabia, and the king of these united races was the Kepheus of the Perseus legend. His name is probably connected, not only with the Greek κῆπος, 'a garden,' but also with the Greek Kebos (κῆβος), of which there is a form Kepos (κῆπος), the Latin Cephus, meaning 'a long-tailed ape.' This is the Indian Hanuman, the sacred ape which, to the present day, lives in the mango-groves or gardens attached to the temples in which he is worshipped. He is the Tamil ape, Kapi, who became the sacred ape of Southern Arabia, Phœnicia, and Egypt, called in Egypt by his Tamil name, and worshipped as the six and eight creating apes called Keftenu, the Western or Phœnician (keft) apes, and Uetenu, or the apes of the green (uet) land of India, to the east of Punt, the Egyptian name of Arabia. Thus Kepheus, the ape king or creator, was the sacred long-tailed ape who was originally worshipped by the Turanian Gonds as Maroti, the tree and

---

[1] See diagram of the path of the pole, with the age of each successive polar star, as given in J. O'Neill's *The Night of the Gods* (Quaritch, 1891), ' Polar Myths,' p. 500.

wind-god, and he represents the anthropomorphic creative power which succeeded the winds as the turner of the stars round the pole, and thus he is the god embodying the earliest conceptions of the individual creator, while, in his astronomical aspect, it is his constellation which furnished the two earliest of the polar stars, and we see how, in the evolution of guesses as to the explanation of the mysteries of creation, the creating tree-ape of the matriarchal tribes of the South preceded the mother-bird of the Northern Finns. It is these creating apes who are said, in an Egyptian inscription, to adore Rā in the language of Uetenu, or India.[1] The queen of Kepheus was Kassiopæia, the shining (Skr. *kash*), the moon-queen of the sun-ape. Andromeda, their daughter, the crescent-moon, the wife of the fish-sun-god, appears in pictorial astronomy as the maiden with the fish swimming close to her. This is the fish seeking to devour her, and in another form of the myth of her deliverance, where she is called Hesione, Hercules saves her, and slays the fish by leaping into its mouth,[2] thus showing that the myth belongs to the same series as that which, in Hebrew mythology, tells how Jonah (*Yonah*), the prophet-dove of the sons of the fish-god, was swallowed by the whale, and thus became the spirit inspiring the prophets who adored the sun-fish Salmanu as the god of wisdom.

In the Perseus myth we also find a wonderful instance of historical evolution proved, like that of the history conveyed in the forms and uses of the cross, by mythic symbols. For Perseus, the fish-sun-god, is shown by his receipt from Hermes of the Harpe to be a direct descendant from the discovering (ἕρμαιος) fire-god. The original weapon or wand of Hermes was the fire-drill, which became the magic staff (ῥάβδος) of Rhadamanthus, and the earliest form of the Zend

---

[1] H. Brugsch, *Religion und Mythologie der Alten Ægypter*, p. 153; see also p. 152.

[2] *Encyclopædia Britannica*, Ninth Edition, Art. 'Andromeda' and 'Perseus,' vol. ii. p. 22; xviii. p. 560.

Baresma, the Hindu Prastara, the rain (*bar*) staff. But its Greek name as the special staff of Hermes, the Caduceus, shows that the origin of the conception of the sacred magic wand was the stem of the tree of life, for Caduceus is connected with the Greek Cados (κάδος), meaning 'a jar or pail.' And just as the Greek Rhabdos is reproduced in the Zend Baresma 'long as a ploughshare, thick as a barley-corn' cut off the tree of life,[1] so is the Greek Caduceus, the staff and jar, reproduced in the Sanskrit Droṇa, the cask or jar, called in the Brāhmaṇas Prajā-pati, or the supreme God Ka,[2] which holds the Soma, the sap or germ of life, in the Soma ritual. That the Caduceus was a symbol of vegetable growth is proved by Comte Goblet d'Alviella who, in his *Migration des Symboles*, points out that it is described by Homer as tripetalous (τριπέτηλος),[3] and in this aspect it can be compared with the Celtic shamrock and the three-leaved plant with which Sigmund restores Sinnfiotli to life in the Nibelungen Lied.

I have shown that the Caduceus, in its trident form, is a direct descendant of the Gond god Pharsi-Pen, the female (*pen*) trident (*pharsi*), formed by inserting an iron trident representing, according to the *Song of Lingal*, the male-god with his two wives, into the mother-jar, formed by the stem of a female bamboo.[4] These converging authorities leave little doubt that the original single-stemmed magic wand became, in the later Trisūla, called the Caduceus, a tree-symbol, representing to the observing races, who made God the physician their supreme god, the growth from the seed of the dicotyledonous tree which had been, in earlier mythology, the sheath or jar of the female bamboo, in which

[1] Darmesteter, *Zendavesta Vendīdād Fargard*, xix. 18, 19 ; S.B.E. vol. iv. p. 209.
[2] Eggeling, *Sat. Bräh.* iv. 3, 1, 6 ; iv. 5, 5, 11 ; iv. 5, 6, 4 ; S.B.E. vol. xxvi. pp. 318, 408, 410.
[3] Homer, *Merc.* 530 ; D'Alviella, *Migration of Symbols*, English Translation, p. 227.
[4] *The Ruling Races of Prehistoric Times*, Essay iii., pp. 193, 229, 230.

the original male-rod, the first form of the Pharsi-Pen, is still kept by the Gonds. This trident Caduceus shows how the plant, as it grows from the seed, throws out first the two primary leaves or cotyledons, between which the plumule which becomes the male stem grows.[1] The two primary leaves became the encircling snakes of the Caduceus, rising from the hanging roots, the Echis or Ahi, the village-snake, the fertile ring of cultivated land surrounding the village-grove, and the Nāga, or cloud-snake, of the rain-worshippers. They curl round the stem like the roots and they grow with it from the seed, the jar of the mother-sap or germ which was originally the tree-trunk, whence the seed drew its life. The Caduceus, with the two snakes' heads converging over the central staff, is therefore a form of the Trisūla called by the Buddhists the Vardhamana, meaning the growing-sign, and it represents not only the process of vegetable growth, as observed by the early tree-worshippers, but also the year of three seasons, and the father-god with his two wives or mothers, the fostering primary leaves.

This trident Caduceus, when Hermes discarded it for the Harpe, became, in Hittite, Phœnician, and Greek mythology,

---

[1] This shows that the first studies of plant-life were made from the examination of the growth from seed of dicotyledonous trees, and not of the monocotyledonous cereals or grasses. Thus the generalisation which made the Trisūla the symbol of the plant of life belongs to the age which made the oak of the Druids the mother-tree of the sons of the barley ; its acorn, called by the Basques Zi, or the seed of life, their mother-seed, and which, in Syria, made the olive-tree first, and the fig-tree afterwards, the mother-tree of the fruit-growing gardening races. It was among these people that the race of physicians, the observant and experimentalising students of nature ($\phi\acute{v}\sigma\iota\varsigma$) arose, who sought the means of curing diseases in the study of febrifuges such as the *Cheironion Centaurion*, said to have been found on Pelion by Cheiron, the Centaur, of remedies for stomachic complaints, such as catechu, obtained in India from the Khadira-tree (*Acacia catechu*), the tree of the sacred fire-drill and sacrificial stake, and the Bel (*Ægle marmelos*), one of the totem-trees of the Indian Bhars or Bhārata (*The Ruling Races of Prehistoric Times*, Essay iii., pp. 166, 294). It was these people who, in their thirst for knowledge, studied the phenomenon of the growth of plants and embodied their botanical knowledge in their religious symbols.

the staff surmounted by the full and crescent moon as in the following examples taken from Comte Goblet d'Alviella's book referred to above, and in these symbols the growth represented by the Caduceus is that of the crescent from the full-moon.[1] The change marked an alteration in the measurement of time from the reckon-

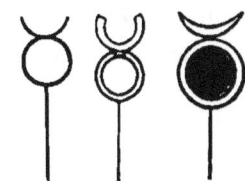

Panic.   Greek.   Hittite.
Fig. 130. Fig. 129. Fig. 232.

ing by seasons denoting the stages in the growth of the tree to one based on the lunar phases, and the Hittite symbol with its cow's horns, representing the crescent-moon, tell us, like the Europa myth, of the transfer of the original conception from Asia Minor, the mother-land of the *Bos frontosus* and domestic ox, to Greece with the coming of Kadmus, the Eastern (*kedem*) arranger. But when this earliest Harpe, the creating symbol appropriated by the fire-god, was transferred to Perseus, the change was accompanied by a fresh development of distinguishing attributes, for he carried, besides the Harpe and winged sandals, the wings of the mother-bird, attached to the Caduceus and the heels of Hermes, the wallet ($\kappa i \beta \iota \sigma \iota s$) and the cap of invisibility, the helmet of Sigurd. In the wallet, a necessary accompaniment of the sun-god, who had been the seed-god, and who had become the sower of the germs of life, we find a reversal to the Kados or jar of the Caduceus, but in the change the wallet is closed, showing that it is the seed receptacle which became the 'Mystica vannus Iacchi,' the winnowing basket of the Bacchic processions, containing the first-fruits, the seed-grain of the future year, and also the $\kappa i \sigma \tau \eta$, the chest or jar containing the sacred barley-cake, the Drōn of the Zendavesta, the seed of life tried with the fire. Thus, these symbols, together with the $\kappa \upsilon \kappa \epsilon \acute{\omega} \nu$, the cup of the Eleusinian mysteries, the Soma cup of the Hindu

---

[1] D'Alviella, *Migration of Symbols*, pp. 227, 231 ; Perrot et Chipiez, vol. iii. p. 232 ; Overbeck, *Kunst Mythologie*, p. lxxxvi. fig. 6 : Perrot et Chipiez, vol. iv. fig. 353.

Soma sacrifice, are the offshoots connected by a succession of changing and related rituals with the original Drona of the Hindus and the magic wand of still earlier beliefs, which became the Caduceus or staff of the sacred jar Kados.

It was as the descendant of Perseus, the bearer of the sickle-shaped sword, or crescent-moon of Eastern Hittite mythology, that Hercules came into the world as the son of Alk-mēnē, the moon (*mēnē*) bow (*alk*, ark), and he was born at Tiryns, the capital city of the Mycenæan kingdom of Argos, founded by Perseus. Additional proof of the Eastern origin of his worship is shown by his connection, in Greek mythology and ritual, with the Tyrian Hercules called Melkarth attended by the divine dogs, the four dogs of the Euphratean god Bel or Bil, Merodach or Marduk, the fire (*bil*) calf (*marduk*), the relationship between the Tyrian and Greek hero being shown by the name Cynosarges, or dog's yard, given to the shrine of Herakles at Athens, where dogs were sacrificed to him.

His worship, which changed into the cult of the fire-drill as the rain pole and mast of the crescent-moon, the heavenly ship of Delphi, became the national worship of Greece during the age of the traditional banishment of the anthropomorphic Heracleidæ, the early fire and sun-worshippers; and this age of mystical religion was the era of the rule of the Dorians, the twin-races, the Spartan and Argive sons of Kastor and Polydeukes, who, like Apollo, Artemis, and Athene, were sexless gods of heaven, without children. This belief in the sexless father-god, arising out of the reverence for the fire-drill, became that which made the polar star first and the sun-god afterwards, the manifestation of the unseen father of life who dwells in the seed as its invisible germ. It was brought to Greece by the Phœnician Semites of the Palestinian coasts, whose principal ports were Tyre, the city of the rock (*tsor*), and the still earlier Sidon, the city of the fish-god, their religious capital being the city of Gebel, the Akkadian Gebil or Bilgi, the fire-god, called by the Greeks

Byblus, the city of the Papyrus ($\beta \dot{v} \beta \lambda o \varsigma$), that is, of the records written on it. It was there that the oracles of God were preserved in the memory of the hereditary priests who, in a later age, revealed the doctrines received from their fathers in the collection of sacred writings from which the books of Sanchoniathon, meaning 'what the god Sakkhun hath given,' were compiled. Sakkhun is probably the Hindu god Sakko, the Akkadian Sukh-us, the rain (*sukh*) god whom the Phœnician sons of Tur, the Hindu Tur-vasu, brought with them from India, first to their settlement at Turos (Bahrein) on the Persian Gulf, and thence to the Mediterranean coasts. This period of Dorian rule, the rule of the Palestinian sons of the land of Dor, and of the worship of the fish-god, was the age when the religion of the Greeks was dominated by the mysticism of the Semites ; when the sons of the earthly fire, the earlier Heracleidæ, called the Phlegyes, were deposed from power, and when the people were ruled by the despotic kings, whose fortresses and palaces, arranged for the Oriental seclusion of the women of the household, survive in the ruins of Troy, Mycenæ, and Tiryns, and in the foundations of the Akropolis of Athens, the city of Athene, the Semite virgin mother-goddess, looking down on the sea-girt Salamis the Semite island of peace (*salem*).[1]

This Semitic age was that which followed the death of Jason, killed, when mourning for Glaukē, the vanished sun-goddess, by the rafter of the prow of the Argo, under which he was sitting, and which fell on him. This was the rafter formed from the oak-tree of Dodona, the parent-tree of the sun-worshipping sons of the seed-grain, the growers of corn-crops, whose rule was followed by that of the trading merchants, the Semite sons of the olive-tree and the fish-god. They, again, were ousted from power by the return of the Heracleidæ as worshippers of the wine-god, Dionysus or Iacchus, who took the place of the earlier god of the Northern

---

[1] *The Ruling Races of Prehistoric Times*, Essay vi., pp. 519, 520, 533, 558, 559.

honey mead, and of Apollo, the beautiful youth, the young
sun-god of a more joyous age than that of the depressing
Semitic gloom. This was the age of the birth of the new
sun-god Achilles, the conqueror of the trading Semites of
Troy, the Dardanian sons of the antelope (*dara*).[1] These
hated and despised foes were represented in Greek historical
mythology by Paris, whose name denotes the avaricious but
plausible traders, the Panis of the Rigveda. Their religion,
as well as that of their successors, the worshippers of
Dionysus, the wine-god, and Apollo, the god of the lyre, was
utterly opposed to the Northern belief in the sun-god, the
armed knight riding on the grey-cloud horse, who killed the
winter frost demons and the serpent of the dark cave of the
winds, the gaoler of the light and the rain, brought prosperity
and gladness to the earth by his yearly triumph over its
oppressors, the darkness and the cold, and who, though his
victory was followed by his decline and temporary death,
caused by the wounds inflicted by his annually reviving foes,
yet returned again to life as the young sun of a new year to
wage his never-ceasing warfare with the enemies of light and
heat. The theology of the Semite-Kushites which succeeded
this Northern creed, seems to have passed into the mystic
phase indicated by the Delphic Trisūla, through the influence
of the meditative and experimentalising physicians, represented in Grecian mythic history by Æsculapias, Cheiron the
Centaur, Jason, and Medea, who depicted the sun-god, not as
the warrior riding the sun-horse, but as the physician ruling
the star-ship who healed diseases, put to flight the carrion birds
of death and ignorance, and created wealth in his annual
voyage. But this conception, which was embodied in the
Jason and Argo myths of the twin races, had been thrust out
of sight, and made to give way to the rigid rules of sacerdotal
law, as Semitic mysticism, metaphysical subtlety, and forma-
lism developed, and became of more account than practical
discovery. This iron rule of priestly law was accompanied by

[1] *The Ruling Races of Prehistoric Times*, Essay iv., pp. 365, 366.

despotic government, which came to be considered necessary, not only for the interests of trade, which prospered under a strong central power, but also for those of religion.

But the Northern conceptions of the master-smith and the sun-knight, though they assumed in Southern lands forms in which they were scarcely recognisable, survived persistently in the beliefs of the Northern nations. They appear in the myth of Lohengrin, the knight carried by the swan in the moon-boat, who distributes the Holy Grail, the water of life and the blood of the gods to the righteous throughout the world. This conception of the Holy Grail travelled, as I have shown, to India in the days of the bird-myth, in the symbol of the two Soma caskets, brought to earth by the divine mother-bird, the snow-bird Shyena,[1] and that this myth was originally one in which the bird who heralds the seasons, and marks by its migrations the intervals between one season and another, is shown by the myth of Lohengrin, whose boat is drawn by the moon-swan, who became the measurer of time, when the seasons indicated by the changes in the growing trees ceased to be thought adequate time-reckoners. But when the spring stork and swallow, who brought to the North the spring-sun, disseminating in April showers the germs of summer and autumn life, became the rider on the sun-horse, and the physician travelling in the moon-boat and star-ship, he was, according to the creed of the believers in the immortalising power of the Elixir of Life, which he bore in his bosom, fed, like the mother-earth, with the holy water and blood, the water which infused the germs of life into the seed and the blood which gave generative power to the virgin earth, and made her the willing mother of the sons whose fathers were united to her by blood-brotherhood.

The whole story is completely told in the myth of Sir Galahad in the *Morte d'Arthur*, and it is proved by the identity of leading incidents in the two stories to be a variant

---

[1] *The Ruling Races of Prehistoric Times*, Essay iii., pp. 248, 302.

form of that of Sigmund, the master-smith, and Sigurd, the sun-knight. When Galahad was chosen to be the holy sun-knight, he, like Sigmund, who drew the sword of light out of the mother-tree of the Volsungs, proves his birthright to the office by drawing the sword which no one else could move from a rock overhanging the Usk or water (*uiske*), the parent river of the sons of Camelot. The rock in which the sword was fixed, and the churchyard stone from which Arthur, the ruling sun-god, drew the sword, which proved his right to rule,[1] both take us back to the primæval legends telling of the birth of life in the creation of the fish-god, and of the light registered by gnomon, the sacred prophetic stone. For the river-rock, whence Galahad drew the sword, is the mother-rock of the sons of the rivers born from the mother-mountain, the Hindu Adrikā, meaning the rock whence the river-parent of the sacred fish, born from the seed of the creating god Vasu, sprang to life, and it was these sons of the rivers and the fish-father, the eel, the Hindu Indu, the Finnic Il-ja,[2] who used the gnomon-stone, the Beth-el, or prophetic stone, the parent of the church consecrated to God[3] as a means of registering the passing time chronicled by the sun.

It was before Galahad drew the sword and began the adventure of seeking the Sang-real, the germ of life, which was to be the seed whence the future year was to be born, thus completing the cycle of the achievements of Arthur and his sun-knights, that he was placed in the Seat Perilous, the seats assigned to the foremost of the sun-knights, which was found to be inscribed with his name. After having drawn the sword of light, he received the invulnerable shield destined for him and marked with the red-cross, the cross of the fire-god of Syria, recording the blood-brotherhood

---

[1] Malory, *Morte d'Arthur*, Bk. i., chap. iii.; Bk. xiii., chap. v. The Globe Edition, pp. 29, 351.

[2] *The Ruling Races of Prehistoric Times*, Preface, pp. xli. xlii. Aind or Indu, the name of the eel-totem of the Kharias, is the root of the Sanskrit Indra, the rain-god of the Rigveda, meaning the sap (*indu*) of life.

[3] The original gnomon-stone has now become the church-spire.

between the sun, the heavenly fire, and the earth.[1] The appurtenances of the sword mark the myth as having in its circuit back to the north traversed Euphratean and Indian lands, for the hilt of the sword, which, according to another story of its acquisition, Galahad alone could withdraw from its scabbard, was formed of two ribs, one belonging to *the serpent-mother of the Indian sons of the tree*, and the other being *the rib of the fish of the Euphrates*, the divine *fish-god Ia*, who came from India to Eridu, while the scabbard was made of *serpent-skin*. The girdle of the sword was made of the hair of the virgin-sister of Sir Percivale, the virgin-star and moon-mother, the Vestal virgin who guarded the national household fire.[2]

After he had been girt with the sword, Galahad received from a knight, who had been sent by God to give it to him, the white horse, the horse Gráni of the Sigurd myth, on which he was to start on the quest of the Holy Grail.[3] Mounted upon it, he came to the hill-chapel, the Mount Hinda-fjall of the Sigurd myth, with the sacred mound on its summit, where he saw the Holy Grail borne by four angels, the fourth of whom placed the holy lance—the spear with which St. George, the red-cross knight of the upright right-angled cross, pierced the dragon or rain-cloud—upright in the holy vessel, containing the Grail or water and blood of life, the mist wreathing the top of the mountain, and it was there that in the mound sun-temples the gnomon lance was fixed. By the miraculous powers of the blood of the Grail on the point of this lance, which had originally been the magic wand, the maimed King Pelles was made whole again.

Galahad finally received the Holy Grail in the spiritual city of Sarras, the Tyre of the Phœnician Semites, the town of the rock (*tsor*) called Sarra and Sara by Ennius and Plautus, the holy land of the East, where the invulnerable shield of the rising sun was made, and whence the myth, in

---

[1] Malory, *Morte d'Arthur*, Bk. xiii. chaps. i.-v.

[2] *Ibid.* Bk. xvi. chaps. iii. iv. v.   [3] *Ibid.* Bk. xvii. chap. xiv.

the form it takes in the Arthurian cycle, had travelled to England. When he arrived at Sarras with his companions, Sir Bors and Sir Percivale, they were all cast into prison by the king, who was a heathen, like the King of Maushil, who, in the Arab legend, slew Gherghis or St. George, the sun-god of the three seasons of the Syrian year, three times. While in prison, the three knights were fed with the Holy Grail sent to them by God, but towards the year's end, the King of Sarras fell sick and sent for the prisoners to beg their pardon. After receiving it, he died at the end of the year, showing that he represented the year-god of the year measured by three seasons, preceding that of the sun-horse measured by the equinoxes and solstices. After his death, the people chose as their king Sir Galahad, the youngest of the three knights, the last of the three seasons of the year, the youngest brother of the three brethren of the fairy tales, who, while his brethren fail, is always successful in all his undertakings. He was the knight of the sun-horse, who had finished in the sun-ship his journey to his Eastern kingdom, whence he was to rise again as the new-born sun, who ruled for a year, and on the last day of the year he died after receiving the Holy Grail.[1]

We thus see that the story of Sir Galahad and the quest of the Holy Grail is an unmistakable variant of the myths of Sigmund and Sigurd, the sun-knight, riding the grey-cloud horse, and that the myth of Arthur and his knights of the Round Table is one telling of the sun-knight, who rose to power, warred with and conquered evil and lawlessness, and obtained, as the crowning achievement of his rule, the Elixir of Immortal Life, which was after his defeat and death to give reviving life to the next year's sun. Also the whole series of the incidents of the life and rule of Arthur, who was according to one myth cast up as a babe by the sea waves, and who is thus identified with the fish-sun-god, are shown, by the retirement of his Queen Guinevere to Almesbury on

[1] Malory, *Morte d'Arthur*, Bk. xvii. chap. xxii.

the death of Arthur,[1] to belong to the sacred series of stories
of the Stonehenge temple of the sun-horse, as Almesbury is
only a mile and a half from Stonehenge ; also, as the year
marked by the position of the gnomon stones at Stonehenge
is one that began with the summer solstice, it is clear that
the temple is not one which, in its final form, was consecrated
by a people who began their year, like the Northern nations,
with the winter solstice, or with the equinoxes, like the
people of Syria and Asia Minor ; but that it was a temple of
those whose creed was that of the Indian Turvasu-Yadavas,
the sea-faring trading race who worshipped the fish-god Ia,
who in India, Persia, Egypt, and Arabia began their year at
midsummer. They, when they came to Greece with the
Minyan leaders, who appear in Greek legends as Pelops,
Peleus, and Pelias,[2] and instituted annual games in honour of
the dead fathers of the barley-growing race, such as those
celebrated at Olympia and Stonehenge, also brought with
them their year beginning with the summer solstice, the time
when the year of Athens and the Olympian Elis began. It
was these people who, in their commercial voyages, first went
to Britain to obtain tin and copper from the Cornwall mines,
and who established themselves as the ruling race of the
island, who took their religious ritual from the Druid sons of
the oak-tree. These Druids started from Persia, where they
worshipped the supreme god at Hu, the Zend Hu-Kairya,
and looked on life as generated from the snake's egg of the
Indian Kushika mythology, and thence passed through Asia
Minor and Greece to Western Europe, taking with them the
medicinal and botanical knowledge, which distinguished both
the Druids and the Greeks of the age of the Jasonic myth.
In England they were known as the Cymri, who succeeded
the Goidelic or Gaelic Celts, the race who were the original

---

[1] Malory, *Morte d'Arthur*, Bk. xxi. chap. vii.
[2] See for legends of Pelops and Peleas, *The Ruling Races of Prehistoric
Times*, Essay vi., pp. 522-532, 555, 556 ; *Encyclopædia Britannica*, Ninth
Edition, Art. ' Olympia,' vol. xvii. p. 765.

worshippers of the sun-horse, whom they looked on as the maker and consecrator of the holy wells, and who succeeded the earlier believers in spells and witchcraft. This interpretation of the evidence entirely agrees with the national traditions of the Cymri, who say that they came, like the Irish Firbolgs, who tell the same story, from the summer land called Defrobani, or Greece, ' where is now Constantinople, by way of the Hazy Sea ' or German Ocean, exactly the route assigned to the Argo in its Northern voyage.[1] They were thus the Minyan crew of Jason's Argo, born from the oak rafter of Dodona, which supported the hall of Pelias, and we have thus a complete and congruent story proved by Greek, Indian, and Northern myths and ritual, and by the evidence of the worship of the sun-horse on the English hills, and in the ruins of Stonehenge, showing how the creed of the sons of the sun-horse and the ash-tree became one which was associated in India and Persia with the year beginning with the summer solstice, and how this was brought by the South-eastern maritime races to Greece, whence it came to Stonehenge by the agency of the Minyo-Semite Mediterranean traders, who also took with them the reckoning of the wheel-year of twelve months and 360 days. The memorials of their calculations they set up in stone at Stonehenge on the site of the great temple dedicated by their predecessors, the Goidelic Celts, to the sun-horse, and Weyland, the master-smith of the Iberian races, and they also instituted the annual games which they brought with them from Greece. In the myth of the Holy Grail and the worship of the oak-tree, which the Cymri brought with them as the teaching of the mother-bird who brings the rain, the bird-mother Khu, who became through the phonetic decay of linguistic changes, the Zend and Druid god Hu, and that which they found in Western England associated with the mythic history of the sun-horse, we see a story which is not only a variant of the Hindu Soma

---

[1] *Encyclopædia Britannica*, Art. ' Celtic Literature ' by Professor Sullivan, Ninth Edition, vol. v. p. 300.

bird-myth, but also of that which embodied the Hindu
teaching as to the origin of life. This they believed to be
generated from the Su or soul of life in the rain, the heavenly
Soma which distilled and diffused through the earth, the
germs of life, and thence disseminated them through the
trees, the mother-grasses, and the barley growing from it, and
through the living beings nourished on the barley and on the
flesh of the animals who fed on the sacred Kuṣha grass.   In
both the Indian Soma myth and that of the Holy Grail, the
blood of the god of life was borne through the air, the home
of the life-giving moisture in the heavenly casket, the rain-cloud,
the vessel of the Grail, whence it descended on the earth after
being, in the original myth, released from bondage by the
lightning flash of the thunder-god, the spear which impregnated it with vital and healing heat.   It was the distilled
essence of this sap of life, re-born in the heavenly vessel, the
mother-tree, which was given as the drink of healing and
immortality to the worshippers of the Hindu Soma, the
barley-growing race, who in Greece and India drank of the
cup of the barley-mother in the cup mixed with barley in
the Eleusinian mysteries, and in the Soma cup of the Indian
ritual, and with this cup they ate the bread of life, the barley-
cake offered at the Soma sacrifice to Vishnu, the year-god,[1]
the sacred cake of the Parsis, of the Eleusinian mysteries, and
of the Mandanite Sabaeans.

This original Finno-Dravidian story, telling of the miracles
wrought by the fire of life and the water of immortality,
was altered by the sons of the horse in accordance with the
individualising spirit of their religious conceptions.   The
rain-bird and rain-cloud, the messenger of the hidden power,
who concealed the germs of life in the seed of the barley-
growing agriculturists, and the egg of the Finns was the
servant of the God who ordained the laws of the invariable
succession of natural phenomena in their appointed order.
It was this unseen spirit who became, in Northern belief, the

---

[1] Eggeling, _Ṣat. Brāh._ iii. 4. 1, 15 ; S.B.E. vol. xxvi. p. 88.

man-god, the knight of the sun riding on the cloud-horse, the brave and beneficent self-ruling warrior trained to observe the laws of righteousness, and keeping perpetual watch over his thoughts and actions. He was not the law-giver who, hidden in his unseen solitude, lays down once for all, irrefragable laws of guidance, but his son and messenger, the law-reformer, who works out his yearly task of redress-ing wrongs and defeating the machinations and attacks of the powers of evil. During his year of consecration he is clothed in the glowing light, the invulnerable coat of mail, which is proof against the weapons of his enemies, who only succeed in compassing his annual death preceding his resur-rection, when he is bereft of its protection. But while he is girt for war he can with the sword of light and his spear, the fire-drill and gnomon, force the rain, the changing seasons, months, and days, to obey his will.

This Northern picture of the righteous warrior, who conquers and controls his originally evil nature, not by unsexing himself according to the practice of the fire-worshippers of the East, but by self-education, and who proves his victory over self, and his right to bear the sword of true and just dealing by remaining always chaste and temperate, is one that is totally foreign to the Southern imagination. For in the South, where the individual was an almost unrecognised atom among those which, when united, formed the State, the maintenance of orderly concert among all members of the confederacy, was thought to be the primary necessity of national existence. Individual effort was held to be impious and irreligious until its motives and grounds of action had been examined by the national council, and had received the imprimatur of their consent; and when the new motive power was brought into action those who sanctioned or used it were bound to watch its effects on the welfare of the State, and to ignore the individual whenever individual interests and those of the State came into actual or apparent conflict. In a well-ordered State of this type,

such as those of Sparta, and the original Zend confederacy,
every child born and accepted by the community as a future
male or female citizen was carefully trained from its earliest
years in obedience, self-abnegation, and in the belief that
their first and most important duty was to obey the constituted authorities, to think of the State as the first object of
their thoughts, and to die rather than refrain from doing
their utmost to fulfil its requirements, and to preserve its
unity and safety intact. In this programme the individuality of the component atoms was only recognised in the
distribution of the State offices, to which the successful toiler
might be eligible. For in the State, as altered from its
original Southern Republican despotism, by the amalgama-
tion of the Northern races, who believed in the privileges of
family and birth, and the transmission of hereditary qualities,
the eligibility to offices of power was restricted to certain
families and clans.

The two tendencies of these different conceptions of
national and personal duty are reproduced in the two
dominant divisions of organised religious belief. In one, the
work of the divine author of righteousness is thought to be
best carried out by individual prophets and divinely appointed
messengers of truth, the ordained clergy, while in the other
the education of the faithful is intrusted to the united
exertions of despotically governed bodies of specially trained
teachers, the Buddhist monks and nuns, the mediæval friars,
and the later Jesuits. But even in societies intended to be
organised on these lines the individualising spirit crept in
among the Egyptian and Syrian hermits, and the Hindu and
Mohammedan Bairagis and Faquirs. It also entered into
those monastic societies which regard the inmates of their
monasteries as a body of united brethren, each of whom is
engaged in working out his own salvation, instead of following
the teaching of the Benedictine motto 'Laborare est orare';
and this same spirit of selfishness underlies the religion of
Protestants, who think that man's first duty on earth is to

take care of his own soul. But while this and other forms of selfish isolation and love of personal distinction are the besetting sins of Northern individuality, despotism, tyranny, and stagnation are those of Southern collectiveness; and in studying the varying results attained by these two motive powers, and by the manifold influences of the past in human history, we find everywhere evidence that the web of national life is in all lands and climes woven of threads spun from Northern, Southern, North-eastern, and North-western intellects, forming apparently diverse but fundamentally similar patterns persistently reproduced by the 'roaring loom of time.'

ESSAY IX

HISTORY OF THE WORSHIP OF IA OR YAH, THE ALL-WISE FISH-SUN
GOD, AS TOLD IN THE MYTHOLOGY OF THE AMERICAN
INDIANS, SCANDINAVIANS, FINNS, AKKADIANS, ARABIAN,
ASSYRIAN, AND SYRIAN SEMITES, IRANIANS, HINDUS,
CHINESE, AND JAPANESE.

In the preceding Essay on the mythology of the Northern
nations, I have shown that Northern beliefs in the divine
creator started from the totems of the hunting tribes, who
traced their descent from animals, and I have also in previous
Essays shown that this totemistic faith culminated in the
worship of the fish-god, the last national totem-parent who
succeeded the sun-antelope, the father-god of the barley-
growing races, who originally, like the Hindu Kshatriyas, who
wear at their investitures the skin of the spotted deer, called
themselves the sons of the deer.[1] But in the Scandinavian
mythology, the evidence as to the evolution of the totemistic
creed, which made animals the national parents, is obscured
and half-obliterated by the anthropomorphic theology
originating in the Northern attribution of supreme creative
power to the god who miraculously produced fire from the
flint, and the fire-drill and socket, who became the master-
smith of the Kabiri. In the Hindu caste genealogy, the old
animal totems still survive, but they are so much mixed up with
totems derived from Southern tree and plant worship, and
with the still more puzzling aberrations engendered when castes

[1] *The Ruling Races of Prehistoric Times*, Essay iii., pp. 284-302 ; iv.
374-376, 403.

ceased to be brotherhoods claiming a common descent, and
became associated trade-guilds of potters, oilmen, weavers,
and other names denoting a community of function, that it
is impossible to frame a satisfactory account of totemism
from caste genealogies without resorting to evidence less
altered by national history than that to be found in India.
It is only among the North-American Indians, who say that
they were brought to America by a man-fish, who became
to the Mexicans the god Teo-cipactli, whose full name was
Hue-hueton-cateo-acateo-cipactli, meaning ' the fish of one
flesh,' that we find the totemistic faith and ritual still preserved in a less altered form than elsewhere ; and it is only
in their mythology that we can trace the stream of national
totemistic tradition to its original source. That the North-
American Indians belonged to a Northern race, who made
their way to the Southern Hemisphere, both in America
and Asia, is proved by the absolute identity between the
national system of relationships of the Iroquois and Indian
Dravidians, shown in the tables of consanguinity in Morgan's
Ancient Society, to co-exist with the form of marriage which
he calls Punuluan. This I have shown to be a union between
alien races, in which the bridegroom received the bride into
his clan by making blood-brotherhood with her, and marking
the parting of her hair with vermilion, a rite still preserved
by all Hindu castes, except those who tie the hands of the
wedded pair together with Kusha grass, or unite them by
tying their clothes.[1] The identity of strain in the American,
Indian and Hindu stocks is still further proved by the existence in both countries of tree-totemism, in which tribal clans
trace their birth to a tree. and finally to plants, such as the
reed and the ear of corn.

But though totemism gave birth to isolated tribes with
shifting alliances, it never produced a national religion,
common to a permanent tribal confederacy. This was intro-

---

[1] Morgan, *Ancient Society*. Tabular Statements, pp. 420, 447 ; *The Ruling
Races of Prehistoric Times*, Essay ii., pp. 56, 57 ; iii. pp. 174, 175.

duced by the fire-worshippers and rain-worshippers, the first of these national creators tracing the descent of the confederated totemistic clans to the fire-stone, producing the heat necessary to sustain life, and the second to the mother-bird of the Northern agricultural races, who brought the spring rains which melted the snows of winter. That these two national parent-gods are worshipped by all the North-American Indians is proved by the elaborate reports published by the Bureau of Ethnology of the Government of the United States. These prove that the chief gods of all North-American Indian tribes are Tunkan (*Ingan*), the stone-god, to whom daily prayers are said, and Wakinyan, the thunder-bird, the god of war, to whom national sacrifices are offered.[1] The stone-god was originally the fire-making flint, which in the Gond *Song of Lingal* made fire for the Gond immigrants into Central India from the North-eastern Himalayas, before they learned to make fire by the fire-drill from the forest agricultural tribes they found settled in the country.[2] This god, who, with the mother-bird, united the totemistic hunters and the agriculturists of the North into the consolidated nucleus of a nation, is the god called by the Akkadians, Zends, and Hindus, the Shu-stone, or the Salagramma, the begetter of life, found by the fire-god Adar in the mother-mountain.[3] It is worshipped as the black stone of the Caaba at Mecca, and is the origin of the stone-gods of Arabia, and of all the Semitic races. The thunder-bird shows, in its Indian name Wakin-yan, that the cult was imported into America after the age which called the mother-bird the frost (*shya*) bird, the Shyena of the Rigveda, and the

---

[1] *Publications of the Bureau of Ethnology*, vol. x. : Mallery on 'The Pictorial Writing of the American Indians,' p. 32. The sacred stone is the stone on which all Northern kings were crowned, the Scottish stone of Scone, which has become the Coronation Stone of the kings of Great Britain, and the stone still preserved at Kingston, or the stone of the king on which the Saxon kings of Surrey were crowned.

[2] *The Ruling Races of Prehistoric Times*, Essay ii., pp. 48, 49.

[3] *Ibid.* Essay iii., p. 144.

Saena of the Zendavesta, for it means the bird which brings
to earth the Waka, or the mysterious germ of life, also called
Takoo Wakan, or the living soul,[1] and this discloses a much
more metaphysical frame of thought in the directors of the
national faith who gave the name, than that which appears
in the realistic name of the frost-bird, the bird that brought
down the rivers the irrigating waters released by the melting
of the mountain snows in spring. That the conception of
the mother-bird as that which bears the germ of life was
connected with the sorcery and magic introduced by the fire-
worshippers, who called their priests the Magi, is proved by
the Ojibewa magic-songs, which represent the thunder-bird as
flying into the arch of the sky, whence he inspires the Midē,
or god-possessed priests.[2] This belief in the mother-bird was,
as I have fully proved, of Finnish origin. It arose when the
Ugro-Finns joined the Northern animal totemists, and it
culminated in the worship of the mother-bird of the Kushite
race Gaṇḍhārī, she who wets (*dhāra*) the land (*gan*), and laid
the egg whence the hundred Kaurāvya, otherwise called
Kuṣhika, or sons of Kush, were born. But the original faith
in the stone fire-god of the boulder-stone as the chief fire-
maker, has, like the original belief in the mother-bird who
brought the spring, been altered by contact with the Southern
cultivating races, who worshipped the mother-grove first, and
afterwards the mother-tree, and it is this Southern influence
which has converted the national fire-stone into the sacred
pole, which the nomad Indians carry about with them as
their most sacred symbol, and keep in the consecrated
tribal tent, oriented to the rising sun when the tribe is
stationary, or in the direction in which they are going when
they are on the march.[3] This pole, among the Omahas

---

[1] *Publications of the Bureau of Ethnology*, vol. xi.: Dorsey, 'A Study of
Siouan Cults,' p. 366.

[2] *Ibid.*: Mallery on 'The Picture-Writing of the American Indians,'
p. 239.

[3] *Ibid.* vol. xi.: Dorsey, 'Study of Siouan Cults,' pp. 390, 403, 413.

and Ponkas, sons of the corn-mother, is made of two pieces
taken from the stems of their mother-trees the ash, the
Yggdrasil of the Edda, and the cotton-wood. This proves
the magic-pole to be a direct descendant from the primæval
sacred fire-drill and socket, which, in Hindu ritual, is made
of the Khadira-tree (*Acacia catechu*), the socket, or mother of
fire, and the Pipal-tree (*Ficus religiosa*), the fire-drill. But the
Omaha worship of the white buffalo-hide, as the totem equally
sacred with the pole, is even more significant, as it is also
connected with the making of the sacred tribal fire. Whenever a permanent village of earth lodges was established
among the Osages and Kansa, seven sacred fire-places,
whence the household fires could be lighted, were consecrated on the west side of the tribal circle, and among
the Osages, seven also on the east side, while the Kansa
consecrated six. The sticks of these fire-places on the
west side were laid by the chief of the Tsishus, or peace
section of the tribe, and those on the east by the chief of the
Panhuka, or war section. Each fire was laid with four sticks
placed in the form of St. George's cross, the outside ends
pointing to the four quarters of the heavens, and the sticks
were all laid sunwards in the order of the deasil, so that the
right side of the circling sun was next to the points of the
cross. In laying the fire of the Tsishu, or peace section, the
first stick laid was that pointing to the west, the birthplace
of the young buffalo-bull, the north and east sticks were
dedicated to the grey and large buffalo-bull, while the south
stick was dedicated to the buffalo-cow, the mother of the
western calf.[1] Thus the story told by the laying of the
peace fire is that of the migration of the buffalo cow-mother
from the South, to meet the buffalo-bulls of the North and
East, of whom the grey buffalo of the East, the bearer of the
white buffalo-hide, was to be the father of the buffalo-calf of
the West, as the laying of the sticks shows that the cross not

---

[1] *Publications of the Bureau of Ethnology*, vol. xi. : Dorsey, ' Study of
Siouan Cults,' pp. 381, 523, 525.

only represented the four quarters but also the circuit of the
sun round the heavens, it also proves that the people who made
this fire-cross their national symbol worshipped the evening sun
of the West, the sun of the Egyptian god Ptah, always depicted as a mummy or dead god. His name, and that of the
Hebrew Japhet are both derived from the verb, *patah* to 'open,'
and mean ' the opener.' This was the sun sacred to the
Semites of Syria, originally the sons of Dan, whose land was
consecrated to St. George, and who began their day, as the
Jews still do, at six P.M., the time of the equinoctial sunset.
This Southern race, who measured time by the equinoctial day,
and looked on the sunset as the time when it began, must
have been one which lived, like the Indian Dravidian founders
of villages, close to the equator. Thus they, like the Hindus
of the present day, thought that the sun of day was their
bitterest enemy, who, unless its heat is tempered by copious
and seasonable rains, burns up and destroys their crops, and
makes the land barren of life. They were the race who
measured time by the twins Night and Day, and who first
divided the day into minutes and seconds, using the duodecimal system of notation, whereas the Akkadians, whose
system of notation was decimal, and based on the lunar months
of gestation and the five fingers of the hand, divided it into
watches marked by the dropping of water through the hole
in the water-clock bowl, a species of water sand-glass, and by
noting the positions of the sun's disc through the day. The
Tamil population of Malabar have from time immemorial
divided their day into 60 portions of 24 minutes each, called
a Nāliga,[1] thus producing, by a reversed system of notation,
the same number of 1440 minutes in the day as is given by
our reckoning of 24 hours of 60 minutes each. It was from this
use of the duodecimal system that the method of reckoning
by fours called gundas, used by every coolie and tradesman
throughout India, originated. These Indian Dravidian rice-
growers have, like the peace section of the American Indians,

[1] Simcox, *Primitive Civilisation*, vol. i. p. 547.

always worshipped the buffalo, and sacrifice it yearly at the autumn festival of the Dasaharā, held in September-October, to the fathers of the race, and they still dance the dances called by the American Indians, whose tribal doctors are priests of the buffalo, the buffalo dance. The most archaic form of this dance among the American Indians, that of the Hidatsa, Mandans, and Winnebagos, is danced by the last four times in the month of May and early in June, and is called by them the dance 'instituted by women,' showing that it came down from the matriarchal times when women ruled the villages in partnership with the men, who were called their brothers. The time when the dance is held is that of the Indian Soma festival of Juggernath to the sun-god of the summer solstice. It is danced in a long tent from 50 to 100 feet long by 20 wide. The dancers, four men and a number of women, enter it from the east. Each woman brings in a handful of fine earth, with which they make two mounds like truncated cones, 4 inches high and 18 inches round. They place these between the eastern door and the fire, which is about 15 feet from the door. The men lay their head-gear and the claws and buffalo tails they carry on these symbolic representations of the two mother-mountains of the mother-goddess and the father-god, which appear in Hindu mythology in the two birth-hills of the Bhārata race, consecrated to the father and mother of fire, the bull Nanda, and Rā-dhā, the maker (*dhā*) of the sun-god Rā, and situated near Mathura, the sacred city of those who obtain fire by rubbing (*math*).[1]

The dance is led by the men, who roar and tramp wildly, like buffaloes ; and the women follow in single file, dancing a very peculiar shuffling step, in which, as described in the reports to the Bureau of Ethnology, ' the feet are kept nearly straight and the heels close together,' an almost exact description of the step danced by the Ooraon girls of Chota Nagpore, in the figure representing the treading down of the

[1] *The Ruling Races of Prehistoric Times*, Essay v., pp. 452, 453.

rice after it is sown. The buffalo dance among the Hidatsa
and Mandans is followed by an orgy, in which only
unmarried women join, similar to those of the village dances
of the Ho-Kols in Chota Nagpore, but whether this orgy
follows the Winnebago dance, the American Government
Report does not say.[1]

We have now to return to the laying of the Osage fires and
to that laid by the Panhuka or warrior section of the tribe
who had become amalgamated with the peace-loving buffaloes.
In this fire the eastern stick is laid first, and the east wind
and dark-horned deer is invoked. The stick of the south is
next laid to the white mother-deer without any horns, that to
the west to the deer that makes a lodge, that of the settled
buffaloes, and that to the north, and the north wind called
the pine wind, to the deer with grey horns, answering to
the grey buffalo of the east of the peace section.[2] This deer
sun-god of the fighting race of the North-east is an exact
counterpart of the Northern god Frey, the god of seed (*frio*),
who fights, in the Edda, with deer's horns. He and his twin-
sister Freya, who wears hawk's plumage, which she lent to
Loki, the fire-god, are the children of Njord, the god of the
pole-star of the North, who was sent to the Æsir from Vanir
the land of love (*vana*), which I have shown in Essay viii. to
be Asia Minor, in exchange for Hœnir, the sun-god, the
horse of light. Their mother was Skadi, the dweller on the
mountains, the Akkadian Istar, daughter (*tar*) of the mountain (*is*), who skates on snow-shoes, and who is the daughter
of Thjassi, the eagle or vulture-god, the mother-bird of the
sons of Mount Ararat and the Euphrates, the snow-bird of
the Zend rain-god Thraetaona.[3] Here in American mytho-

[1] *Publications of the Bureau of Ethnology*, vol. xi. : Dorsey, 'Study of
Siouan Cults,' pp. 427, 505, 513.

[2] *Ibid.* p. 381.

[3] Mallet, *Northern Antiquities*, Bohn's Edition, The Prose Edda, Part I.
23, 24, 37 ; Part II. 2, pp. 418, 419, 429, 460, 461 ; *The Ruling Races of
Prehistoric Times*, Essay iii., p. 247.

logy we have the deer-god of war united with the warrior
thunder-bird, and the same deer-god in the Edda is united
with the hawk, the mother-bird of the mining races of the
West worshipped in Egypt as Hat-hor, the mother of the
hawk-headed Horus, and the mother-bird of the Hindu
agricultural and mining races, the Kharias and Kharwars; and
of them the Kharias, who are the older and ruder race, worship
as one of their totems the sheep, the mother of the Jewish
race.[1] This myth of the Edda, which makes the deer-god
and the hawk-goddess parents of the warlike races worshipping
the pole-star, is exactly reproduced in the American-Indian
myth which makes the thunder-bird of war the dark-horned
deer of the East and the white doe of the South, the totems
of their warriors. But this hawk mother-bird, who is the twin-
sister of the deer of the East, is also the Greek Kirke, the
sorceress and the hawk ($\kappa\iota\rho\kappa\acute{o}s$). The sorceress is the patron-
goddess of the cats, and the hawk-mother Freya is the
goddess drawn by two cats, the European counterparts of
the two tiger-mothers of the Gond Turanians, the race who
brought sorcery into India.[2] These cats again appear in
Egyptian mythology as the cat-goddess Bast, who is a form
of Hat-hor, and the goddess to whom the Aten, the carp,
called in India the Rohu, or sun-fish, is sacred.[3] This
divine fish is not the original sacred fish of the sons of the
rivers, which I shall show to be the eel first, and afterwards the
dolphin, the mother-fish of the maritime races, but the fish
sacred to the believers in the sanctity of the sacred pools of
living water whence the rivers rose, a sanctity extended,
where the doctrine reached the thirsty lands of the arid countries bordering on the desert, to the reservoirs for storing
water, the artificial lakes in which the lotus, worshipped both
in India and Egypt, was born, the Indian sacred tanks, like
that worshipped in Mathura, and called Rā-dhā-kund, the

---

[1] *The Ruling Races of Prehistoric Times*, Part I., Preface, pp. xxxvii.,
xxxix; Risley, *Tribes and Castes of Bengal*, vol. i. p. 461.

[2] *Ibid.* Essay iii., p. 193.   [3] *Ibid.* Essay ii., pp. 125, 126.

tank (*kund*) of the giver (*dhā*) of the sun Rā, the sun-mother.[1] This is the home of the sacred carp.

This analysis shows us that the original myth, telling of the confederation of the collection of variously-named American Indian tribes called by the trappers the Sioux, and by themselves the Dakotas, a name meaning the joined people, proves by the history of the tribal fires, that the united races were the sons of the buffalo of the South and West, and of the deer of the East and South. The two races who worshipped the equinoctial sun rising in the due East and setting due West. They were united, according to the evidence of the myths of the Edda, in Asia Minor, the birthland of the goddess Artemis, who was both a bear-goddess and the deer-goddess Elaphia, and the birthland of the St. George's Cross consecrating their tribal fires.[2] Asia Minor was, as I have shown, the home of the myths of the sons of fire, generated by the fire-drill and socket. It was there that the Indian matriarchal cultivators, sons of the buffalo cow, met with the mining races of the North-east, sons of the pine-tree, invoked in the prayer of the Osage Panhuka to the pine-wind of the North and of the Tauric Artemis. The united tribes became the barley-growing Iberians, the Ibai-erri or people (*erri*) of the rivers (*Ibai*), the race who instituted the maritime commerce of Asia Minor, and made all the people of the sea coast worshippers of the great mother Cybele of the pine-tree of the North, and of the fish-mother Myrrha or Smyrna, who became the Babylonian bi-sexual queen-goddess Semiramis, called in the male form Samirdus, who invented weights and measures and the art of silk weaving. She became Poseidon, the Thunny fish and the lotus flower, to whom black bulls were sacrificed yearly, the representatives of the earlier buffalo bull killed as their totem sacrifice by the Indian Dravidians, and this god of the black bulls became, in another transformation, the black bull Pashang of the Zendavesta, the

[1] *The Ruling Races of Prehistoric Times*, Essay v., p. 460.

[2] *Ibid.* Essay iv., p. 360.

barley-eating god Pashān of the Rigveda, who gave his name
to the Greek Poseidon.[1]

It was these Iberian barley-growing sons of the buffalo
whose migration is recorded in the historical genealogy of the
sons of Shem in Genesis, and the American Indian ritual
of these tribes enables us to place before our minds a most
graphic picture of their march southward, headed and directed
by the sacred pole. We can see how the sons of Shelah,
meaning the spear-pole in Hebrew, started under his guidance
in their tents from Armenia, called Arpachsad, the land (*arpa*)
of the conquerors (*kasidi*), Shem's son, and how they, the
Hebrew or Iberian sons of Eber, were like the American
nomad Indians[2] guided by the mystic pole, the fire-drill and
socket of the fire-worshipping Magi, to India, as the sons
of Joktan, and to the Euphratean countries of Nahor, the
father river (*Nahr*), as the sons of Peleg.[3]

The historical evidence furnished by these myths proving
that the Asiatic people, who traced in their national history
their descent from the deer of the North-east and the buffalo
of the South, disagrees with the verbal evidence of the
American Indians, who have the same national totems.
They all say that they entered America from the North,
and therefore the original myth that they brought with
them, assigning the origin of their household fires to the
buffalo and the deer, must be one which told their history
before they came to America under the guidance of the
fish-god, whose mythic history, as I have traced it, shows
him to be a mythological descendant of the totem buffalo-
god.

The correctness of the conclusion, which derives the
historical myths of the American Indians from Asia, is
irrefragably proved by the cosmogony of the Mexican Sia, as
told in the admirably graphic report on this tribe furnished

[1] *The Ruling Races of Prehistoric Times*, Essay i., p. 24 ; iii., p. 254, 286;
v., 437-440.
[2] See p. 236.　　　　　[3] Genesis x. 21, 31.

244 THE RULING RACES OF PREHISTORIC TIMES

by Mrs. C. Stevenson, who, with her husband, lived with the
Sia, and so completely gained their confidence as to learn
from them the innermost secrets of their creed and ritual,
called Wakanda-thatica or Stories of the Mysteries of God
(*Wakanda*) by the Sioux, which are so carefully guarded
that they are not told even to the initiated except after
fasting and prayer. The awe with which these stories are
regarded cannot be better illustrated than by quoting the
words with which the Sia describe the recitation of their
cosmogony. They say : 'The hour is too solemn for spoken
words; a new life is to be given to us.'[1] The Sia, who
call themselves the sons of Ût'set, the corn-mother, are a
hospitable race of artistic potters, living on the Jemez and
Salado rivers, tributaries of the Rio Grande in Mexico. The
district peopled by them and their immediate neighbours,
the Zuni and Tusaya, is called the great farming district of
Poshai-yänne, the last of their creating-gods. They, like
the Zuni, who thought that the religious standard of
Christians would be sufficient, if they had more time to
devote to religious duties,[2] are most strict in their religious
observances, and, like the Jews, they make the ritualistic
code binding on all members of each family. Thus, Mrs.
Stevenson tells of one Zuni father, who disowned his son
because he had stayed away from one of the religious cere-
monials. Also they, like all other Mexican Indians, place
their children under the protection of the water-god by
baptizing them; a custom which I have proved to be
universal among the Iberian races, who called themselves the
sons of the rivers, and which, beginning with the baptism by
dew and fire, at the Spring festival of the Palilia, produced
the elaborate baptismal ceremony initiating the partakers in

[1] *Publications of the Bureau of Ethnology*, vol. xi.: Dorsey, 'A Study of
Siouan Cults,' p. 369; Stevenson, 'The Sia,' pp. 9, 16, p. 67.

[2] This remark is based on a reminiscence of a saying of a Zuni or Tusaya, I
forget which, I read in one of these American Reports, which I have not now
by me, that 'the Christians would be good people if they had more time for
religion.'

the Soma sacrifice, the ablutions of the Greek penitential
ritual, and of the penitents in the Eleusinian mysteries.[1]
The Sia also, like the Ugro-Finns, look on the wife as
guardian of the household hearth, for the husband takes up
his abode in his wife's house. Their chief is the Tiämoni,
whose sceptre is a crooked staff, similar to that assigned in
Egyptian pictorial mythology to Osiris, god of the barley-
growers. He holds his office for life, inheriting it, among the
Zuni, from his mother, while he is elected by the Sia alter-
nately from the corn, coyote, and reed clans. Besides the
Tiämoni and his vicar, the Sia have also a war-priest and his
vicar, who rule jointly with the Tiämoni and arrange the
national hunts. They are thus like the more nomad Indians
of the North, divided into peaceful cultivators and warlike
hunters of the deer. They also reproduce the constitution
of the Hindu Dravidians and Greek Spartans under which,
the king, the law-giver and judge, ruled at home, and his
joint king, the Dravidian Sena-pati or commander-in-chief,
provided for the national defence.

The three corn, coyote, and reed tribes all indicate, as will
be seen from the national history, distinct periods in the
nation's growth. The earliest represented by the coyote, or
wild-dog tribe, is that in which the dog, the totem of the
Northern hunting races, brought, as he is said in Sia tradition
to have done, fire to earth from the lower world, the home of
the stone-god. The dog was, in Ojibewa tradition,[2] given
with the drum, the sacred rattle and tobacco to Minab'osho,
the first man of the Creation-myth of their Mide-wiwin or
national priests, who were taught medicine by the bear
Manido.[3] The drum and sacred rattle are the musical instruments played at the dances of the Mundas and Ooraons

---

[1] *The Ruling Races of Prehistoric Times*, Essay iii., 217-232, Preface,
xliv.-xlviii.

[2] *Publications of the Bureau of Ethnology*, vol. x.: Mallery on the ' Picture-
Writing of the American Indians,' p. 492.

[3] *Ibid.* p. 234.

of Chota Nagpore, and the tobacco marked the age of the incense offerings, when the first temples were built in honour of the fish-god, the totem parent of the sons of the rivers, when he came down from the mountain heights to his shrine in the plains, where he was worshipped with the incense which reproduced the mountain clouds and mists, the dwelling-places of the soul of life at the sources of the parent-rivers. It was Minab'osho who taught the sacred otter, who, in the Nibelungen legend, is one of the earliest sacred triads it speaks of. That of Fafnir, the encircling snake; Regin, the god of the twilight and rain (*ragna regn*); the heavenly smith, and the Otter. They were the three sons of Hreid-mar, the keeper of the nest (*hreid-r*) of the mother-bird, the storm-bird. The wise Otter who learnt wisdom from the totem fish he ate,[1] was slain by Loki, the wolf fire-god, the Northern form of the fire-dog who became the sacred dog of the Parsis, who are called in the titles to the second Mandala of the Rigveda Medah Saunaka, the Medes, sons of the dog (*shvan, sunah*). The mother-dog of fire is Saramā, the bitch of the gods of the Rigveda, mother of the twin four-eyed dogs;[2] Sarameias and Sharvara, who became the Greek Hermes, the dog of the gods, and Kerberos, who guarded the river of the dead in the West. Saramā is another form of the mother of fire, called in the Rigveda Mātarishvan, the mother of the dog (*shvan*).

---

[1] In Chinese mythological ritual, the otter is said to sacrifice fish in the first month of the year, December-January, and is thus recognised as a sacrificial priest who was frequently a member of the early mythological triads. Thus Devapi is the rain-priest, in the Hindu triad, of the parents of the national kings, with Shantano, the ancestral king-husband of the mother-river Gungā, and Vahlika, the father of the fire-worshipping Takkas of Balkha, the Northern conquering race. We shall find towards the close of this essay, another triad with a sacrificial priest, that of the Zend Frashavstra, the Hindu Prashastri or teaching-priest, Jamāspa, and Vīstāspa, who introduced the teachings of Zarathustra. (*The Liki, The Yüeh Ling*, Bk. iv. § 1, Pt. I. 8; *The Royal Regulations*, Bk. iii. § 2, 25; S.B.E. vol. xxvii. pp. 251, 221).

[2] Rigveda, x. 14, 10, 11.

The next stage in the national growth is that in which the sons of the reed were born. This corresponds in the Nibelungen legend to the stage in which Sigurd having slain Fafnir and Regin, rides to the top of the mountain of the deer (*hinda*) called Hinda-fjall, and wakes from her winter sleep Brunhilda, the mother of the springs (*brunnen*). It is the age in which, as we shall see presently, Ūt'set, the mother of corn, left the nether earth in which she had lain as the seed whence the corn-mother, the Greek Demeter, the barley-mother, was to be born, and made her way to the upper air on the top of the mother-mountain, whence the parent-rivers rose, by the help of the growing reed, the mother of the Kushite race, the reeds growing round the mountain lakes which sent their waters to the plains by their children, the rivers. It was then that she intrusted the star-bag to the beetle, the Egyptian symbol of life, who took the bag to the upper earth, but on the way let the stars escape, leaving only the Pleiades, the three stars of the belt of Orion, and the Great Bear for Ūt'set, to place in the sky. It was after this was done that the twins, who, as we shall see, exactly correspond with the twin-brethren of Asiatic and European mythology, were born. The whole story is one which tells how the Kat'suna, the men with masks, whom Ūt'set took with her to the upper air, and who traced their descent from totems, looked on the Pleiades, the belt of Orion, and the Great Bear as their parent-stars ; and these stars are, as I have shown, those which measured the year of the matriarchal races of the South, the corn-growers of Asia Minor and of the worshippers of the pole-star.[1] The last age told of in the Sia birth-story, was that in which the society organised by the twins and divided into separate clans, such as the Hindu castes of the potters, oil-men, weavers, fishermen, etc., the American-Indian societies of the snake, the ants, and the knife, united by community of function and ritual, and by common guardian gods of each

---

[1] *The Ruling Races of Prehistoric Times*, Essay ii., pp. 123-133 ; iv. 357-365, 401-403.

social guild, became the united sons of corn dwelling in the land of Poshai-yänne, the sun-god, who, after his marriage, is slain like the sun-god of each year, and who rises up again by the help of the feather of the eagle, the Phœnix-bird which brings the young sun to life from the dead sun.

The Sia story of the origin of this society of confederated artisans and husbandmen, traces the creation of the earth and its inhabitants, to Sūs-sistinnako, the parent-spider, who is an exact counterpart of the Hindu Kīrat or Kirttidda, the Spinner, the constellation of the Pleiades, called the Krittakas, which also means the spinners, and was the mother of Rā-dhā, the maker (dhā) of Rā, the sun-god, while her father was Vrisha-bhānu or Bhrika-bhānu the ray (bhanu) of rain (vrisha) or fire (bhrigu), the Greek βροχή, meaning rain accompanied by thunder.[1] Thus, according to the Hindu myth, the thunder-god, and the constellation of the Pleiades were the parents of the mother of the sun-god, the pool whence the parent-rivers rose, and the Pleiades were, as I have shown, the stars which ruled the year of the Southern Hemisphere, divided into two periods of six months, each measured by the motions of the Pleiades in November and April.[2] The mother of the sun was originally the void of darkness, the Phrygian goddess, Baau, the Akkadian Bahu, and this, in the Sia myth, is the circular web Sūs-sistinnako, spun and divided into four compartments by a cross of meal, the St. George's cross, denoting the earth in Chinese. He placed in each of the two Northern divisions of this cross a seed, and then sat down in the South-west quarter of the web, answering to the Indian Peninsula and Archipelago, where the Pleiades were worshipped as mother-stars, and which lies to the west of America. He then sang, and the two seeds shook like rattles, accompanying his song, and from them, after a while, were born Now-ūt'set, the mother of the West, the mother of the buffalo race, who

---

[1] *The Ruling Races of Prehistoric Times*, Essay v., p. 451.
[2] *Ibid.* Essay ii., pp. 123 ff.

lighted their fire with the Western stick, used also to light
the fire on the Hindu altar,[1] and U͂t'set, the mother of corn
and of the deer-race of the East, who lighted their fire with
the East stick. These three parent-gods exactly correspond
with the Gond-god Pharsi-pen, the female (*pen*) trident
(*pharsi*) the central bisexual-god Pharsi-pot, and his two
tiger-wives, Manko Rayetal and Jango Rayetal, the outer
prongs of the trident.[2] The creation by music is analogous
to the Gond story of Lingal, the father corn-god, of the
' threshing-floor of Gonds,' another form of the sexless father-
god the Ashēra or pole, which was, as we have seen, sacred to
the Indians. Lingal is said to have played on his lyre with
eleven strings, and to have thus made Rikad Gowadi and his
wife, the parents of the forest-races, dance the national dances,
and give their daughters as wives to the cultivating Gonds
from the North-east, who made fire with flints.[3]

The two Sia-mothers of the West and East, inspired by
Sūs-sistinnako, made the moon to rule the night, of black and
yellow stone, turquoise and red stone, and the sun to rule the
day, as the fiery globe we see in heaven, made of white shell,
blue turquoise, red stone, and an abalone shell, or in other
words made it to rise in the red-dawn, behind the mountain,
born of the primæval ocean of the East, the home of the
shell, and to pass through the blue sky of day, to the sea in
the West, a picture exactly similar to that imaged in the
story of Manu's creation. For in this, Manu, the thinker,
the father of the dwellers on the new earth, sowed in the
ocean of the flood, clarified butter, sour milk, curds and whey,
and from this seed the mother-mountain Idā was born, first
as the mother of the sheep, and afterwards of the cow-born
people. The picture of the creation of the earth, as set
forth in these stories, is clearly that of the mother-mountain,
the rock yielding the fire-stone, of which the moon and the

---

[1] *The Ruling Races of Prehistoric Times*, Essay iii., p. 166.
[2] *Ibid.* Essay iii., p. 193.
[3] *Ibid.* Essay ii., pp. 48, 49, 89.

sun were made, the parent of life rising out of the shell-yielding ocean, and this is reproduced in a more elaborate form in the account of how Sūs-sistinnako made the earth before Now-ūt'set and Ūt'set made the moon and sun. He divided the earth into the same six regions as are reckoned in the Akkadian cosmogony, each region having in its centre, the spring of life rising at the top of the mother-mountain, which was not, as in the primæval conception, the fire-yielding rock, but a plateau covered with productive soil, whence the mother-tree grew ; a mountain similar to the mountain Himinbjorg, in the Edda, on which was the mother-ash-tree, Yggdrasil, with the well Mimir (*memory*) at its roots. The Sia trees were the spruce in the North and the pine in the West, the mother-trees of the American-Indian deer-race, and of the worshippers of Cybele, and the Finn sons of the bear in Asia Minor and North-eastern Europe and Asia ;[1] the oak, the mother-tree of the Druid sons of the tree (*dru*) in the South, and the aspen, the mother-tree of the Armenians of Iberia and Georgia in the East.[2] This land of the East was the mother-land of the deer-race, that of the first Akkadian mountain Ararat, the wet-mother of the fertilising floods sent down the Euphrates, which was super-seded in later mythology, when the sons of the bird (*khu*) had become the sons of the tortoise (*kush*) by the mountain called Khar-sak-kurra in the Western Himalayas, whence the Indian rivers rose. One of the two last regions of the world in the Sia and Akkadian cosmogony was the Zenith, likened by the Akkadians to an upturned boat supported by the pillars of the world, the mother-mountains of the four regions, and bending like the Egyptian heaven-mother Nut, over her consort Seb or Geb, the convex earth. Under this convex earth the tortoise was the Nadir, the home of the dead to the Akkadians, and of the first created beings to the Sia. This nether earth was lighted by the moon and stars, the

---

[1] *The Ruling Races of Prehistoric Times*, Essay iii., pp. 263, 264.
[2] *Ibid.* Preface, p. xxi. note 2.

latter being made by the mother-sisters of white crystal, and not scattered about, as they afterwards were, in the upper world, when Ūt'set took the stars with her, as I have already related. The three regions called in Akkadian cosmogony the firmament, the nether-earth, and the upper air or middle region, dividing the other two, are called by the Sia, Hu-waka the creating (*hu*) waka (*author of life*), the heavens, Ha'arts, the parent nether-earth, which Sus-sistinnako first created, and Tinia, the middle plain, to which, as we shall see presently, Ūt'set, came when she passed through the crust which separated it from Ha'arts, the nether world, by the help of the growing reed. In these names we can recognise the creating Druid and Zend god Hu, whose name is, as I have shown, a form of the original Finnic Khu, the mother-bird, the thunder-bird ; also the Etruscan supreme god Tinia, who was in their cosmogony, as in that of the Sia, the god of the atmospheric region, who speaks in the thunder and descends to earth in the lightning.[1] He is thus apparently the god of the maritime races, who, from this centre land of India and the Persian Gulf took the worship of the fish-sun-god both to Etruria in the West, where the Etruscan language is closely connected with the Ugro-Finnic-Akkadian, and to America in

---

[1] *Encyclopædia Britannica*, Ninth Edition, Dennis's ' Etruria,' vol. viii. p. 367 ; Leland, *Etrusco-Roman Remains*, chap. i. p. 18 ff. The name of Tinia, the creating place of the thunder-god, in the Sia dialect, and the thunder-god in Etruscan, may be compared with the Akkadian *tin* 'perfect' denoted by the character $\bigwedge$ indicating the stem, sprung from the two parent roots (*Transactions of the Ninth International Congress of Orientalists*, vol. ii. ' Akkadian Affinities,' by the Rev. C. J. Ball, pp. 698, 699) ; also with the Chinese *din*, *t'in* 'heaven' reproducing the Akkadian *dim*, another form of *tin* meaning the wells or springs of Ia, called I-dim, or the house (*I*) of the spring, whence the thunder-god drew the rain, and whence grew the Ash Yggdrasil with its roots in the well Mimir and the Urdar-fountain, where the two mother-swans fed. This was the fount of the ancient stone-pillar (*Urd-r*), the father of fire (Mallet, *Northern Antiquities*, ' Prose Edda,' 16, and *Scandinavian Mythological Doctrines*, pp. 413, 490, 491), whence gushed the water which created the tree-stem. It was from *tin* also, meaning hand, that the Akkadian *ti*, life, the Chinese *t'en*, god, were formed.

the East. There these three regions are an epitomised form of the six for the Tinia or middle region, the home of the thunder-god is that of the four regions of the four quarters above, which are the Hu-waka or heavens, and beneath them Ha'arts, the nether-earth, to which the sun goes at night.

The Sia parent-tree of the zenith was the cedar and a variety of the Druid oak, the tree of the Nadir, and these trees are both significant, as the cedar was the tree used for the pillars of the earliest Akkadian temples at Girsu, and was originally made sacred in Syria, as the tree of the incense-worshipping sons of the river-reed, still used in producing incense to propitiate the thunder-god of the Kansa Indians.[1] Also the belief in the oak as the national parent-tree is one belonging, like that of the cedar, to the Mediterranean zone, for it is the sacred tree of the national Greek temple at Dodona, consecrated also to the mother-doves, the Peleiades or Pleiades, who succeeded the northern raven as the mother-bird in the age of the worship of the fish-god, when the first temples were built, and when the mother-constellation of the matriarchal races was changed from that of the Spinners to that of the Doves. These were the birds sacred to the Babylonian Semiramis, the bi-sexual fish called Smyrna or Myrrha in Asia Minor. The fish parent-god, and the dove-stars sacred to her, were made guardian-gods of married pairs, as is shown in Kimah, the Hebrew name for the Pleiades, and the Assyrian Kimtu, both being derived from the root *kamu*, to tie. They were the stars which consecrated the union of the married pairs of the Kushite race, who were united by tying their hands together with bonds of Kusha grass; the mother-stars of the Hebrew race, the doves released by Noah to return with the olive-leaf, consecrating the olive-tree, the tree sacred to the race who extracted oil with the oil-press, and divided civilised men

---

[1] *Publications of the Bureau of Ethnology*, vol. xi.: Dorsey, 'Study of Siouan Cults,' p. 385.

into classes united by community of function. It was these married pairs, united under 'the sweet influences of the Pleiades,' who reproduced in the family the original joined races born, according to Hindu tradition, from the union of the daughters of the Pleiades, the Southern mother-stars, with the turners of the Northern fire-drill, the father-stars of the Great Bear, which make the stars revolve round the pole, and these bear-stars became, when the bear-race was united with that which made the Pleiades their mothers, the stars of the seven antelopes.[1] Thus the Sia reproduction of the Akkadian cosmogony tells us that the national birth-story of the potters, the guild to which the Sia potters belonged, the ruling race descended from the Greek father-god Peleus born of the potter's clay ($\pi\eta\lambda\delta\varsigma$), and of the oaks and doves of Dodona, was framed after the Hindu cult of incense had reached Syria,[2] and had there been adopted by the worshippers of the fire-cross of St George, the symbol of the earth, used in the creation of Sūs-sistinnako.

It was after the creation of the earth of the six regions that Now-ūt'set and Ūt'set made the first sun and moon of the nether land of Ha'arts, and when these had been made, Ūt'set, the mother-goddess of the day, as Now-ūt'set was goddess of the night, created the Chaska cock of dawn, the sacred sun-bird of the Munda mountaineers of Chota Nagpore. She afterwards made the altar as a human figure, with the legs crossed like the Hindu earth-altar, made in the form of a woman. The crossing of the legs, which is like the crossing of the lines of the rain-god and the goddess-mother of the Māghada fire-worshippers on the Hindu altar,[3] seems to be a reminiscence both of the sitting attitude of the Buddhas, and of this symbol of the transverse cross. It denotes, as I have shown in Essay VIII., the path of the sun at the solstices when it rises and sets, not in the due east and

[1] *The Ruling Races of Prehistoric Times,* Essay iii., pp. 287-289 ; i., p. 24 ; v., p. 509.
[2] *Ibid.* Essay iii., pp. 299-301.   [3] *Ibid.* Essay iii., pp. 163, 168.

west as at the equinoxes, but in the north-east and north-west at the summer, and in the south-east and south-west at the winter solstice. This story, therefore, tells of the reckoning of the year by the solstices, and this solstitial year was, by all the Northern European and Asiatic nations who used it, measured, like the year of the American Indians and Esquimaux, from the winter solstice.

It was after the making of the altar that Ūt'set, under the guidance of Sūs'sistinnako, created the race of intelligent beings, the sons of the thinking mother of corn, the American form of the Hindu Manu, the thinker. They were divided into the Pai-ä-tämo, or men of Ha'arts, the nether earth, the spirits of the dead fathers, who lived in holes in the ground. The two first men she created were Quer'ränna, the moon-man, and Ko-shairi, the flute-player, the Hebrew Jubal, son of Lamech, whose name is a Semitic form of the Akkadian, Hindu and Gond god, Lam-ga, Linga, and Lingal, who, in the Gond cosmogony, tamed the savage primæval father and mother by playing on his musical bow. He was the sun-messenger between the original sun made by Now-ūt'set and Ūt'set, and the Kat'suna, the men with masks, the people of the northern totems who were the third race created by Ūt'set after the Kopishtaia, the cloud, lightning and thunder people, the race who believed in the water-elves and nymphs of Greek mythology. The final contest between Now-ūt'set and Ūt'set arose after the creation of the three first ruling races of civilised men, the Pai-ä-tämo, the cultivating race of the matriarchal age, ruled by Now-ūt'set, the buffalo-mother of the West; the Kopishtaia, the fire-worshipping race who reverenced the thunder-god; and the Kat'suna, the northern sons of the deer and other northern animal totems. Ut'set invited her sister to see the sun-rise, and called to the bird Shu'ahkai, a small, black bird with white wings, the crow mother-bird of the northern sons of the deer, to place itself between the rising sun and the two sisters, so as to make its consecrating rays fall on the face of Ūt'set, and to prevent them from illuminating the face of Now-ūt'set. This means, in other words, that Ūt'set made the northern mother-bird the creatrix of the deer-race, who worshipped the dawning sun of the East, instead of the Spinners, the Pleiades, stars of the Southern cultivators, who also, as the sons of the buffalo, worshipped the setting sun of the West. When Ūt'set was consecrated by the rising sun, her followers fell on those of Now-ūt'set, but Ūt'set, after a time, proposed to stop the battle, and that she should wrestle for victory with Now-ūt'set. In the contest Ūt'set won, and then, in Northern fashion, killed Now-ūt'set and cut out her heart. Another form of this story is to be found in that of Hadding, the northern hairy (*hadd-r*) sun-god, who is, as we shall see, the sun-man of Ūt'set, the pupil of Wagn-hofde, the Wain-head, or the Pole-star, who went down to the nether world at the winter solstice to get the plant of life, Angelica, after he had wedded Ragnhild, the maiden of the twilight (*ragna*), the sun-maiden of the dying or setting sun of the buffalo-race, the sun of the autumnal equinox, which ushers in the time of harvest and the sowing of cold-weather crops in India, and begins the Jewish year. He, before he brought to earth, at the winter solstice, the young sun of the future year, the plant Angelica, was imprisoned by the winter-frost of Loki, the fire-wolf, who ruled the North under the name of Mid-Odin, before the mother-bird of the deer-race, the raven who inspired Odin. During this age Odin had abdicated the government of the world owing to the adulteries of his wife Frigga (the seed, *frio*), goddess of the matriarchal age when matrimony was unknown, preceding that of the deer-god Frey, and the hawk mother-bird, his sister Freya. The imprisoned sun-god, by Odin's help and advice, slew the fire-wolf, and ate its heart.[1] It was the exact counterpart of this hairy sun-god of the North, who had become the father of fire by killing the fire-wolf and eating its heart, who became the sun-god of the creation of Ūt'set, after she had united the

---

[1] Elton and Powell, *Saxo Grammaticus*, Bk. I. pp. 24, 29-31, 37, 38.

buffalo and deer-race by eating the heart of the buffalo-mother. For it was after the defeat and death of Now-ūt'set that, in consequence of the rise of the waters which flooded the nether earth and made it uninhabitable, that Sūs'-sistinnako placed the sun in heaven as the hunter-sun, dressed in deer-skins, with fringed skirt, deer-skin leggings, and a kilt with a snake painted on it. He wore moccasins of deer-skins, embroidered with red and yellow beads and blue turquoise, the sun and moon-colours. He carried a bow in his right, and an arrow in his left hand, and a quiver of cougar-skin hanging at his back. Like the Ya'ya, or symbolical sheaf, representing, as I shall show presently, Ŭt'set in the Sia ritual, he wore eagle and parrot-plumes on each side of his head, *and the hair round his head and face is red as fire, and quivers as he moves.* This is a complete picture of the red-headed, hairy (*hadd-r*) sun-god of the North, Hadding, after the sun-god had been united with the deer-god Frey, the son of Njord, the north pole, or pole-star. The yellow line round his mask shows him, as Mrs. Stevenson says, to be the sun- and rain-god, that is, the god who succeeded the fire-god. His path through the heavens, as ordained by Sūs'-sistinnako, was from east to west, and his daily journey was divided into three stages, for his breakfast, dinner, and supper, the three strides of the Hindu sun-god Vishnu.

This hairy sun-god of the sons of the deer, who begins the day by lighting his fire in the east, is a clear reproduction of the hunter-god Orion, the Wild Hunter of the North, who succeeded the Pleiades as ruler of the year, divided into the three seasons, indicated by the three stars in his belt. It was this star which, as I have already several times proved, hunted or drove the stars round the pole, indicated by the pole-star, which remained immovable. He was also the Akkadian god Dumu-zi and the Egyptian Smati-Osiris, both of whom as the star Orion, launched their year-bark on the seas of time in the month of November, the month when the Pleiades

year began, and this boat is the moon-hare, the constellation
Lepus at his feet. This he hunts through the lunar phases
of the twelve lunar months assigned to Orion the hunter, both
in Hindu and German legend. It is after his twelve months
hunting that he rests, as I have shown in Essay vii.,[1] twelve
days, according to the Rigveda, in the house of Agohya, the
pole-star, and according to German acted legend, that he
spends twelve days in revelling, and ends his revels by killing
the deer, the old year he has hunted. In considering the
evidence as to the home of the genesis of this conception of the
year, the constant recurrence in the numeration, to the number
twelve, must be noticed, and it must be remembered that, as
I have shown in the beginning of this Essay, that the use of
twelve, as a factor in measuring the night and day, is of Tamel,
or Dravidian origin. Hence we see how the deer-hunter
became the Hindu hunting-god Mriga-sirsha, the antelope
(*mriga*) head (*sirsha*),[2] who killed the dying year with his
three-knotted arrow. The three stars in his belt, which became the three-knotted cord of the Hindu Brahmins, and
that called kamberiah, worn, with its three knots, by all the
Dervishes of the East, who represent the Sufi or mystic
doctrines of Mohammedanism, derived through the Zends of
Persia from India, the original home of the three-knotted
cord.[3] These stars and knots, as I have shown in Essay iv.
of this work, represent in Brahmin chronometry the three

---

[1] *The Ruling Races of Prehistoric Times*, Essay vii., p. 22.

[2] But see *The Ruling Races of Prehistoric Times*, Essay iv., pp. 401, note
2, 402, where I have proved that Mriga originally means the animal that
wanders round, and that from it, is formed the Zend *Mēreghā*, a bird, the
Hindu *Murghi*, the domestic fowl, and it is used to mean a bird in Rigveda i.
181, 7, hence it was originally the mother-bird, then the deer, and last of all
the antelope, its meaning in modern Sanskrit.

[3] O'Neill, *The Night of the Gods* (Bernard Quaritch), vol. i. p. 127 :
Encyclopædia Britannica, ninth edition, vol. vii., Art. ' Dervishes,' p. 113.
These Dervishes, like the Zends, who made their cord of six strands instead
of three, as the Brahmins did, reckon six pillars (*Erkian*) of spiritual life.
These which reproduce in a metaphysical form the six strands of the girdle,
the paired three are, (1) The existence of God, (2) His unity, (3) the angels,

seasons of the year of the early corn-growing races,[1] before
a fourth season was added for the autumn fruits.

It was during the reign of Orion, the hairy sun-god of the
North, that the Kat'suna, sons of the northern national totems,
gathered round the mother-mountain of the eastern sun.
Their rulers were the sons of the deer-sun, who appear in
Hindu ethnological history as the Kshatreya, or warrior
races who assert their descent from the deer, by wearing at
their initiation skins of the spotted deer,[2] and it was these
people who symbolised the earth as the tortoise-boat floating
bottom upwards on the mother-ocean. This tortoise-mother
was, as I have shown in Hindu mythology, the mother-
mountain Idā, raised from the waters by her father and
husband, the god Manu, the thinker, to be the mother of a
new race of men, born from the earth, regenerated by its
baptism in the ocean.

The Sia story-teller looks at this legend from a different
point of view, and tells how the children to be born of Idā,
called here Ūt'set, the corn-mother, were driven from the
nether earth by the flood of the rising ocean, ascended from
the nether earth of the fathers, to the heavenly plateau of
the corn-growing race, emerging through the reed placed by
Sūs'-sistinnako, on the mesa, or central mound of the lower
world, and reaching, like Jack's bean-stalk, to the entrance of
the upper world. He tells also how the new race changed
their totems, from those of the deer of the East and the
buffalo of the West, to the wolf of the East, sacred to the wolf
sun-god Apollo, the god of day, the sun-god Rā, and the bear
of the West, his twin sister Artemis, the bear-mother, the

(4) the prophets (5) the day of resurrection or rebirth, and (6) good and evil
through God's predestination, a reproduction of the creed of Ahura or Asura
Mazda, the one supreme and only God, who fixed the unchangeable sequence
of the laws governing natural phenomena. The whole subject should be considered with reference to the numerous instances of the use of six as a sacred
number, which I have brought forward in these Essays.

[1] *The Ruling Races of Prehistoric Times*, Essay iv., pp. 402, ff.
[2] *Ibid.* Essay iv., p. 403.

goddess of night, and the stars that circle round the pole.
The reed, which they ascended was the mother-reed of those
in which Kavād, the father of the wise (*kavi*), Kushite kings
of the Zendavesta and Bundahish was found; the reeds
encircling the mountain tarn whence the parent rivers of the
corn-growing Iberians, the people (*erri*) of the rivers (*ibai*),
descended to water the earth and to bear their sons and the
produce of their lands, to the mother-ocean. This reed is,
in the Zendavesta, the golden pipe bringing the waters of
the ocean through the mountain to become the clouds and
mist which wreathe its top, wherein the soul of life dwells,
before descending to the earth in the rain and the running
waters of the irrigating streams.[1] Ut'set found the way
through the reed, carrying the stars in a sack, while the procession was closed by the turkey, who, in Sia ritual, was the
mother-bird of the women, who, in the national sacrifices,
carry wands with turkey plumes, while the men's wands have
eagle plumes. When the growth of the reed was barred by
the earth, Ut'set called Tsika, the male locust, to make a
path for them. The locust denotes the stem of the reed
making its way irresistibly through the earth as the plumule
of the growing tree, which, with its two cotyledon leaves
formed the plant reproduction of the Dakota fire-cross, the
cross called by them Sus-beca or the dragon-fly.[2] This Latin
cross of St. George became, when it was changed from the
fire-cross, to be the symbol of the growing plant, the trident
called by the Buddhists, Vardhamana, or the growing one.[3]
This is the trident still worshipped by the Takkas of the
Punjāb, as the three gods of the three seasons of the year,
Shesh Nāg, Takht Nāg, and Bāsuk Nāg, the three prongs of
the Gond trident of Pharsi-pen, and that of the people called

[1] *The Ruling Races of Prehistoric Times*, Essay iii., p. 144, 145.

[2] *Publications of the Bureau of Ethnology*, vol. x.: Mallery on the ' Picture Writing of the American Indians,' p. 725.

[3] Comte Goblet d'Alviella, *The Migration of Symbols* (A. Constable and Co.), pp. 237, 241.

in the Zendavesta, Keresavazda, or the people of the horned (*keresa*) club (*vazda*).[1] The locust, as the stem of the plant and of the symbolic cross, was followed by Tuopi, the badger, the borer of the earth, the totem in Sia ethnology of the people of the South, the first cultivators ; then by the deer of the East, the sun-animal, the elk of the North, and the buffalo of the West, sacred to the vanquished Now-ŭt'set, and, according to the Osage myth, the first father of the corn-mother. The sons of the animal totems, the mask-wearing Kat'suna, who came to the upper earth with Ŭt'set, were first the people calling themselves Iberians, that is, the people (*erri*) of the rivers (*ibai*), the sons of Eber, who became the sons of the barley-seed and of the eight-rayed star, denoting to the Akkadians and ancient Chinese, god and seed.

Ŭt'set gave the star-bag to the beetle, the sacred scara-bæus, or symbol of life, in Egypt, and he was followed by the turkey, the mother-bird of the American corn-race, born not from the local mother-mountains of the four regions, but from the central mountains in which they were united. When the turkey, the American substitute for the mother-bird of Asia, had passed through the door leading to the lower waters, it was closed. Then Ŭt'set, as I have already mentioned, found that the only stars left in the beetle's star-bag were the Pleiades, the mother-stars of the southern sons of the buffalo-mother, the three stars of Orion's belt, the parent-stars of the earliest corn-growing races, who measured their year by the daily revolutions of the stars led by Orion round the pole, the central revolving fire-drill, only visible in the pole-star, and by the lunar phases. These three stars were the parent-stars of the Kat'suna. The remaining stars of the beetle's bag, the seven stars of the Great Bear, the guardian of the polar circle, whose number seven measures the fortnightly lunar changes, are those which became the stars of the night-mother, Artemis, the mother-bear, the parent stars of the next age of the Twins.

[1] *The Ruling Races of Prehistoric Times*, Essay iii., pp. 190-193.

It was in the far north and not in the south, the mother of cultivation and of the buffalo race, that the Sia entered the upper earth in their escape from the encroaching ocean, thus clearly showing that the myth told of the migration of the buffalo race from the coasts of the south to the inland north, where it met with the deer race, the people of Asia Minor and Southern Greece, who changed the name of Artemis, from that of Arktos, the Braurian bear-mother to Elaphia,[1] the deer-goddess of Elis and Olympia, the mother-goddess of the race of the Olympian worshippers of the solstitial sun, who divided the year into four seasons marked by the solstices and equinoxes. The Sia, in their new northern settlements, at first only ate grasses, whence they developed Indian corn, just as millets and barley had been developed, in Asia Minor and Mesopotamia. They there changed the genera of their totem animals, only the badger remaining as the totem of the South, while the deer, buffalo, and elk of the East, West, and North, were replaced by the wolf of the East, the Greek god Apollo, the sun-god of day, worshipped in Argos as a wolf, the bear of the West, his twin sister Artemis, and the cougar, or tiger of the North.

Thus the age inaugurated by the emergence of the corn-mother into the light, and to the worship of the rising sun of day instead of the setting sun of night, was that of the Twins, called in the Rigveda Ushasā-Naktā, Dawn and Night, the children of Saranyu,[2] who, as I have shown, were

[1] I do not mean to say that these people spoke Greek, or gave Greek names to their gods. The names they gave in the then spoken tongue of Asia Minor, were translated by the later Greeks, just as the Tamil names in India, were subsequently translated into Sanskrit.

[2] *The Ruling Races of Prehistoric Times.* Essay iii., pp. 210-218. These sons of Saranyu, who were also children of the wolf and the bear, were also children of the fire-dog Saramā, the bitch of the gods, and Mātarishvan, the mother of the dog, for Saranyu and Sara-mā are both forms of the root Sar ; Saramā was the Sia coyote, who stole fire from heaven. See Essay vi., p. 510.

the same twins as the gods called by the Greeks, Apollo and
Artemis, and the people by whom this worship of the twin-
gods of light and darkness was introduced, were the Turanian Gonds, the sons of the tiger, and the worshippers of the
growing plant, the trident, whose two mothers were the
tiger-goddesses, Manko Rayetal and Jango Rayetal. It was
they who ruled India, north of the Godavery, before the
Kushika, and their worship of the tiger is commemorated in
the name of Vajjians, or sons of the tiger (*vyaghra*), given
to the confederated Mallis, or mountaineers, and Licchavis,
sons of the fire-dog (*lig*), who ruled the north-east provinces
of the old Gond kingdom in the days of Buddhist history.
It was this new race who looked on the bear, the eater of
honey, the food of the prophet-race,[1] as the inspirer of magic,
and it is the bear which is the Manido, or inspired prophet of
the Ojibewas, the great medicine animal, who, in one of their
hill stories, brought back the dead to life.[2] These followers
of the bear-inspired prophet were thus a race who studied
the growth of plants and their uses, and who looked on the
Supreme God, not only as the father of life, but also as the
Great Healer and Physician, the medicine man. It was under
the inspiration of his prophets, and the kings they advised,
that society, in the kingdoms formed by confederated villages
and provinces, ceased to be a conglomeration of tribes descended from totem ancestors, and became divided into
castes united by community of function. This change was,
as I have shown, the result of the assumption of government by the Kushika kings descended from northern tribes,
the Gond sons of the tiger, and the fire-worshipping sons of
the dog.[3]

---

[1] *The Ruling Races of Prehistoric Times*, Essay iii., pp. 208, 209.

[2] *Publications of the Bureau of Ethnology*, vol. x.: Mallery on 'The Picture
Writing of the American Indians,' pp. 234, 245, 255, 256.

[3] *The Ruling Races of Prehistoric Times*, Preface, p. lviii. lix. Essay ii.,
p. 87; iii., pp. 310, 311.

Under this new arrangement of totems, the shrew was made totem of the Nadir, and the eagle of the Zenith. Ut'set asked the badger, wolf, cougar, bear, shrew, and eagle to harden and drain the earth, raised from the ocean, but they, as belonging to the Northern hunting, and not to the Southern cultivating race, failed, and the task was completed by a woman and man of the old Kapina, or Spider Society, the farming sons of the Pleiades and the buffalo. They repeated again, in the drier mountain-lands of the North, their earlier conquests of the moist rice-lands of the South, when they made rice, what it still is, the staple food of all the coast people of the Western Pacific, from Ceylon to Japan. But the Sia Kapina did not grow rice, but spun cotton, a dry crop, over the land, *thus showing that they came from India to America, for cotton is an indigenous plant of India, first used for weaving purposes in India and China, whence it was brought to America by the immigrating races.*

They also made a latticed road of wood towards all the quarters of the earth. And these stories tell in mythical fashion, how the weavers and carpenters of American history became like the weavers of India, who called the mother-stars the spinners,[3] and the Takkas or carpenters of the Punjab who worshipped the growing tree, the leaders in the progress of the agricultural communities, hitherto composed only of farmers and herdsmen.

It was to the leaders of this new society of the growers of the dry food crops of the North which superseded the rice crops of the South, that Ut'set assigned the supremacy, when she left the upper earth, and committed her image, the sacred ear of corn, to the care of her votaries as the Ya'ya or Iarriko, the national god-symbol.

[1] It is the mythology of these artisan rulers which is preserved in the saying in the Rigveda, that the Twins (*yama*) spun the first web in which men clothed themselves, the web of Time (Rigveda, vii. 33, 10, 11), and it was these ruling weavers of the age of the Twins, who made Penelope, the spinner of the web (πηνή) the wife of Odusseus, the wandering sun-god. See *The Ruling Races of Prehistoric Times*, Essay iii., pp. 210, 211.

The story then goes on to tell of the matriarchal rule of the six women, who were left by Ut'set in charge of the six divisions of the world, and this division of the world into six kingdoms, recalls the sanctity of the number six, venerated by the Twin races, who believed in the divinity of pairs, made the six Pleiades their mother-stars, and called themselves the Ashura or worshippers of six (*ash*) gods.[1] They were the people called Khati, in Assyria and India, a name which, like that of the American-Indian Dakotas, means the joined (*khat*) people ; the Hittites, who worshipped the Hittite and Cypriote six-rayed star ✳, showing the pole of the Zenith and Nadir, intersecting the four quarters made by the solstitial sun, the sun that measured the year of the American Indians, and also that of the races who began their year with the summer, and not with the winter solstice, like those Greeks who measured time by the Olympian calendar, the year introduced into Elis by the twins Castor and Polydeukes, the Zends and Hindus who celebrated the Soma sacrifice at this season.

The state of society, ruled by the six matriarchal queens, is depicted in the Sia story by telling how the men left the women, and went to the other side of the river when the women quarrelled with them, because they insisted on sacrificing children to prevent a too-rapid increase of the population, a reminiscence of the age of the yellow race, the Hittite Semites, whose ritual is represented in that of the Haranite Sabæans, described in Essay VII.[2] These were the people who made the original human pair, Adam, the father of the red race, and his mate, their parents, and who, therefore, no longer called themselves the sons of animal totems, and it was they who offered children and human victims, in the burnt human sacrifices of the Arabs and other Semites, and in the children thrown into the mother-river Gungā, by the Hindus, and

---

[1] *The Ruling Races of Prehistoric Times*. Essay iii., pp. 287-289.
[2] *Ibid.* Essay vii., pp. 55, 56.

who, as the Mexicans used to do, ate these sacrificed human victims,[1] thus showing that the sacrifice was a direct continuance of the primæval totem offerings. It was also this race which introduced the Meriah sacrifices of the Khonds.[2] It was during this separation, that is, when the rule of the world was divided between the northern patriarchal and the southern matriarchal races, that the women gave birth to the Skoya or giants, who were said to eat children, that is, to feed on sacrificed human victims. These cannibal children of the matriarchal women were the offspring of the marriages by capture between the Northern invaders and their alien wives, the Southern women. They were the Hindu Rakshasas, also called Ugra, the Hindu form of the Ogres of Northern mythology, who warred with the Southern agriculturists.

It was the sun-god who put an end to this period of anarchy, and permanently united the patriarchists of the north with the matriarchists of the south, by becoming, by Kochinako, the yellow virgin of the north, daughter of the spider-mother, and mother of the yellow race, the father of the dwarf twins, Ma'asewe and Uyunyewe. They were the dwarf childless and therefore sexless gods, who in all the mythologies of Europe and Asia, became the parent-gods of the yellow artisan races, the measurers, who in Asia Minor called themselves sons of Minos, meaning the measurer, and deified the judge Danu of the Jews and Akkadians, as the brother of Minos, Rhada-Manthus, he who divines by twirling (*math or manth*) the rod. They were, as I have shown, the twins Day and Night, born in Asia Minor on the river Xanthus, meaning the yellow river, in India, on the Yamuna, or river of the twins (*yama*), the children of Saranyu, the mother of corn (*sar*), who became the Zend, Yima, the tiller of the garden (*vara*) of God, the parents of the yellow gardening race, who, in the belief of all Asiatic nations, conquered the forces of destruction and death, by tilling and

---

[1] Prescott, *History of Mexico*, Second Edition, vol. i. p. 70.
[2] *The Ruling Races of Prehistoric Times*, Essay iii., pp. 275-277.

cultivating the land, and measuring time, so as to adapt their industry to the exigencies of each revolving season. In the Sia story, the twins, when they had attained maturity, went to their father in the east, the sun of the deer races by the rainbow bridge called Bifrost in the Edda, built for them by their grandmother, the Spider. Their father put them in the sweating house, as the Greek barley-mother, Demeter, put her nursling, Demophoon, and Thetis, the moon mother, her son Achilles, the young sun-god in the oven and the fire. When the twins were tried by fire and matured, as the seed is in the heated earth, their father put them in a room filled with animals, the totems of the primæval age, which they tamed. He then sent them forth to conquer and civilise the world, giving each of them a bow and arrows, and three ' rabbit sticks,' the three ' paridhis' or sacred twigs, placed in a triangle on the Hindu altar, to symbolise the three seasons of the year of Orion and the moon-hare.[1] This sacred triangle became, when doubled by joining the apices of the two triangles together, the Dorje of Vishnu ⧓ the figure representing the path of the sun from north-east to north-west at the summer, and south-east to south-west at the winter solstice. These three lines of the triangle had also, as I have shown, become to the nations who looked on the year as the succession of the seasons of sowing, growing and ripening, the three prongs of the trident of the growing tree, the parent of Soma, or the divine sap, and this when doubled, became the Assyrian thunderbolt of two tridents united by a common staff.[2]

These six rabbit-sticks were, as we shall see, those of the year of the rabbit, which began each Mexican cycle of four years, thirteen of which make up the great cycle of fifty-two

[1] *The Ruling Races of Prehistoric Times*, Essay iii., pp. 164-168.
[2] Comte Goblet d'Alviella, *The Migration of Symbols* (A. Constable and Co.), fig. 44, p. 97.

years, which opened with the lighting of the national fires in
the culmination of the Pleiades, the stars of the Spider, in
November. Also, the rabbit is, in the Chinese Zodiac, the
sign that corresponds to our Libra; and Chinese astronomy has
a further peculiarity, which marks it as having been evolved
from a time reckoning, which began, like that of the Twins,
with Night and Day, for the signs in their Zodiac proceed in a
retrograde direction contrary to the course of the sun, and
they were first used as signs of a horary Zodiac, marking the
hours of day and night.[1] I have already shown that the
duodecimal measurement of the hours of the day and night
originated in Malabar, and these six time-sticks also belong
to the same duodecimal notation. They represent each a
half year of six months, corresponding to the two periods into
which the Hindus divided their year, beginning with the
winter solstice. The Devayāna or six months, of the
lengthening day, when the sun went from south to north,
and the Pitriyāna, or six months of the decreasing day, after
the sun, at the summer solstice, turned from north to south.

The American Twins, who represented a year of twelve
months, divided into two periods of six months each, the year
which succeeded that of Orion, were led by Ma'asewe, the god
of Devayāna. They first attacked the wolf of the East, their
mother in Greek mythology, killing him, not with their arrows,
but with their sticks, the second and fourth stick thrown by
Uyuyewe, the twin of the autumn sun, or those denoting the
three months that closed each period of six months. The
first of these wounded the wolf, and the last crushed its
stone-armour and killed it. When the victory of the new
year of four seasons, the year of the gardening race, over that
of three, the year of Orion, was achieved, Ma'asewe cut out
its heart, thus imitating the Northern Hadding, who was
also the ruling god of the winter solstice, who ate the wolf's
heart.

---

[1] Prescott, *History of Mexico*, Second Edition, vol. i. pp. 96-106 ; *Encyclo-paedia Britannica*, Ninth Edition, vol. xxiv. 'Zodiac,' p. 793.

They next went to the north to get wood for their arrows, and there they slew the totem cougar, by throwing him off the bridge, the Milky Way, by which he was leading them to the forest of the north, and they then made the North Pole Star, the god Njord of the Edda, the father-god of the worshippers of the right-angled upright cross, the ruler of the new year of four seasons. The pole-star worshipped by the race who introduced the year of four seasons, was probably the star worshipped by the Omahas as the female red-bird.[1] This star of the bird Vega, the star of the Vulture in Lyra, was that worshipped by the Egyptians as Ma'at, the goddess of the unchangeable laws of nature, and it was the pole-star from 10,000 to 8000 B.C. The sanctity attached to the constellation by the early Turanians, who made Danu, the judge, the father-god of the artisan races, is shown by the name of the Weaving Sisters, given by the Chinese to three stars in Lyra, which was just, as I have shown, the constellation of the Vulture.

After the north, the conquering twins went to the west, the land of the bear, and mother-bird, the hawk Kirke and Freya, to get feathers for their arrows and Uyunyewe's arrow, and not his stick, as in the case of the wolf, slew the bear Ma'asewe, cut out its heart, as he had cut out that of the wolf, and kept it.

They had now made their arrows, which were useless when they attacked the wolf, deadly, and had accomplished the task assigned in the Hindu Brāhmaṇas to Vishnu, also worshipped as Kṛishṇa, the black antelope, who, we shall see, was one of the twins' victims. Vishnu, the dwarf-god of time, who grows with the time he reckoned and measured, was also like the American dwarfs a twin god, for he was born in the Yamuna, the river of the twins (*Yama*), and he, like them, made the arrow, called his Vajra, or thunderbolt. This we are expressly told in the Brāhmaṇas was the year of three

---

[1] *Publications of the Bureau of Ethnology*, vol. xi. : Dorsey, 'Study of Siouan Cults,' p. 379.

seasons, symbolised in its point, the fire-god Agni, the slain fire-wolf, the feathers, the Soma or soul of life, born of the mother-bird, the hawk-daughter of the western bear, and the connecting piece (*Kal mala*), the shaft gained from the woods of the North, guarded by the tiger.[1] When the Twins had slain the bear and obtained the control of the year of three seasons, their next task was that assigned in Hindu mythology, to Krishānu, the rainbow-god, who wounded the mother-bird who was bringing Soma to earth, and brought down one of her feathers, which became the Palāsa tree (*Butea frondosa*), the sacred tree of the Brahmins and Mundas.[2] In telling this story, the Sia cosmogony relates how Ma'asewe, who had incorporated the hearts of the eastern wolf and the western bear, and had thus become the ruler of the sun's course from east to west, the gates of heaven, as we shall see in Arabian mythology, killed the young of the mother-eagle of the South, the Soma bird, the Bindo bird of the Gond *Song of Lingal*, which brings the rains of the south-west monsoon at the summer solstice, together with their parents. He thus took the place of conqueror, hitherto assigned to Uyunyewe, the god of the summer solstice, and the Pitriyāna, in the quest of the arrows. This shows how, in the new reckoning of time, the opening of the year was transferred from the summer solstice, beginning the six months sacred to Uyuny-ewe, the time of the Hindu Soma festival instituted by the Twins, to the winter solstice beginning the six months of Ma'asewe. The feather of the slain bird, which became in India the Palāsa tree, is in the Sia story the Piñon nut-tree, planted by the squirrel, which grew up, like Jack's beanstalk, to the eyrie of the eagle of the South, and it was by this tree that the Sia Twins descended to earth as the sons of the squirrel, the name by which all Hindu forest tribes, formed of the amalgamated matriarchal Dravidians and the patri-

---

[1] Eggeling, *Satapatha Brāhmana*, iii. 4. 4, 14-17 ; S.B.E. vol. xxvi. p. 108.

[2] *Ibid.* iii. 3. 4, 10 ; S.B.E. vol. xxvi., p. 78.

archal Mundas and Gonds call themselves.[1] It was when the mother-bird was slain and deposed, that the mother-tree became in her place and in that of the mother-mountain, whose deposition is told of in the next adventure, the parent of the united patriarchal and matriarchal races.

On the return of the Sia Twins to earth, their next task was to go down to the Nadir and destroy the Skoyo, or giant woman of the fire-mountain, by burning her in her own fire. She was the Cyclops, the fire-socket, with one eye, slain by the Æolic Apollo, the wind and thunder-god, called Bronchios the roarer, who succeeded Apollo Lycæus, or the wolf (*lukos*) Apollo, and buried him under the burning mountain.[2] Her Sia slayers took from her abode the bows and arrows of the rain and thunder-god which were, as I have proved a few pages back, the double Dorje, or united triangles of the year-god of the solstitial sun. This sun-god henceforth became, with the pole-star, the ruler of heaven, to the united worshippers of the St. Andrew's cross of the solstitial sun, and of the Latin cross of St. George, the rain-god of the people of Asia Minor and Syria. The union between the worshippers of the pole-star and of the solstitial sun is commemorated in the Egyptian five-rayed

star of Horus ✕ , the son of the hawk-mother, while

the sanctity of St. George's cross is attested, by its being adopted, for ages before Christianity arose, as the sign which was to render futile the witch-craft of the fire-worshipping children of Maga, the mother of the earthly fire, of the totem race, of the wolf-god. The combined solstitial and equinoctial sun henceforth became the sun-god Rā, of the four seasons, the god of the sons of the seed, the Sia corn-clan, symbolised by the eight-rayed star, the god who ordained the perpetual law of the invariable sequence of natural

---

[1] *The Ruling Races of Prehistoric Times*, Essay ii., pp. 48-50.
[2] *Ibid.* Essay iii., pp. 176, 213 ; vi., 504, 505, 515.

phenomena. This sun-god, who was also the rain and thunder-god, who brought rain and heat in their appointed regular seasons, became the Hindu Pururavas, the eastern (*puru*) roarer (*ravas*), the father of the Purus, the city (*pur*) builders of Hindu legendary history, and of all those who lighted their sacred fires from the Hindu earth-altar, dedicated to the sun of the solstices, with the fire-drill of the Soma-worshipping sons of the sun-horse called Pururavas, and the socket named from his wife Urvashi, the ancient (*ur*) creatrix (*vashi*).[1] They were the race who, as I have shown from the Sia story of the sons of the squirrel and the Hindu story of the Palāsa tree, worshipped the Soma tree containing the sap of life, as the child of the rain-bird and fire-mother.

The last task of the twins was the slaying the antelope of the Zenith, the Hindu antelope-star, slain with his own three-knotted arrow, the three seasons of his annual life, the god Marīchi, meaning the fire-spark, of the Rāmāyana, who was slain by Rāma, the plough-god, as a deer, and who went to heaven as one of the tail-stars of the Great Bear, when that constellation, called by the Hindus the seven antelopes, became the ruler of heaven, instead of the single antelope Orion, or, in other words, when the seven days of the fifty-two weeks, into which the lunar-solar year was divided, became the fundamental unit used in measuring time, instead of the single day and night. The Sia Twins were led to the antelope, which lived on the top of the mountain of the Zenith *looking westward*, as we are specially told in the Sia story, by the mole, the counterpart of Tuopi, the badger. Ma'asewe shot him from under the earth through the mole-hole, and thus slew the antelope-sun of the Syrian Semites, descended from Terah, the antelope, who measured time by the west, or setting sun of the autumnal equinox. The mountain where the antelope died was the central mountain

---

[1] Eggeling, *Sat. Brāh.* iii. 4. 1, 22 ; S.B.E. vol. xxvi. p. 91 : *The Ruling Races of Prehistoric Times*, Essay iii., pp. 165-168.

called Hinda-fjall, the deer (*hinda*) mountain, in the Nibelungen Lied, of which I have several times spoken, on which
Sigurd, the rider of the sun-horse, woke Brunhilda, the goddess of the springs (*brunnen*), from her winter sleep.
Ma'asewe, the god of the year of the winter solstice, which
superseded that of the autumnal equinox, as it had previously
done that of the summer solstice, cut out the heart of the
antelope he had slain, and scattered it to the four quarters of
the heavens, which were thus consecrated to the new year.
He decreed that in future, the antelope, that is, that the sons
of the deer, should eat grass, which was to be developed into
their national totem of the corn-seed, and not devour men
and animals, as was the custom in totem-worshipping days.

This story of the antelope is clearly a Sia reproduction of
the Asiatic legends of the antelope-god, the Semite Terah,
the son of Nahor, the river (*nahr*) Euphrates, father of
Abram, the sun-god of the corn-growing races, the Akkadian Dara called Ia, the house (*I*) of the waters (*a*), and the
antelope (*dara*) of the deep, the antelope-father of the Hindu
Brahmins, whose skin every Brahmin must wear at his initiation. It was the deer and the antelope which led the corn-
growing races from the mother-mountain, down the fertile
slopes of the river valleys, and proved them to be well
drained and suited for corn-lands, by feeding there on the
favourite antelope-grass *Poa lynosuroides*, called Kusha, or
grass of the sons of Kush, which will not grow on swampy
wet soil. This grass became that of which they made their
'prastara' or magic rain-wand, and its juice was mixed with
their Soma cups of the seed of life (*Su*), which, with the
sacred bread, were substituted for the animal totems, as the
sacramental meal of the race regenerated by baptism in the
running waters of life, drawn from the parent rivers.[1] It was
these Kushite-Semite sons of the antelope who, as I have
shown, used the constellation Hydra, called by the Akka-

[1] *The Ruling Races of Prehistoric Times*, Part I., Essay iii., pp. 163, 164,
206; Preface, xliv-xlviii.

dians, 'the divine foundation of the prince of the black antelope,' as the ruling star of the second season of the year of four seasons beginning with the summer solstice, which succeeded the year of Orion.[1]

The Sia dwarf-twins, after their victories, went round the earth attending rejoicings to celebrate their success, and found in their travels the honey whence mead, the sacred drink of the Hindu twin Ashvins and of the gods of the Edda, was made from the blood of Kvasir (the leaven) by the Northern dwarfs.[2] This was the age of the intoxicated priests and prophets who sought inspiration in strong drink, the age, as I shall show presently, preceding that of the Zend reformer Zarathustra, when Bhang, the Persian Hashish, was substituted by the sons of the fish-god for strong drink, as the giver of the seeing-eye, which enables the inspired teacher to see into ' the life of things,' and when the drinking of intoxicants was forbidden as a degrading habit. The twins gave their sacred honey-drink, the Hindu Madhu, to the two families who received them hospitably—one among the Oraibi, and one in the Kat'suna or totem village. They told them to leave their homes and take refuge in a neighbouring round-house, the round houses which, in the Bronze Age, replaced the long houses of the Neolithic stone period. Those who had refused hospitality to the twins were turned into stone, and their spirits went up to heaven as that of the Piñonero, the Canada jay, hated by all hunting Indians. This was the bird of the Piñon-tree, down which the twins had come from the eagle's nest.

It was after the destruction of the race who worshipped animal totems, that the Sia twins again went up the rainbow-bridge to their father, the sun, who placed them in the Sandia mountain, the sacred central mountain of the next race, the Kushites, gathered round the mother-mountain of

---

[1] *The Ruling Races of Prehistoric Times*, Essay iv., p. 370 ; see the whole dissertation on these two years, pp. 357-372.

[2] Mallet, *Northern Antiquities*, ' Prose Edda,' Part II. 3, 4, pp. 461-463.

274 THE RULING RACES OF PREHISTORIC TIMES

the East, who formed the association called by the Sia those
of the ants. Their seven societies venerate the number seven,
the fundamental number used in reckoning the year of weeks
as more sacred than the six of the twin races, who used the
duodecimal system of the Malabar sons of the buffalo,
and the sacred seats forming a degree of the circle, in Babylonian notation.[1] Only the last four of the ant societies
learn the songs for rain, taught by the gods of the year of
four seasons, and thus these seven societies seem to be connected with the seven Lumasi of the Akkadians, the seven
stars, which, as I have shown, mark the years of three and
four seasons.[2] These ants were the Vamsa of the Rigveda,
'who broke down the walls in which cloud-demons imprisoned
the light,'[3] or who, in other words, disentangled the history
of the year, and the Myrmidons of Greece, sons of Zeus, the
father-ant, and of Peleus, born of the fine potter's clay ($\pi\eta\lambda\delta s$)
of the race of artistic potters, to which the Sia belong. They
were the conquering race, who, as we shall see presently when
dealing with Arabian mythology, overcame the ape-gods, who
ruled the winds, and drove the stars round the pole in the
early Hindu cosmogonies, and who led Jan-shah, the lord
(*shah*) of life (*jan*), to the kingdom of the Semites, whose
sacred river dries up every seven days.[4] It was they who, in
the final form of the Kushite political system, substituted
for the six kingdoms indicated by the six-rayed star of the
Cypriotes and Hittites, a kingdom of seven united states, with
the seventh province assigned to the king in the centre, and
the six kingdoms allotted to his most trusted subordinates,
round it. These are the seven kingdoms of India of which
the centre is Jambu-dvipa (Central India), and the seven
kingdoms of Iran, of which the centre is Khorasan, the
heritage of the Arabian and Persian Janshah, the king of life,

[1] *The Ruling Races of Prehistoric Times*, Essay iv., p. 383.
[2] *Ibid.* Essay iv., pp. 355-364.   [3] Rigveda, i. 151, 9.
[4] Burton, *Arabian Nights* (Nichols and Co.), vol. iv., 'Story of Janshah,'
pp. 282-285.

who was son of the king of Kābul, the birth-place of the Kuṣhite
race, ruling the seven kingdoms of India, and of the daughter
of the king of the seven kingdoms of Khorasan.[1] It was these
united fourteen kingdoms which represented, in mythologic
ritual, the dawning lunar-solar year to be measured by the
twenty-six annual phases of fourteen days each, making the
year of fifty-two weeks, and three hundred and sixty-four
days, the year of the fish sun-god. It was he who ruled the
central corn-land, the home of the teeming population of
agricultural ants, who, according to the reminiscence of the
ant-age recorded by Herodotus, were said to dig up gold in
India.[2] This ancestral corn-land was that known to the Sia
as the land of Poshai-yänne, the sun-god or sun-fish, who met
his death by being stabbed after his marriage, like the sun-
god Sigurd, with a spear, after which he was thrown into a
lake, the constellation Aquarius, wherein the fish-sun died, and
whence he was born again as the sun-eagle of the new year.[3]
Whether the name Poshai-yänne reproduces the Chinese
Fo-sho, the form to which they have changed the Indian
Buddha, or not, there is no doubt whatsoever that the Sia
Poshai-yänne exactly corresponds with the all-knowing fish-
god of India, whose characteristics were transferred, as I
have shown in Essays IV. and VII.,[4] to the historical Buddha,
the great religious teacher of the sixth century before
Christ. He was, as I have proved, only the last successor of
a long line of previous Buddhas, or knowing ones. The first
of these was Sumedha, the sacrifice (medha) of Su, that is,
the living offspring of the Su, or sap of life, the life-reviving
rain which feeds the rivers, and is distributed by them to the
receptacle of the regenerating sap, the tree or plant of life,
the mother of the human race. It is she who gives life to

---

[1] Burton, *Arabian Nights* (Nichols and Co.), vol. iv. pp. 274-277; *The Ruling Races of Prehistoric Times*, Essay iii., pp. 145, 146, 253.

[2] Herod. iii. 102, 103.

[3] *Publications of the Bureau of Ethnology*, Stevenson's 'The Sia,' pp. 66; *The Ruling Races of Prehistoric Times*, Essay iv., pp. 376-387.

[4] *Ibid.* Essay iv., pp. 395-399.

the grain, the seed, the son of the god of life, and through him secures the continuance of human life on earth. The visible sign of this soul of life, residing in the running water whence the sap was first-born, was the fish, the eel, which I shall presently deal with more fully, and who showed by his migrations that he knew the secrets of the changes of the the seasons, and thus became the symbol of the Il-ja,[1] or El, ultimate source of all life on earth, the all-wise god who gave creating and regenerating power to the waters of the rivers, whence it was distributed through the universe. Hence the fish-god, whose heavenly messenger and prophet was the fish-descended sun rising from the waters of the ocean, became the great Buddha, that is, he who knows (budh) the teacher of the world. Each of these Buddhas was born according to the legend of the mother Māyā, meaning in Sanskrit, supernatural wisdom, the goddess Māga, the magic-mother of the Magi, under the national Sal-tree, the tree of life before the life-giving grain, and reputed to issue from the right side of his mother, the side turned by the sun towards the points of the solstitial cross in its annual circuit. He was thus the god Sāl-manu, the fish-god, who knew all sciences, and brought wisdom to mankind, the fish-god, god Ia of the Akkadians, having, as I shall show directly, the same sacred number forty. He was, in the Sia story, born of a virgin-mother, made pregnant by eating two Piñon nuts, the fruit of the tree planted by the squirrel, down which the Twins had come from the eagle's nest, and which gave its name to the jay, into which the Kat'suna, or men with masks, had been transformed. This story is almost exactly the same as the Hindu legend in the Mahābhārata, telling how Jārasandha, the first king of the united Kushikas and Māghadas, was born as the child of the two queens of the Māghada king, son of the mother, Māga, each of whom, when

---

[1] *The Ruling Races of Prehistoric Times*, Part I. Preface, pp. xlii, xliii, where I show that the Finnish Il-ja, meaning *eel*, is derived from the goddess-mother of the three seasons, Iṛu, Ilā, Iṛā, or Iḍā.

made pregnant by eating a mango-stone given them by the moon-god (*Chandra*), bore half a child, the parts being miraculously united by an old woman called Jārā, one of the Rākshasas, the Sia Skoya, meaning old age, or lapse of time. The general of Jārā-sandha was Hansa, or Kansa, the goose (*khans*), the son of the doe-antelope, who, as we have seen, preceded Poshai-yänne, called Kalānemi, and of Ugra-sena, the ogre, or Skoya, who took Mathura, the sacred city, and ruled it, when Krishṇa, the young sun-god, who became Vishnu, the fish, was born.[1] This story, as told by the Sia and the Hindus, shows how the moon-goose, who was the nurse of the fish-sun, became Prajā-pati, the moon-nurse of the Buddha, another form of the fish-sun, ruler of the gardening race, born of the fruit-tree, and succeeded the race descended from animal totems. And this coincidence is again confirmed by the Buddhist history, which tells how, when Kaṣhyapa, the tortoise-king, father of the two Māgadha queens, was ruler of the earth, his capital, Kaṣhi (Benares), where the two queens, mothers of the ruling god of the gardening race, were born, was ruled by the Kiki, the blue jay, answering to the Sia Piñonero, or Canada jay.

Poshai-yänne was first the boy who lighted cigarettes for the Tiämoni, who answered, as I have shown, to the Patesi or priest-king of the Akkadians, and of all the people of South-western Asia, and for his advising priests. In other words, he ripened their parent-grain, as the young sun-god. He excited their envy by wearing successively bracelets of turquoise, flat shells, red and black stones, emblems both of the sun and moon god, and to win these the Tiämoni staked successively the four regions of the North, West, South, and East. Poshai-yänne won the North at the game of forty pebbles, the number forty being sacred to Ia, the Akkadian fish and antelope god : the West at the game of six cubes, the dice thrown by the Pāṇḍavas, the sun-princes, reputed children of the sexless sun-god, Pāṇḍu, who, like the Sia

---

[1] *The Ruling Races of Prehistoric Times*, Essay v., pp. 462, 463.

twins, killed the sun-deer.[1] He won the South and East at
the game of four sticks and four mounds of sand, in which
the winner had to guess correctly the stick or mound in which
the pebble or stone of life was placed, a reminiscence of the
mother-tree and the mother-mountain. He then similarly
won the Zenith and Nadir in exchange for bracelets of large
turquoise, and of white and pink beads, and the victory made
the sun-god, who was first the god Rai of the Wends, the
god Rāhu and Rāma of the Hindus, Rā of the Egyptians,
and who became the Hindu god Vishnu, and Sal-manu, the
fish-god, ruler of the land.

The land which he ruled was that of the four mounds or
mother-mountains of the pastoral races, and the four sticks,
the forest-clad river valleys of the corn-growing sons of the
rivers, and of the growing plant or tree. This is the land
depicted in Essay III.,[2] in the diagram of the countries of
Irān and India, the two united seven kingdoms watered by
the Ganges, Jumna, Indus, Oxus, and Euphrates. The rivers
in this primæval geography represented, as shown in the
diagram, the four quarters North, East, South, and West,
the quarters won by Poshai-yänne, the sun-god of the sons
of the rivers, while the remaining intermediate quarters, representing the other four sides of the land of the eight-rayed
star, the divine seed, were assigned to the Gond Turanians
of the North-east, the worshippers of the pole-star, the
Māgadhas fire and sun-worshippers of the South-east, the
Dravidas or tree-worshippers of the South-west, and the
Irānian sons of the sheep and cow of the North-west. This
eight-sided figure forms the figure of the tortoise-earth
resting on the primæval ocean, as it is depicted by Varāhamihira.[3] It is reproduced in the eight-sided sacrificial stake
of the Hindus, the altar of god on which the sacrifice of man,
said in the Brāhmaṇas to be the true sacrifice, is continually

[1] *The Ruling Races of Prehistoric Times*, Essay iii., 262, iv. 343, 367, 368.
[2] *Ibid.* Essay iii., p. 220.
[3] Sachau's Alberuni's *India*, chap. xxix., vol. i. pp. 295-303.

offered up to the gods. This sacrifice of man is, as I have
shown, the Soma sacrifice in which the sacrificer cleanses
himself from sin in the baptismal bath of regeneration.
This he entered clothed in the Jarāyu or afterbirth, the skin
of the black antelope, and emerged as the child of god, born
of the life-reviving water, and thus became the twice-born
son of the cow-race, and of the golden-horned bull of heaven,
the sun, showing his new parentage by living entirely on
milk, called Vratā, or fast-milk, during the performance of
the sacrifice.[1] The eight tribes dwelling in these eight
divisions of the united kingdoms, round which the fish-hill
sun revolved during the year,[2] were a reproduction of the
Gond division of the nation into the four immigrant or
superior tribes who became, in the geography of the tortoise-
earth, the four races of the triangles of the north-east,
south-east, south-west, and north-west, the Greek cross
consecrated to the solstitial sun rising and setting in these
quarters, and the four aboriginal or inferior races of the
river-triangles, the Maltese cross formed from the Latin
cross of St. George consecrated to the pole-star and the equinoctial sun. This conception of the mountain-land, the
eight-sided dwelling-place of the sun-god parent of the sons
of the seed was not one which belonged exclusively to its
Hindu authors, but one which was diffused by them through
the whole world, and which formed the foundations of the
Semitic belief in the all-wise god, the fish-born creator of
life. It is symbolically reproduced in the octagonal dome of
the Rock of Jerusalem, and the sacred stone of the convent-
hall of the Bektāshi dervishes, which has eight sides, with a
candle or eye in its centre,[3] thus showing that it denotes the

---

[1] *The Ruling Races of Prehistoric Times*, Essay iii., p. 221 ; iv. p. 367 ;
Preface, p. xlv.

[2] See the diagram of the cow-fish of the Hittites, the goddess Ashteroth,
the moon-cow united with the father-fish (*The Ruling Races of Prehistoric
Times*, Essay iii., p. 172).

[3] J. O'Neill, *The Night of the Gods* (Bernard Quaritch), 'The Cardinal
Points,' pp. 167-169 ; 'The Bethels,' p. 12S.

sun shining on the eight quarters of the world known to the maritime traders of the East. It is still more distinctly depicted in the Japanese enclosure of the deity Haya-Susa-no Wo-no-Mikoto in the place of Suga, the reed-parents of the sons of the rivers, in the holy (*idzu*) land of Idzumo, the holy one. This home of the sun-god of the sons of the reeds is surrounded with 'the eight-sided, holy quarter's fence,' and these eight sides or regions are, in Chinese mythology, the eight-sided 'fang' on the back of the divine tortoise. This land, again, is, in Japanese mythology, as in that of the Sia, the island-home of the creating twins sent by the Kami (upper) gods to 'make, consolidate, and give birth to' the land of Japan. Their names are Izanagi and Izanami, the soliciting (*izanau*) male (*gi*) and female (*mi*), and thus they, like the Greek Apollo and Artemis, are of different genders. They created the land by churning the ocean with the divine spear, the magic pole of the American Indians, as the Hindu gods churned it with Mount Mandara, the first conception being derived from the creating fire-drill, the second from the creating mountain which collects from the ocean the rain which is to water the earth and make it fruitful. It was from the curdled ocean-foam, dropped from the spear of the Japanese twins, that the island-home of life, Onogoro, meaning the 'self-curdled,' was formed round the north pole. It was on this island that they raised the eight-sided Suga, or reed-palace of the sun-god, on one pillar supporting the roof, the rotating pole, and fenced in with the fence of eight clouds, and it was in this palace that the creating high (*taka*) god, Takehaya-Susa, placed the first man, Ashi-Nadzu Chi, the father-reed (*ashi*) stroker. This churning with the spear, the father of the father-reed, is exactly similar to the symbolical churning in the Sia rain-ritual, where the priest stirs the holy water in the cloud-bowl, into which To'chanitiwa, used to produce suds, has been dropped with the sacred reed, the Baresma, or rain-reed (*bares*) of the Zendavesta, 'long as a ploughshare, and thick as a barleycorn,' till suds, the curdled ocean-foam of the Japanese story, have been produced. This is sprinkled on the altar and all who take part in the festival, beginning with the women and ending with the white bear consecrated to the sun of the North and West, and the parrot, who is placed on the top of the altar as the mother-bird of the forest-races.[1]

It was the creating twins, a brother and sister, like the Japanese, and not two brothers, as the Sia twins, who, according to the Peruvian myth told by Garcilasso de la Vega, were sent down by their father to the marsh-reeds, the birthplace of the sons of the rivers, near Titicaca, and he gave them a golden rod, the magic pole which was to show them, by entering the ground at one push, where they were to establish the court of the Inca children of the sun. It was these twins who were further, according to Japanese theology, the Passive and Active Essences, the fire-drill and the fire-socket, who were developed from the three supreme Kami, or heavenly gods who dwell in the pole-star, the three weaving sisters of the Chinese in Lyra, and who are called Ame-no-Minaka Nushi, lord of the awful centre of heaven, Taka Mi-Musubi, and Kami Mi-Musubi, the ineffably-begotten height (*taka*), and the ineffably-begotten Kami, or upper god.[2] These stories, when translated from their mythical form, and compared with the conceptions of the twin fire-sticks, the twin-children of Saranyu, Dawn and Night, whose father was Vivasvat, the god of light, with two (*vi*) forms (*vas*), and the children of the wolf-mother, Leto, Apollo, the

[1] *Reports of the Bureau of Ethnology*, vol. xi. Stevenson, 'The Sia,' pp. 82, 83.

[2] J. O'Neill, *The Night of the Gods*, vol. i. 'Axis Myths,' pp. 31, 32, 35, 37, 62, 63, 169, 224. This island-myth is also, as Mr. O'Neill points out, reproduced in the Greek myth of the birth of the island of Delos, where the second avatar, or birth of Apollo and Artemis, took place. It was the centre of the Cyclades, so called from the κύκλος, the circle or wheel. It was raised by Poseidon by his trident, symbolising the growing plant, from the ocean, and was called Ortygia, the island of the quail (ὄρτυξ), which is shown by its Sanskrit form Vartika to be derived from *vart*, 'to turn,' as the quail by his migrations shows the turning of the year.

god of day, and Artemis, the goddess of night, tell how the
year of the three seasons, the three Kami of the Japanese, the
three stars in the belt of Orion, who, in his moon-boat, dragged
the stars, her attendants, round the rotating pole, as the three
weaving sisters of the Chinese in the polar constellation turned
it, and made the fire-drill of heaven, only visible in the pole-
star, revolve in the course of the day and night. It was they
who marked the passage of time, which creates all things by
slow but sure processes, and who raised from the mother-ocean
the eight-sided tortoise-mountain, ruled by the sexless twins,
who created it. Their offspring are the sons of the seed, born
from the sun-god who ripened the grain, and their dwelling-
place is the tent or mountain of heaven, supported by the invisible magic-pole which grinds out the heat and rain, which
creates and maintains life on earth. This tent is again symbolised in the cottage supported on the leg of the chicken, the
cock of the rising sun, the Chaska cock of Ut'set, in which
the mother-witch Baba Yaga sits in Russian fairy tales.[1]

It was these twins who built the heaven palace of the
pole-star, the eight-sided island altar of god, the home of
the eight tribes, sons of the mountains and the rivers, who
regulated the computation of time, and who changed the
year from one of three seasons to one of four, who inspired
the divine prophets, the leaders of the people and the
founders of the national ritual. It was they who, according
to the Brāhmaṇas, established the Soma sacrifice of the sons
of the sun-horse and the moon-cow. But before proceeding
further with this question, I must first point out that this
widely distributed myth of the generation of the tortoise
island earth from the sea is one which could only have
emanated from a maritime people accustomed to long coast-
ing voyages. These led them to believe that the whole
inhabited earth was bounded by the sea, and this is the
conception of the Hindu story of Manu and the birth of his

[1] J. O'Neill, *The Night of the Gods*, 'Axis Myths,' vol. i. p. 225; Ralston,
Russian Fairy Tales, pp. 74, 75, 76, 144.

daughter and wife, Ida, the mother-mountain, for he was led
to the centre of the earth, where she was to be born by the
horned fish, the mother-dolphin, the Ma or mother-ship,
which carried Ia, the fish-god, from India to the Persian
Gulf, and who in astronomical mythology became the constellation of the ship Argo. These maritime races were
originally the sons of the Indian rivers, and those of
Asia Minor. It was among the forests on their banks,
which supplied boat-building materials, that they learnt
how to make rafts first, and river and sea-boats afterwards, and gradually acquired the arts of propelling their
creations by poles, oars, and sails, and it was they who
first adored running water as the receptacle of the soul of
life. This had been, in the earlier mythology, brought with
the showers of spring by the mother-bird, the stork, who
was in the belief of the sons of the rivers re-born in the eel,
which became their totem parent during the age when the
mother-stork and storm-bird were superseded by the mother-
hawk of the mining races. It was from the mother-
hawk, the bird with the piercing eye-sight, and the eel, that
the ruling, metal-working, farming, and pastoral tribes of
Central India, the Kharias, Kharwars, Mundas, Rautias,
Lohars, Gualas, Santals, and Ooraons trace their descent, and
it is their eel-parent-god Aind, Indu, or Induar, who has
become Indra, the rain-god of the Rigveda, who succeeded
the earlier Sukra, the wet (*suk*) god, whose name, Ush-ana,
proves the storm-bird to have been originally Uk-ko, the
Ugro-Finnic storm-bird of the American Indian races.[1] The
close similarity between the original names of the bird Khu or
Khur and the fish Kha or Khar proves their common origin,
while their adoption by the Akkadians and Egyptians as the
sacred Khu and Kha proves that the Finnish totems of the
bird and the fish were adopted as parent-gods throughout
the whole of South-western Asia. But we have seen in the

---

[1] *The Ruling Races of Prehistoric Times*, Preface xli. xlii.; Essay iii., pp.
146-148.

Japanese myths that the twin-parents of the sons of the
mother-hawk and the mother-fish were the three Kami, the
year of three seasons, and a similar conclusion is to be
deduced from the mythology of the Finnish eel. This name,
in Icelandic āll, in German aal, is in Finnish Il-ja, born (*ja*)
of Il. Il, again, was the Basque Iru, the three, the year of
three seasons, the mother-mountain Iḍā, Iṛā, or Iḷā, forged by
the eternal forger, Il-marinen, the stars of the Great Bear,
the mother-stars of the race who made the Great Bear, the
seven antelopes of the Hindus, the seven bulls (*haptoiriñgas*)
of the Zends, the guardian of the pole-star and the revolving
pole, and divided time into periods of seven days. Il-marinen
was the second god of the triad in which Väinä-möinen, the
god of moisture, was the first, and Uk-ko, the storm-god and
storm-bird parent of the Esthonian mother-goddess Linda,
the bird, was third. The triad, in short, was the air sun-god,
who sent down the generating dew brought by the third
member, the bird of spring, and from the sun-born dew the
sacred eel was born according to the primæval mythological
creed of the sons of the rivers. It was the eel which told no
less surely than its mother, the dew bird of the annual
changes of the seasons, for it was in autumn, when the birds
migrated and went to their winter haunts, that the old eels
left the rivers to go down to the sea, and in the spring that
the young eels came up the river with the spring-stork. It
was the eel born of the dewy mist, the casket holding the
germ of life and the bird of the mother-mountain, the harbinger of spring, who was in Egyptian mythology, Nunet,
the vulture wife of the fish-god, Nun, who became the
Elohim of Semite theology. Their progeny were the
antelope sons of the deer-sun of the East, rising behind the
mother-mountain, and of the mother-dolphin, which led the
Grecian priests from Crete to Delphi.[1] She was the Hindu
Makara, who was first the alligator Mugger, the storm-
cloud and the dragon-ring of stars, surrounding and guarding

---

[1] *The Ruling Races of Prehistoric Times*, Essay vi., pp. 517, 518.

the pole, as the constellation Draco, called in the Rigveda
Simshumāra, which took the Gonds over the waters of the
flood to the tortoise earth. This name, Simshumāra, was
afterwards transferred to Capricornus, the goat-fish, also
called Makaram, the Alligator, the Makkar of the Babylonians, in which were situated Al-gedi and Deneb Al-
gedi, two of the stars called the Ten Kings of Babylon,
marking the sun's course from February to November in the
lunar-solar year of thirteen lunar months.[1] It was the constellation of the Goat-fish, and that of the Bull (vrishabha),
Taurus, containing Rohinī (Aldebaran), the mother-cow,
which is said in the Rigveda to draw the chariot of the
Ashvins, the twins ruling the year,[2] and they were in the
year of the fish-sun of thirteen lunar months, the constellations marking the autumnal and vernal equinoxes. This
Makara, the alligator, dolphin, and goat-fish, became in his
final avatar, the fish-father-sun, Pradyumna, the foremost
(pra) bright one (dyumna), the son of Krishṇa, the black,
antelope sun, and Rukmini, the moon-goddess, who established
the year of the fish-sun nursed by the moon, which I have
described in Essay IV. pp. 374-387. This year was in Hindu
history that of the Pāṇḍavas, who ended their thirteen
months of years of exile by conquering the Kaurāvyas, sons
of the mother-bird, who had gambled with them as the
Tiämoni did with Poshai-yänne. It was these sons of the
fish-god who came to Mexico, bringing with them the computation of time, on which their cycle of the rabbit of fifty-
two or four times thirteen years was founded. It was the
year of Kadrū, the thirteenth wife of Kashyapa, the tree
(dru) of Ka, and Kashyapa was the father of the two Māghada queens, who ate the mango stone whence Jāra-sandha
was born. It is this year of the sons of the nut who
worshipped the fish-god Yah, which was that represented by

[1] The Ruling Races of Prehistoric Times, Essay iv., p. 384.
[2] Rigveda, i. 116. 18; The Ruling Races of Prehistoric Times, Essay iii.,
222-224, 250, 258, 259, 268, 269, 284, 286.

the thirteen children of Jacob, the gnomon-stone set up at Luz, the place of the almond tree or stone (*luz*), the Hebrew counterpart of the Hindu mango-stone and the Sia piñon nuts, the mothers of Poshai-yänne. It was the sons of the sun-fish and their predecessors, the sons of the sun-horse, who instituted the elaborate form of the Hindu Soma sacrifice, with the baptismal consecration of the sacrificer, and it was they who substituted for the sacrificial feasts on the flesh of totem ancestors, the sacramental water and barley cup given at the Hindu Soma festivals and the Eleusinian mysteries, and to this was added, by the sons of the fish-god, the fish eaten both at the early Christian sacrament and on fast-days and Fridays, sacred to the mother-goddess of seed (*Fri*), with the sacred bread, and this holy fish is still the food of those Hindu tribes who refuse to eat flesh-meat.[1]

It is this Hindu festival to Soma, called Varshābhu or the rain (*varsha*) plant, which is reproduced in the Sia festival to the rain-god of the snake society, which is held, not, like the Hindu festival, in the summer, but at the beginning of the new year at the winter solstice. Its ritual has been most carefully described by Mrs. Stevenson, who attended the festival as an eye-witness,[2] and photographed its ceremonial as well as all the ritualistic apparatus.

The snakes worshipped are the snakes of the six quarters of the world, formed by the six-rayed star of the twin races, the Nāga or hooded snakes of the Hindu Nagas whose image is painted on the Sia medicine-bowl.[3] These six gods, and the five of the Histian or knife society of the totem races, sons of the cougar of the North and the bear of the West, who founded the sacrifice of slaying with the knife,[4] form the

---

[1] *The Ruling Races of Prehistoric Times*, Essay iv., pp. 374-376.

[2] *Publications of the Bureau of Ethnology*, Stevenson, ' The Sia,' pp. 76-86.

[3] *Ibid.* Stevenson, ' The Sia,' Plates xv. xvi. pp. 82-84.

[4] Note the Hindu distinction in the Brāhmaṇas between (1) the primæval races who killed the victim by a blow on the frontal bone, the men of the club ; (2) those who stabbed it with the knife behind the ear, the barley-cutting fathers, the age of the Sia Histian ; and (3) those who killed it, in the way

eleven gods of generation of the Hindu twin races, the gods
of the five seasons of the year to whom animal victims alone
were offered, and those of the six seasons of the year who
received offerings of the fruits of the earth when the animal
victims were offered to the totem-gods.[1] These people of
the knife society are the Sia form of those called by the
Dakota, or joined tribes, the Isanti or Santi, from Issan, a
stone knife. They are divided into the Wahpeton, or men
of leaves, the woodland sons of the cougar or tiger, and the
Sisseton, or men of the prairie, the sons of the bear, while
the whole confederacy of the Dakotas, including the Santi
and other tribes, are called the Sioux, a corruption by the
trappers of Nadowessi, meaning the rattlesnake people.[2]

It was the Gond sons of the tiger and the Pleiades in
India and the sons of the sun-antelope who worshipped the
Great Bear as the seven antelopes, and who counted eleven
gods of generation, who introduced the sacramental drinking
of the Soma or sap of life. I have already shown how this
cup was, in India, filled with the sap of the parent fig-tree,
Kusha-grass, barley, and the water from a running stream,
and, in Greece, with water, barley, and mint, and this totem
cup of the barley-born races is almost exactly reproduced in
the Sia sacramental bowl filled with water on which corn-
meal and corn-pollen are sprinkled. This is consecrated by
the six stone pebbles, the parents of fire, the sacred number
of the twin races, dropped into the medicine-bowl by the
Yanit-siwittänñi while the eight consecrating stanzas are
sung. These answer to the eleven Samidheni, or kindling
verses of the Āpri hymns of the Rigveda and of the Brāhmaṇas,[3] sung at the animal sacrifices preceding the Soma offer-

thought to be orthodox when the ritual of the Brāhmaṇas was framed, by
choking it, by keeping its mouth closed or strangling it with a noose.
Eggeling, *Sat. Brāh.* iii. S. 1, 15, ; S.B.E. vol. xxvi. pp. 189, 190.

[1] *The Ruling Races of Prehistoric Times*, Essay iii., pp. 265-267.
[2] *Publications of the Bureau of Ethnology*, vol. x.: Mallery, ' Picture Writing of the American Indians, p. 272.
[3] Eggeling, *Sat. Brāh.* i. 4. 1, 7 ff; S.B.E. vol. xii. p. 102, note 2.

ing. In the Sia ritual, the sacred water is twice given to the worshippers, beginning with the women, in an abalone shell, first, after the fourteen gourds filled with the sacred water have been, during the singing of the fifth stanza, poured into the medicine-bowl by the Yanit-siwittänni, and before the corn-meal and pollen have been sprinkled on it while the sixth stanza is being sung; next, after the six pebbles have been dropped in, during the seventh stanza, and after the suds already described have been produced by the reed and sprinkled on the audience, and the sacred dances have been danced by the men and women dancers. In this ritual the use of the abalone shell, like the conch shell of the Hindu Vishnu, shows that it was framed by a maritime people, and the confession of sins which preceded the Greek Eleusinian mysteries, and which does not appear in Mrs. Stevenson's account of the Sia, is shown by Prescott to be part of the Mexican ritual.[1]

The sprinkling of the corn-pollen, or the life-giving water, shows that the ritual dates from the age when the worship of the date-palm, the tree of life of the Babylonians and the trading Shus of Western India, the tree which only fruits when the flower of the female tree is fertilised with pollen from the male tree, succeeded that of the earlier Syrian and Hindu fig-tree. This mother-tree was, in India, that of the maritime traders, worshippers of Balaram, son of Rohinī, the red cow, the star Aldebaran, whose cognizance was the date-palm, and it was the mother-star Aldebaran which was worshipped by the Sabæans of Southern Arabia, while the date-palm was the parent-tree of the Zend neophytes who made their Kösti, or sacred girdle, of the fibres of its leaves. This was the race which succeeded the Khati or Hittites, the rulers of India, who worshipped the ass as their totem, for Balaram is represented as slaying the ass, and thus making the red-cow-star the national guardian-star. These new rulers, the sons of the sun-horse, called also the Ikshvāku, or

---

[1] Prescott, *History of Mexico*, Second Edition, vol. i. chap. iii. pp. 56, 57.

sons of the sugar-cane (*iksha*),[1] were the Yadavas, who
succeeded the Turvasu, or people whose god (*vasu*) is the
revolving-pole (*tur*), and whose bright god (*deva*), was Ya,
that is, the fish-sun, the Ia of the Akkadians and the Yahveh
of the Jews. It was his followers, the sons of the seed, the
eight-rayed star, and of the palm-tree, who became rulers of
the maritime commerce of the Indian Ocean and the Mediterranean, which had been founded by the twin-races the
Tur-vasu, who, as I have shown, derived their system of
measuring the twenty-four hours of the twins Day and Night,
from the people of the Malabar coast who divided them into
sixty Naligas of twenty-four minutes each. They were the
people called Tursena and Tyrrhenians in the Mediterannean,
and Tursha by the Egyptians, who introduced the Dravidian
custom of common meals throughout all the countries on the
Mediterannean coasts, including Carthage.[2] It was these
people who diffused the worship of the barley and the
mother-sheaf, a custom preserved among the Sia in their
Ya'ya, or image of Ut'set, worshipped at the rain-festival
of the Sia snakes, which she left behind her when she went
away from the earth, before the birth of the twins. The
Ya'ya were made of an ear of Indian corn, crowned with
eagle's and parrot's feathers, and placed in a basket woven
with cotton-wool. They were renewed every four years, the
term of the national Mexican cycle, and the seeds of the ears
were sown as the sacred national crop. This symbol proves
that the corn-growing races were sons of the parrot of the
forests, and of the mountain-eagle ; and the cotton is an
Indian plant, as the parrot is an Indian bird. The name
Ya'ya, or I'arriko, is similar to that of the Hindu Yayati,
father of the twin sons of Devayāni Yadu and Turvasu, and
the name of Yadu, the holy (*du* or *duga*) Ya, is certainly
that of the fish-god Ia.

---

[1] *The Ruling Races of Prehistoric Times*, Essay v., pp. 446, 466 ; Essay iv.,
pp. 404, 405.

[2] *Ibid.* Essay iii., pp. 293, 297, 298 ; Arist. *Polit.* ii. p. 11 ; vi. p. 5.

The order in which the Sia place the six Ya'ya is also most significant. Three, the three Hindu gods of the three seasons, who are especially worshipped as the Upasads at the Soma sacrifice, are placed on the altar, two in front to represent the twins, who are always coupled in Hindu ritual, while the sixth, the tiger-god Pandhāri of the Gonds, is placed on the tail of the cougar or tiger, painted on the ground, in front of the other five. This sixth god was, towards the end of the Sia rain-ritual, taken up and put in the place of the first parent Ya'ya, the midmost of the three, the central prong of the Gond trident-god Pharsi-pen. This substitution of the young for the old god is a Sia reproduction of the Hindu change of belief, which made the sexless sun-god Pandu, whose children's fathers were Dharma, the god of law, Vayū, the wind, Indra, the rain-god, and the Ashvins or heavenly twins, the father-god of the regenerated race, born of the Pandavas. Besides these noteworthy resemblances to Hindu ritual, which I have here adduced, there is also, in the Sia ritual, a reproduction of the Syrian worship of the Latin cross of St. George, acknowledged as a sacred symbol throughout the ancient world, placed on the breasts of Buddha and Apollo, the sun-gods, used as the ideogram of Anu, the sky-god, in cuneiform writing, and also as the Egyptian hieroglyph for Ptah, the sun-god of the western or dying sun.[1] The Sia cross was crowned with eagle plumes, tied with white cotton cord, and is shown to have been an indigenous symbol, by its use in Mexico, as the sign of the god Tlaloc, or Quetzacoatl, the bearded white god, whose name means 'the feathered twin, or the serpent,' who was worshipped by the Toltecs as their great civiliser,[2] and also by the stone cross, surmounted by the bird found at Palenque. It was guarded by a wolf and a bear with a cub, the animals of the East and West, all of white stone, shewing it to be a

---

[1] Comte Goblet d'Alviella, *The Migration of Symbols*, English Edition (A. Constable and Co.), pp. 12-14.

[2] Prescott, *History of Mexico*, Second Edition, chap. iii. vol. i., p. 50 note.

symbol of the equinoctial sun-god. It was placed in front
of the three dwarf images of Koshairi, the parent flute-player,
the year of three seasons, standing behind a shell and guarded
by two wands tipped with plumes of the turkey mother-bird,
and represented the year of four seasons introduced by the
Twins.

The above resemblances between Asiatic, European, and
American-Indian historical myths and ritual, which might
be largely added to, prove most conclusively, as Prescott has
already pointed out,[1] that the American-Indians brought
with them to America, national traditions and rites, which had
first originated in Asia and Europe; that the great national
emigration took place, after the establishment of maritime
commerce in the Indian Ocean, while the Sia ritual proves
that the immigrants from whom they traced their descent
had, before their departure from Asia, celebrated a festival to
the rain-god, very similar in its details to the Soma sacrifice
of India, that they worshipped the mother corn-plant, and
used the fertilising sacred pollen of the Hindu and Babylonian worshippers of the date-palm.

I have shown in Essay III. pp. 323 ff., that the doctrines of
the Hindu Soma worshippers paved the way to the rise of
Jain asceticism and self-torture; and that similar doctrines
were also brought to America, is proved by the ablutions,
penances, vigils, fastings, and flagellations of the Mexican
priests.[2] That this resemblance to Hindu self-torture
extended also to the ceremonies of the nomad American
Indians of the North, is shown in the rites of the Dakota
buffalo sun-dance. This is a much more elaborate ceremony
than the dance, 'instituted by women,' of the Hidatsa, Mandans, and Winnebagas, which I have already described, and
from which the Dakota dance is descended. It is held some
time near the summer solstice, and the first ceremony is the
religious preparation of those among the tribe giving the

---

[1] Prescott, *History of Mexico*, Second Edition, vol. iii. Appendix, p. 343.
[2] *Ibid.* vol. i. chap. iii. p. 56.

dance, who wish to pray to the sun Wakan-tan-ka.[1] This
is begun some months before the time fixed for the dance, and
at the close of the preliminary ceremonies, the neighbouring
tribes are invited, and universal peace proclaimed. A large
prairie is chosen, police appointed to keep order, and a crier
sent round to tell all the tribes where they are to pitch the
upright, conical tents, of the primitive pattern, which must
alone be used at the dance. When the time arrives, during
the first two or three days, the tents are pitched and the
ground prepared. On the fourth day, the chosen men go out
to look for the mystery tree. When they have found one
suitable, they come back and dig a plot in the centre of the
camping-ground, removing all grass and roots. They make
it square, with projecting points at the corners, to indicate the
solstitial path of the sun, rising and setting at the solstices,
north and south of the east and west.[2] They cover the
ground with sweet-smelling creeping grass, like the Hindu
Kuśha grass, strewn on that part of the Hindu sacrificial
ground which is set apart for the spirits of the Fathers, and
also with wild sage, and place the buffalo's skull on the sacred
grass. The chosen warriors then go out to cut down the
mystery tree, riding to it furiously, as if charging an enemy.
When it is reached, the warrior appointed as leader strikes
one blow with his axe on the east side, showing that the rite
is one instituted by the fighting sons of the deer, who make
their fire by laying the east stick first. The second strikes it
on the south, and the third and fourth on the west and north,
and the final strokes are given by a selected young virgin.[3]
It is placed on a litter of sticks, no one, not specially appointed
to do so, being allowed to touch it, and when it arrives on the
ground, it is set up with solemn ceremonies in the centre of
the consecrated spot. Ropes are attached to it, ending with

[1] *Publications of the Bureau of Ethnology*, vol. xi. : Dorsey, 'Study of
Siouan Cults,' pp. 451 ff.

[2] *Ibid.* vol. xi. : Dorsey, ' Study of Siouan Cults,' p. 451, § 146.

[3] *Ibid.* vol. xi. : Dorsey, ' Study of Siouan Cults,' p. 465.

hooks, to be inserted in the flesh of those who have vowed to
show their mastery over themselves, by swinging in honour of
the sun-god. This swinging ceremony takes place, after all
the devotees, both male and female, have undergone the
necessary consecration in the preparation tent, placed to the
east of the pole. This swinging ceremony is a sequel of the
custom of the corn-growing races, of swinging the mystic
basket of seed-grain in the wind to strip it of its useless husks,
to gain for it the power of generation, given by the god of
the air, and to simulate its infant sleep. It is still preserved
in India, in the ceremony of swinging the young sun-god
in August;[1] also the Dakota swinging-penance is exactly like
that annually performed in Bengal by the devotees, who
swing themselves on hooks at the Charak-pooja, while the
preparations for cutting down the mystery tree are very like
those observed in Chota Nagpore, in cutting down the Kurrum
tree (*Naudea parvifolia*), at the barley festival in August,
and in both cases, those who cut the tree must fast.[2] It is all
but utterly impossible that this peculiar form of swinging-
penance should have originated independently, both among
the Bengalis and Dakotas, and when the numerous other
coincidences between Hindu, Chinese, Japanese, and American myth and ritual, especially the measurement of time, both
in India and America, by the Pleiades, Orion, the Pole Star
and Great Bear, are also taken into account, I cannot see how
it is possible to doubt that the American Indians came to
America from Asia, some of them passing through China and
Japan, and some perhaps by direct voyages. The immigration
must have taken place at different epochs, and the incomers
must have continued to arrive, up to a time after the institution of the lunar-solar year of thirteen lunar months, the year
of the fish-sun, and after the religion of the worshippers of
the Soma, or divine sap of life, of the sons of the barley-seed,
had become one, in which personal endeavours after holiness

---

[1] *The Ruling Races of Prehistoric Times*, Preface, p. xv.
[2] *Ibid.* Essay iii., pp. 232, 233.

and the eradication of evil habits and passions, had been added
to the national worship of, and obedience to, the gods who
ruled the times and seasons, and watched over the national
prosperity of those confederated tribes, who followed the rules
laid down in the national ritual. This last epoch was, as is
shown in the Dakota swinging penance, an age when military
powers and courage were held in high esteem, and also one in
which the individualising genius of the north had triumphed
over, and become amalgamated with, the communism of the
south. The union is shown by the combination of the
worship of the seed-grain, and the mother-tree of the agricultural races, with that of the cross of St. George, the emblem
of the worshippers of the pole-star of the north, and with the
worship of the antelope-sun, who guided the sons of the deer,
the worshippers of the rising sun of the east, and not the
western setting sun of the buffalo races. It was from this
antelope-sun that, as we have seen, the fish-sun was descended,
for Pradyumna the Makara, the Hindu fish-sun, was the son
of Krishṇa, the antelope sun-god, and Rukmini, the moon.
The same descent is also shown in the name of the totem fish
of the trading Shus, called Sal-rishi or the Sal antelope
(*rishya*), and the Sal-tree (*Shorea robusta*) is the mother-tree
of the Hindu Turano-Dravidian races, so that the Sal-fish was
the totem of the tribe born from the union of the northern
antelope sons of the barley, with the Dravidian sons of the
tree.[1] The supreme god of these descendants of the goddess-
mother of corn, Sar, worshipped by the Akkadians as 'Sala
with the copper hand,' the wife of Dumu-zi, or Tammuz, the
star Orion,[2] the goddess of the sun-worshipping workers in
metal, who was also the constellation Aquarius, was the Sume-
rian Sal-manu, the divine fish worshipped by the Shus of India
and of the Euphratean Delta. He was the god of the maritime races dwelling on the coasts of the Indian Ocean, the

---

[1] *The Ruling Races of Prehistoric Times*, Essay iii., pp. 284-286; vi.
510-511.
[2] Sayce, *Hibbert Lectures for* 1887, Lect. iii. p. 212.

Napût Apām, or son of the waters, in the Rigveda, who became the Latin sea-god Neptunus. He was the Jewish Solomon, son of Bathsheba, she of the seven (*sheba*) measures (*bath*), the god of the fifty-two weeks completing the lunar year, who was also the god Ia, of the Akkadians, the Yah of the Hindus and the Yaveh of the Jews.[1]

It is his history, which is given in the Arabian and Talmud legend, telling how Sakhr, called in the *Arabian Nights* the King of the White Jinn, the flying spirits of the sun-lit air, dwelling in the north, came and stole the ring of Solomon, while he was absent, and when it was in charge of Aminah the Faithful, his female substitute. This was when the sun-spirits of the north came south with the sun at the winter solstice, to fight the black Jinn of the South, who slay the sun in his winter house.[2] The year is then ruled by Aminah, the female-sun, the mother and nurse of the young sun of the new year, born at the winter solstice, to become the conquering male-sun of the next year, who was the god riding on the sun-horse. This was the sun-mare of Sakhr, which he lent to Bulukiya,[3] the horse Grani, of Sigurd, the sun-god of the Nibelungen Lied, who fights and conquers, at the summer solstice, the evil powers of night and winter. Sakhr is the god worshipped by the Akkadians as Suk-us, a name of Istar, the wet (*suk*) goddess, the Hindu Sukra, the rain-god, before Indra, the eel. He, as the god called Ushāna, is, as I have shown, a southern form of the Finnish Ukko, the thunder-god and thunder-bird, the mother storm-bird. This is the vulture-bird of the Egyptians, Nunet, the wife of Nun, the fish-god, who ruled the year, as the vulture-star Ma'at, a name of Vega in Lyra, the polar star, from 10,000 to 8,000 B.C., when the myth was framed. This mother-bird of the

---

[1] *The Ruling Races of Prehistoric Times*, Preface, p. xxi.; Essay iii., p. 286, 295, 307 ; iv. pp. 340-376.

[2] Burton's *Arabian Nights* (H. S. Nichols and Co., 1894): 'The Adventures of Bulukiya,' vol. iv. p. 263 ; 'Tale of the Fisherman and the Jinni,' vol. i. p. 37, note 6.

[3] *Ibid.* 'The Adventures of Bulukiya,' vol. iv. p. 267.

south, ruling the year in the winter absence of the sun is the Esthonian Pörga Neitsi or hell-maiden, dwelling below the earth, who was beguiled of the ring which Sakhr stole, by the youth who slew the frog-sun by its help, and who could fly while wearing it on the little finger of his left hand.[1] She was also the Dolphin, the fish-mother of the sun-fish, who ruled the heavens while he was borne in her womb. On losing his ring, and thus becoming vanquished by the mother-bird of the pole-star, Solomon, the winter-star of the South, was reduced to beggary, that is, was made powerless, but, after forty days, the number sacred to Ia, the sun-fish of the winter rains, who rained forty days and forty nights,[2] Sakhr fled, throwing the ring into the sea, whence the young sun-fish rose, and thence as in the other stories of the loss of the ring, it was recovered by a fisherman from a fish. This was the magic ring which has become the fisherman's ring, placed as the ring of marriage on the finger of the Pope at his consecration, and broken at his death. It was originally the plain wedding ring, the ring of ten lunar months of gestation, the annus of the Romans, and afterwards became the lunar-solar ring of recurring months forming the year, beginning with the emergence of the sun from Aquarius in November, and his return to it again at the same season. This is the masonic ring bearing a stone on which is inscribed in a circle the two equilateral triangles which formed the thunder-bolt of Vishnu, but in this ring the triangles are not joined as in the thunderbolt by their apices, to form the year of the dwarf

twins, but are interlocked thus,[3]  to represent the

double path of the solstitial-sun. This is the masonic sign

[1] Kirby, *Hero of Esthonia* (Nimmo and Son, 1894), vol. ii. pp. 240 ff.

[2] Genesis vii. 12.

[3] The figures in this diagram are not in the masonic sign, they are merely placed in it to show that it is the sacred symbol of the nine spirits of heaven, the square of the original sacred triangle, the sign of the year of three seasons. The two interlocked triangles are a sacred symbol on monuments of the Bronze Age. Boyd Dawkins, *Early Man in Britain*, chap. x. p. 378.

of the temple-building sons of the fish-god, who completed
the measurement of the year, begun by the triangle of the
three seasons. It is the figure of the tortoise earth, as
described by the Hindu astronomer Varāhamihira, with the
land of Panchāla the Gangetic Doab, ruled by the Srinjayas
or sons of the sickle (srini), in the centre. The sacred land
of the Pāṇḍavas, who instituted the year of the fish-sun of
thirteen lunar months, the corn-land of the Sia-god Poshai-
yänne. This ninth land, added to the eight originally forming the eight-rayed star, completes the number nine, the
number sacred to the Akkadian Igigi or spirits of heaven,
and to the Hindu god Vishnu. In masonry it is the sign of
the Royal Arch, only allowed to be worn by the most advanced
masons. It represents the corn-land in the centre, watered
by the irrigating waters, the home of the fish-sun, and this
land is flanked by the two sacred door-posts, the two St.
Andrew's crosses, denoting it as the sign of the year of the two
solstices, and the two triangles, the six rabbit-sticks of the
Sia twins. This ninth centre-land is the key-stone of the
converging arches of the eight-sided temple which form the
dome, the sacred symbol of wisdom, the dome of the great
Cathedral of St. Sophia, at Constantinople, and the Pantheon
at Rome, from whence was taken the pattern reproduced in
St. Peter's. The young fish-god, when he gained the year-
ring, became ruler of the year of the race, who looked on the
god of light, who directed the course of the sun, which made
the seed to grow and ripen, as a father-god more powerful
than the invisible god of the revolving pole, the god Ka, who
planted in it the germ of life. It was then that he sent his
Wazir, Asaf, the son of Barkhya, the lightning (barukh) god,
to arrest Sakhr, and that the fish-sun, who circled the
heavens in his annual course, deposed the mother-bird, the
pole-star, from the rule of time. It was then that the year
was changed from that of the revolutions of the oil-press, the
year of twelve months, of three hundred and sixty days,
described in Essay VII.,[1] to that of three hundred and sixty-

[1] *The Ruling Races of Prehistoric Times*, Essay vii., pp. 7, 61.

four days, divided into thirteen lunar months, of twenty-eight days each, each month containing two fortnights, or twenty-six fortnights in all, and twenty-six is, as I have shown in Essay VII., the sacred number of the Hindu Kabir-puntis, the Hindu worshippers of Kabir, originally the wise (*kabir*) god of the building races of Asia Minor and Thrace, called the Kabiri.

That this deduction is a legitimate account of the history of the evolutions of these conceptions, is shown in the history of the offspring of the male lion of the West, the lion of the tribe of Judah, and the female wolf, the totem goddess of the East, mother of Apollo and Artemis. Sakhr[1] told Bulu-kiya that their first children were the Jinns, the flying-sons of the mother-bird, the mother of fire, whose father was the fire-dog, the Akkadian Lig, who became the lion, and their next the seven males and seven females of the lunar-solar race, the seven horses and seven cows of the cosmological hymn of the Rigveda,[2] the fourteen days of the lunar phases which make up the year of the fish-god.

This year of the all-wise Solomon was that inaugurated by Hásib Karim al Din in the Arabian story of the queen of the serpents, which contains that of the adventures of Bulu-kiya. This story, the full significance of which I will describe presently, ends by telling how Hásib[3] was made by the Wazir of the king of the seven kingdoms of Irân, to show him the way to the queen of the serpents, who was alone able to heal his master, the year-sun, of his sickness. The queen of the serpents, in Hindu mythology, is Kadru the tree (*dru*) of Ka, the thirteenth wife of Kashyapa, and thirteenth month of the lunar-solar year, who, before she was the mother of the year-sun, of the sun-fish god, was the goddess ruling

---

[1] Burton's *Arabian Nights*, vol. iv., ' The Story of Bulukiya,' pp. 265, 266.
[2] Rigveda, i. 164. 3 ; *The Ruling Races of Prehistoric Times*, Essay vii., p. 7.
[3] Burton's *Arabian Nights*, vol. iv., ' The Queen of the Serpents,' pp. 330-338.

the year of three seasons of the growing plant. She told
Hásib that in order to compass the cure of the king he must
take her to the Wazir's house, where the latter would tell him
to kill her, and cut her in three pieces. Hásib was to refuse,
and to leave the task to the adviser of the king ruling the
era of the year of three seasons. When the Wazir had
killed her and cut her up, the flesh was to be boiled in a
caldron of brass, the sacred metal of all high-caste Hindus,
who will only eat and drink out of brass, and that used to
make all the utensils of the temple of Solmon, Ia, fish-
god, not made of gold. The caldron was to be placed on a
brazier, and the first scum that rose from it was to be put
aside as though the Wazir would tell Hásib to drink it.
The second scum was to be drunk by Hásib, and this,
as I shall show presently, was the Bhang or elixir of wisdom,
which became to the followers of Zarathustra, whose chief
counsellor was Hásib, the Zend Jamáspa, the agent for
securing the power of divine insight, which succeeded the
intoxicating drink, the deadly draught of the first scum,
which had been used by the earlier inspired teachers.
From this draught Hásib gained, like Solomon whose
counterpart he is, all wisdom and knowledge ; learnt, as the
story-teller in the *Arabian Nights* expressly says, how to
predict eclipses, to understand the ordinances of the planets,
the scheme of their moving, and thus to know the new
astronomy which measured the path of the solstitial and
equinoctial sun round the eight-sided enclosure of the
tortoise earth. Hásib when thus possessed of the knowledge
of the fish sun-god, Ia or Salmanu, was to give the first
phial to the Wazir, who had intended to kill him without
telling him it was the second, and thus to cause his death.
He was then to take the three pieces of the snake, the year
of three seasons of the worshippers of the rain-snake Nága,
the Gond Nagur or plough of heaven, to the king, who would
be completely healed by eating it, and thus destroying the
old year of the believers in the inspiration of intoxication.

300 THE RULING RACES OF PREHISTORIC TIMES

As the new-born king, the young sun-god of the new epoch, he made Hásib, the depository of the new knowledge, his Wazir or chief counsellor. This ending of the story of Hásib is exactly like those of the numerous solar legends in Europe, which I have quoted at the end of Essay VI., telling of the revolution in knowledge which took place when the sun-god, circling the heavens in a track marked through the stars, was made ruler of the year. In all these the young sun-prince, who like Peleus, the father of the sun-god Achilles, has been taught all knowledge by the divine physician, the Centaur Cheiron, the sun-horse, kills, or is accessory to the death of, as Hásib was to that of the snake, the evil beasts of the old faith of their ignorant predecessors. The credit of the deed is, when the hero falls asleep after his victory, claimed by the prime minister of the king, whose foes the sun-prince has destroyed. His falsehood, which is punished with his death, is proved by the awakening of the real conqueror, and his return with the tongues of the slain beasts as evidence in his favour.[1]

Hásib is, as Sir R. Burton tells us,[2] an Arab counterpart of Jamásp, meaning the twin (*jama*) horse (*aspa*), the heavenly twins' brother, and prime minister of the Persian king Gūstāsp, called in the Zendavesta Vīstāspa. He thus belongs to the Zend triad of three brothers, who were the chief allies of Zarathustra, the Zend reformer, the high-priest of the religion based on personal strivings after perfection. It was they who, as we are told in the Yasnas, helped him to place Ahura, or Ashura Mazda, the supreme god of knowledge, on his throne in the starlit heaven, called the Garothman, or glittering (*gar*) home of endless lights, the stars among which the sun-god moves, the place where the holiest souls reach the summit of bliss,[3] and this god was, as Assur,

---

[1] *The Ruling Races of Prehistoric Times*, Essay vi., pp. 524-528, 564-567.
[2] Burton, *Arabian Nights*, vol. iv., p. 26, note 1.
[3] Mill, *Yasna*, li. 15-19; Darmesteter, *Zendavesta Yasht*, xxii. 15; S.B.E. vol. xxxi. 184, 185; xxiii. p. 317.

the supreme god of the Assyrians, the fish-god, the guardian
of Nineveh, meaning, as its ideogram shows, fish-town.
These three brethren belonged to the Hvogva clan, the
Sanskrit Shu-gva, meaning the coming (gva) shu, the
begetter, or soul of regenerated life. They were Frashaostra,
the Hindu Prashastri, or teaching priest, Jamaspa, and
Vistaspa. The first was father of Hvogvi, wife of Zarathustra, son of Purushaspa, who learnt the innermost secrets
of the divine law by the inspiration of bangha, the well-known
hashish, a preparation of hemp (*Cannabis Indica*), introduced
as the agent of religious enlightenment by Zarathustra.[1] It
was bhang which became, to the oracular priests and
priestesses of the fish-god, the Akkadian god Kha, called by
them the god of the oracle (*kua*),[2] a more effective stimulus
of divine illumination than the intoxicating mead or honey-
drink, drunk by their predecessors, the worshippers of the
twin-gods, and it inspired not only the servants of the
Assyrian fish-god Assur, or Kha, but also the priestesses of
Delphi. Jamaspa, the representative in the triad of the age
of the twin-gods, which succeeded that of the primaeval
teacher, who was first the village schoolmaster, and afterwards the chief adviser of the national king, is said in the
Yasnas to be the chooser of holy wisdom, and he was the
counsellor of the third brother, Vistaspa, the king who
protected Zarathustra, and who was, through his wife
Hutaosa, the chief of the family of the Naotaras, or people
of the new (*nao*) star (*tara*),[3] that is to say, it was through
their agency that the morning and evening stars, the planets
Mercury, Venus, and Jupiter, were made the rulers of heaven
by the worshippers of the rising sun, illuminating the earth
in his annual course through the stars, instead of the pole-

[1] Darmesteter, *Zendavesta Din Yasht*, 15; S.B.E. vol. xxiii. p. 267,
note 3.

[2] Sayce, *Assyrian Grammar*, Syllabary, No. 442.

[3] Darmesteter, *Zendavesta Abān Yasht*, 68, 98; S.B.E. vol. xxiii. p. 70,
note 1; 77, note 1.

star, the ruling star of the worshippers of the mother-mountain and the revolving pole. This king Vīstāsp, of the new star, was also inspired by the holy Bhang, and it was under the auspices of this king of heaven, the rider on the sun-horse (aspa), advised by his two inspired brethren, that the Athrāvans, the Hindu Athārvans, who in the Rigveda predicted an eclipse of the sun through the turīya, or instrument for measuring the motions of the heavenly bodies round the revolving pole (tur), were sent forth with the Chista, the ark or chest of the law (dīn), to preach the new faith.[1] This ark or chest of the law was the Hebrew high-priest Aaron, meaning the ark or chest, and his sons were the Semitic Kohathites, the Armenian Kahanai, the prophet-priests of the sacred ephod, the sacred shirt of the Parsis, telling the oracles of God issuing from the breast of the Almighty.[2] These prophet-priests were the priests of the sun-and-fire-god Atri, meaning the devouring (ad) three (tri), the god of the three seasons. These three brethren, Frasha-ostra, Jamāspa, and Vistāspa, the divine propagators of the revolution consummated in the days of Vistāspa, making the rider on the sun-horse, the god of living lights who circled the heavens in his yearly course, the god of life instead of Ka, the invisible germ dwelling in the seed, appear in the Bundahish as the three fires, the Atri of the Hindus. These are (1) Frobāg, the generating fire, the Hindu Viru, the Latin Vir, the counterpart of Frashaostra, or Prashashtri. This was kindled on the mother-mountain of the Kushite

---

[1] Darmesteter's *Zendavesta Dīn Yasht*, 15-17 ; S.B.E. vol. xxiii. p. 267, note 3 ; 268.

[2] *The Ruling Races of Prehistoric Times*, Preface, p. xvi. The ephod was sanctified by Gideon after he had destroyed Penuel, the town of the face (pen) of God (el), the sacred triangle, and cut down the Ashēra, Judges vii. 26 ; viii. 17, 24-17. The sacred shirt of the Parsis was, as I show, Essay IV. 406, the direct descendant of the garment made of the skin of the totem ancestor. It and the sacred girdle were the magic dress which consecrated and infused the spirit of the parent-god of the oracle into the child of God who wore it.

race in Kābul ; (2) the fire Gushāsp, or Jamāsp, kindled in Adarbaigan, the petroleum-yielding fire-land to the south-east of the Caspian Sea, where fire-worship first became a formulated creed ; (3) the fire Būrzin Mitro, kindled by Vīstāspa in Khorasan, the central kingdom of the seven kingdoms of Irān. These three fires are said in the Bunda-hish to be united in the Bāhrām fire of the Parsis always kept burning in the temples.[1]

The Bāhrām fire is, as I have shown, the sacred fire of the immortal god formed of the two essences—(1) the female or receptive essence, the Phœnician goddess Baau, the Akkadian Bahu, wife of the southern sun of the winter solstice, the Bohu, or deep of Genesis, the Aminah of the story of Solomon's ring, and the female Swastika of the Hindus. She was the mother-void, or ocean-bed filled with water of life, the origin of the deeps and abysses, the brazen seas of the Babylonian and Jewish temples ; (2) the active essence, the male Swastika, the father-bird Shakr, and the sun-fish-god Solomon, who was first Ram, the sun-god, the Rāma Hvastra, the hissing (*shvas*) wind, and sun-god of the Zendavesta.[2] These were the divine wife and husband, ruling the year divided into the six months of the summer and six months of the winter solstices, the Pitriyāna and Devayāna periods. It was they united who formed the God Jehovah, and who are mentioned in the Jewish ritual for Pentecost, and also in their daily formula addressed to the Deity, 'In the name of the union of the holy and blessed Hu, and his Shechinah, the hidden and concealed Hu, blessed be Jehovah for ever.' The name Hu, the Zend and Druid form of the bird Khu, is masculine, while Yah is feminine, the two forming the bi-sexual regenerating and creating god. It is these two combined essences which are invoked separately as Hu and Yah on the

[1] West, *Bundahish*, xvii. 6-9 ; S.B.E., vol. v. pp. 63, 64.
[2] *The Ruling Races of Prehistoric Times*, Essay v., pp. 469, 470 ; Part II. viii. pp. 173, note 2, 174 ; Sayce, *Hibbert Lectures for 1887*, Lect. iv. p. 264.

seventh day of the Feast of Tabernacles.[1] They formed the
god invisible to mortal eyes, whose existence was only known
through the revolutions of the pole, which he turns while
dwelling in the cloudy canopy of the mountain mist, the
inner chamber or shrine, the naos of the eight-sided palace
of the god of the tortoise-race. These two united divine
essences are the two creating or soliciting (*izanau*) twins of
the Japanese, Izanagi and Izanami, the artificers of the three
Kami, of which I have already spoken, the Chinese Yin and
Yang,[2] the ever-revolving day and night calling on all the
powers of nature to contribute their due share to God's un-
ceasing work of creating, reproducing, and sustaining active
and progressive life.

The middle fire of the three, combined to make the bi-une
creator, the twin (*jama*) fire of Jamāsp, was that which is
said in the Bundahish to have specially helped Kai-Khusrib,
the Pahlavi form of the divine king called in the Zendavesta
Hushrava,[3] king of the Hus or Hushim, sons of Dan, who
united the Aryans in one kingdom, and destroyed the idol-
temples on the Chaēchasta Lake, the modern Urumiah in
Adarbaigan (*Baku*). Hushrava, who is thus identified with
Vistāspa, is the king Sushravas of the Shus in the Rigveda,
who conquered (1) Kutsa, the priest and earthly representative of the god Ka of the Purus, the indwelling soul of the
seed, who was, as I have shown, superseded by the creating
sun-fish-god, the substitute for the seed-life hidden in the

---

[1] *Rabbinical Comment on Genesis*, by P. J. Hershon, 1885, pp. 138, 302 :
J. O'Neill's *The Night of the Gods*, vol. i. ' Axis Myths,' p. 238. This god
Hu, the Hu-kairya, or active (*kairya*) creator of the Zendavesta, and the Hu
of Jewish liturgy, furnish an actual demonstration to be added to that given
by the sacred shirt of the Parsis and the ephod of the Jews, proving how the
Parsi religion developed into the Hebrew worship of Jehovah. The Egypt
out of which the Jews came to Palestine through the wilderness was in the
original myth Persia, the. land of Manna.

[2] O'Neill, *The Night of the Gods*, ' Axis Myths,' p. 38.

[3] We see here an actual instance of the Pahlavi Khu becoming the Zend
Hu, thus proving that Hu is Khu, the bird.

plant, the living son of god. The Purus, again, are the clan
of Zarathustra, the son of Purushaspa; (2) Atithigva, the
coming (gva) guest (atithi), a name of Divodāsa, the solar
god of the twin races before the birth of the fish-sun, whose
name is paralleled with Hvōgva, or Shūgva, the coming (gva)
Shu, the new begetter; (3) Āyu, the son of Puru-ravas, the
thunder-god, and Urvashi, in the form of the moon-goose
swimming on the Plaksha lake, showing that the myth
belonged to the age of Kansa, the moon-goose, of which I
have already spoken. Sushravas was thus the sun-god who
superseded the god Ka and the thunder-god, and the moon-
goose, who measured time by the seasons, and the lunar
phases marking the period of gestation. He was, in the
Genesis genealogy of the kings of Edom, the land of the
red (edom) man, the third king Husham, of the land of the
Temanites, that is, Southern Arabia (Tema), the home of the
sons of Rohinī, the red cow, the star Aldebaran, the men
born of the Akkadian Te, called in the Assyrian Temennu,
or 'the foundation of life,'[1] and its ideogram means 'the lord
of seed.' Thus Jamāspa, the twin-horse (aspa), was the
counsellor of the conquering king, who was first the rider on
the sun-horse, and afterwards the sun-fish, the all-wise ruler
of the sons of the seed, the eight-rayed star, the sign of
god and seed in Akkadian and ancient Chinese. This god
of the Bāhrām fire of the two bi-une essences of the
Phœnician mother-goddess, the darkness or void, the sun of
the West, and the father-sun-god Rai of the East, the god
of the Wends, was the god of the living soul of life, the
Soma brought from heaven by the mother-bird to Kadrū,
the thirteenth wife of Kashyapa, and the final month of the
year of the fish-god, the Ia, or house (I) of the waters (a),
who became the Nun, the sacred fish, the invisible god of
light hidden in the mists, worshipped as the supreme god
by the Akkadians, Egyptians, and Semites, the god whose

---

[1] Rigveda, i. 53, 9, 10; *The Ruling Races of Prehistoric Times*, Essay iii.,
pp. 211, 212, 228, 240, 274.

existence was first discovered by the Indian and Babylonian astronomers and metaphysicians.

This age of the fish-god, worshipped by the Assyrians as 'Sallimannu, the fish,' the god of the city of Temen-Sallim (the foundation of peace),[1] was that of the society described in the *Arabian Nights*, an age some thousands of years earlier than that of the Ommayad and Abbasid Caliphs, when these stories took their present form. Though among them there were many which then first saw the light, the great majority, and the most historically important of the tales are direct descendants of the old stories of the professional historio-graphers and guardians of the national knowledge of the East, compiled before the days of writing, when the national schoolmasters and teachers were thought by the people, and believed by themselves, to be divinely inspired, and when the office was held to be a sacred trust which obliged those who undertook it to be scrupulously careful of the preservation of the exact truth in their historical and scientific deliverances. Under their guidance, history was not biographical, nor were the actors named in it spoken of as individuals, but the stories, giving the summary of the record of events, were narratives in which the composers aimed especially at accuracy, conciseness, and the stimulation of an awakened interest in those who heard them and retained them in their memory for their own instruction and that of future genera-tions. If the Rawi, or professional story-tellers, who told them to the Arabian public, and who were the Arabian and Egyptian representatives of the Improvisatore of Italy, the Bhats of India, and the Celtic bards, or Sce255alds, had invented them in the age of the Caliphs, they would have described the society existing in that epoch of foreign wars and internal disturbances, when men's minds were full of military details and of the spirit of religious enthusiasm, which burned in the souls, not only of the leaders, but of the rank and file of the Mohammedan armies, who propagated by the sword the

---

[1] Sayce, *Hibbert Lectures for* 1887, Lect. i. p. 58.

belief in one God and his prophet Mohammed. But throughout the eight volumes of stories, believed, in Sir R. Burton's
edition, to be those of the earliest collection, and the four
volumes of supplementary stories taken from other Arab
editions, military matters are rarely spoken of, and in only
two stories, those of Omar-bin-al-Nu'uman, and his sons,[1]
and of Gharib and his brother Ajib,[2] are the chief actors
soldiers.

It is perfectly impossible that the ideal society these
stories depict, and in the midst of which the original
authors lived, could ever have been organised in a time of
religious warfare and national conquests, for no people whose
thoughts were filled with dreams of war and religious zeal
could have made a national hero of the peaceful though
adventurous trader who plays the principal part in most of
them. By far the greatest number of these stories tell of
the adventures of shop-keeping traders and artisans, and it
is they and the court officials and grandees who are the chief
actors in these dramatic narratives. In two of the stories,
those of Taj-ul-Muluk and the Princess Dunya, and of
Prince Ardashir and Hayat-al-Nufus, the hero prince wish-
ing to marry the princess whose hand was refused to him,
attains his end by disguising himself as a merchant, instead
of forcing consent by war, as his father proposed to do.
Also, when a prince or man of high birth falls into misfor-
tune, and finds himself an unacknowledged outcast in a
foreign country, he becomes a trader, just as Prince Zau-al-
Makan, in the story of Omar-bin-al-Nu'uman and his sons,
becomes assistant to the man who lighted the fires in the
public baths of Damascus, and Badr-al Din Husan, the son
of the Wazir Nur-al-Din Ali,[3] became a cook and confec-
tioner in the same city.

[1] Burton, *Arabian Nights* (H. S. Nichols and Co.), vol. ii. p. 1 ff.
[2] *Ibid.* vol. v. p. 162 ff.
[3] *Ibid.* 'Story of Badr-al Din Ali and his son Badr-al Din Hasan,' vol. i.
p. 179 ff.

The region spoken of in these tales as that in which trade and traders ruled the land, was the vast extent of territory between the sea-coast of China on the east, the home of the Princess Badur, and the Canary Isles, called Khalidan, on the west, where her husband, Kamar-al-Zaman, was heir-apparent.[1] The connection of the subjects of the Caliphs with the eastern half of this region was, during their rule, so very small, that it would have been utterly impossible for any set of writers, whose minds were only imbued with the impressions generated in that age, to place the birthplace of these stories in the far East in the palace of King Shahryar of the Bánu Sásán, ruler of India and China. The social organisation of the country ruled by the Caliphs, as painted in these stories, which still exists there without any material changes, proves conclusively that their civilisation comes from India. There is little or no mention of settled land-owners holding large estates, or of any distinctions of rank based on birth, and it is thought to be perfectly consonant with propriety that Abdullah, the fisherman, and Ala-ed-din, the son of a poor widow, should, when they were enriched with the gifts of Abdullah the Merman,[2] and those of the slaves of the wonderful lamp and the ring,[3] become sons-in-law of the king.

A society such as this, in which traders are the ruling class, and in which the dominant interest is maritime commerce between India, China, and the islands of the Malay Archipelago on one side, and Syria, Egypt, and North Africa on the other, with its centre in the Persian Gulf, could only have been formed during an age of general peace, when the statesmen were merchants, and their interests were the principal care of the ruling powers. The rule of the Shus, or Shu-varṇa, the Vaishya traders of Western India, who

---

[1] Burton, *Arabian Nights*, 'Tale of Kamar-al-Zaman,' vol. iii. p. 1 ff.

[2] *Ibid.* 'Abdullah the Fisherman and Abdullah the Merman,' vol. vii. p. 237 ff.

[3] *Ibid.* 'Ala-ed-din and the Wonderful Lamp,' vol. x. p. 33 ff.

had been sons of the fig-tree, but who subsequently became
sons of the palm-tree,[1] and who were first the yellow race,
whose clothes were dyed with turmeric,[2] who founded the
commerce of the Indian Ocean, must have exactly corresponded with these conditions. It was they who, as I have
shown, were the people who based their ritual on the worship of the household-fire,[3] as the people of Korea and China
still do; who introduced the cup of mead as the sacred drink
in the sacrifices instead of fermented spirits;[4] who were
clothed in goat-skins, as the Hindu and Akkadian priests of
the goat-god, who watched the solar disc on Babylonian
monuments;[5] who founded the Soma sacrifice as sons of the
fig-tree, first of the Bur-tree (*Ficus Indica*) afterwards of the
Udumbara-tree (*Ficus glomerata*);[6] who instituted ceremonial
shaving and the reverence for the barber-physician;[7] who
reformed the sacrificial ritual by substituting offerings of
milk, curds and whey, and clarified butter for animal sacrifices;[8] who founded the Jain religion of penances aiming at
moral improvement;[9] who wore the girdle of woollen threads
sacred to the sons of the sun-ram;[10] and who showed that
they were worshippers of the sun revolving round the pole
in twelve months, by fixing the date of initiation by the
girdle at twelve years after conception.[11] The ruling
traders, who succeeded this yellow race, were formed from a
union between the original trading sons of the fig-tree and
the race of lunar warriors, the first Indian Kshatriya whose
clothes were dyed with madder, as the sons of the red man,[12]

---

[1] *The Ruling Races of Prehistoric Times*, Essay iii., p. 328; iv., p. 347;
v., p. 448.
[2] *Ibid.* Essay ii., p. 87; iii., pp. 138, 233, 277, 278: v., p. 448.
[3] *Ibid.* Essay iii., p. 328. [4] *Ibid.* Essay iii., p. 208.
[5] *Ibid.* Essay iii., pp. 149, 238; iv., p. 403.
[6] *Ibid.* Essay ii., p. 118; iii., pp. 238, 239.
[7] *Ibid.* Part I. Preface, pp. xliv, xlv; Essay iii., p. 270.
[8] *Ibid.* Essay iii., pp. 327, 328.
[9] *Ibid.* Essay iii., pp. 323-329. [10] *Ibid.* Essay iv., pp. 406, 407.
[11] *Ibid.* Essay iv., p. 403. [12] *Ibid.* Essay v., p. 449.

who were born of the spotted-deer and the palm-tree, and who abhorred intoxicating drinks.[1]

A society ruled by the yellow and red races, which had no hereditary distinction of ranks beyond tribal divisions, and in which individual property in land was almost unknown, is one which could only have arisen in a country where, as in India, land was originally divided among village communities, where all acknowledged householders in the village were partners in the town land, and where the only distinctions of rank originally recognised were those allowing the superiority of kings, and their ministers, appointed by customary modes of election, to rule over the people. The transfer of these fundamental attributes of Indian society to the Euphratean countries and Arabia must have taken place before the system of distinctive castes was rooted there, for no trace of it appears in the *Arabian Nights*.

But while their political organisation marks the Arabian traders as related by birth to the Southern founders of villages, their marriage customs show that they were still more closely allied than the Indian people to the Northern races, for they repudiated entirely the Southern rule of exogamy which is generally observed in India, and which forbade marriage between men and women of the same village or the same sept or clan, for in these stories the marriage of a son with the daughter of his paternal uncle is looked upon as the most desirable of unions. The Arab merchant kings, therefore, ruled a race who continued the endogamous marriage customs of the North, with a preference for the internal policy of the Southern exogamists, the founders of the village city and state, a division of society fundamentally different from the Northern tribe, clan, and family.

The conclusion that they were descended from a race formed, like the sons of the palm-tree, by the union of Northern invaders with indigenous Southerners is confirmed by the

[1] *The Ruling Races of Prehistoric Times*, Essay iii., pp. 149, 332 ; iv., p. 403 ; v., p. 448, 449.

story of King Shahryar of the Bánu Sásán, ruler of India and
China, and his brother Zaman, king of Samarcand, with which
the book opens. These sons of Sásán were the Sassanides,
whose rule over Persia began 226 A.D. But the name, as
that of the race who introduced the Pahlavi language, probably extends very much further back, and it seems to be
connected with the Vedic Sasá, the food offerings of the
ritual in which libations of milk[1] given to Indra were offered
to the moon-cow, and these offerings are, as I have just shown,
the distinctive sacrifices of the Vaishya traders. The meaning of the names Shahriyar, the city friend, and Zaman, the
age or epoch, shows that these offerers of milk ruled the age
of the founders of cities and of the beginnings of mercantile
activity. Both kings discovered the infidelity of their wives,
who bestowed their affections on black men, the dark Indian
Dravidians, called in Arab history the Himyar, meaning the
dark races. These Himyar had founded the maritime and land
trade of South-western Asia, and had introduced their village
system, not only along the coasts, but in all the fertile inland
valleys watered by the rivers of India as well as those of the
Tigris, Euphrates, and Oxus, and it was they who brought
the Indian word Ra'ayat, meaning 'a cultivator,' into Arabic
as Ra'ayá. They were the first rulers and colonisers of the
Himyarite empire of Southern Arabia, who paved the way to
its becoming the great commercial power of the Indian
Ocean, the land called by the Egyptians the Holy Land of
Punt. The two kings were reclaimed to a belief in women's
honour and fidelity by the stories told by Sharyazád, the
learned and accomplished daughter of King Shahryar's Wazir,
whose name, meaning the 'city freer,' was, according to Sir
R. Burton, probably originally Shirzad, lion-born,[2] and this
shows her to belong to the Northern warlike race who call
themselves, in India, Singhs, meaning both sons of the moon

[1] Grassmann, *Worterbuch zum Rigveda*, s.v. 'Sasá.'

[2] Burton, *Arabian Nights*, vol. i., 'The Story of King Shahryar and his
Brother,' p. 13 note.

(*sin*) and the lion (*sinha*), the sons of the lion of the tribe of
Judah, the Yadavas, who completed the reformation of the
Indian ritual begun by the earlier Vaishya, sons of the fig-tree,
and of the twin gods, the Turvasu worshippers of the revolving pole (*tur*). Sharyazad's sister, Dunyazad married Zaman
at the end of the thousand-and-one nights, during which
these stories were told, and the two sisters inaugurated the
age which substituted learned Northern women for indigenous
wives, and made the endogamous marriages of the North to
take the place of the exogamous unions of the South among
the ruling races of the land. A similar transition stage from
exogamy to endogamy is marked in India by the marriage
rules of the Koech Rajbansi, the great agricultural caste of
Eastern Bengal, who are not divided, like the exogamous castes
into distinct septs, but are all the children of Kashyapa, the
legendary father of the Kushite race, born of Gandhārī, the
mother-bird who laid the egg whence all the sons of the
tortoise Kush issued. These, as I have shown, became the
Ants, who measured time by the periods of seven days, the
Myrmidons of Greece, and the Vamra of the Rigveda, born
of the egg-shaped tortoise-earth commemorated in Grecian
legend as Egina, the tortoise island ruled by the fish-seal-
king Phokus.[1]

Though the Koech Rajbansi profess to disallow marriages
between those nearer than seven generations on the father's
side and three on the mother's, they are very lax in the
observance of this rule and prefer to marry neighbours rather
than go away from home to seek a wife.[2] While the exoga-
mist castes marry by the bridegroom marking the partings of
the bride's hair with the red mark of Sindurdan, and thus
making blood-brotherhood with his alien wife, the Koech
agriculturists who are already clan brethren, as well as the
Savars or Su-varna, the race (*varna*) of Shus, sons of the bird
(*khu*), and also of the fish-antelope (*rishya*) Sal-rishi, marry

[1] *The Ruling Races of Prehistoric Times*, Essay vi., pp. 523, 524.
[2] Risley, *Tribes and Castes of Bengal*, vol. i. ' Koech,' p. 494.

by uniting the hands of the pair by the bond of Kusha grass.[1] These customs conclusively prove that the substitution of endogamous for exogamous marriages was begun by the sons of the antelope, whose southward career, led by the Kusha grass, I have so often traced in these Essays. It was they who introduced the custom of calling the dead ancestors, who had always been worshipped by all the Southern sons of the Pleiades, to sit on the Hindu sacrificial ground on the sacred Kusha grass, and eat the rice and barley offered by their descendants who grew the Southern rice and the Northern barley.[2]

The greater laxity as to consanguineous marriages which distinguishes the pastoral Arab Semites from the Hindu tribes, who were both cattle herdsmen and agriculturists, proves that the former people had retained more of the original Northern customs of the race who looked on inter-tribal and consanguineous marriages as a duty, and who were most anxious to keep the blood of the tribe pure from outside contamination. It was this feeling which prompted the writer of the Semite history in Genesis to tell how Rebekah preferred that Jacob should seek a wife among his near relations, both on the father's and mother's side, rather than marry among the Hittite daughters of the land of Canaan, and it was the same reason which gave birth to the law that Persian and Egyptian kings should marry their own sisters.

In order to trace out clearly from local evidence the chronological epoch when the composite city and village building traders of South-western Asia were first ruled by Northern invaders, who had the political wisdom to leave the old institutions intact, and to content themselves with the exaction of tribute as their payment for maintaining law and order throughout their vast dominions, we must turn to the early histories of India and the countries bordering it on the north-west, called in the *Arabian Nights* the kingdoms of Shahryar

---

[1] *The Ruling Races of Prehistoric Times*, Essay iii., pp. 174, 175. 280.

[2] *Ibid*. Essay ii., p. 130 : iii., p. 207 ; iv., p. 401.

and Zaman. In both India and Irān we can trace the growth from the village and province of the ideal State called the Seven Kingdoms in Zend and Hindu tradition,[1] commemorated, as I have shown, in the *Arabian Nights,* in the story of the cure of Karazdán, the king of the Seven Kingdoms of Persia, by Hásib Karim-al-Din or Jamáspa. These were formed of seven united provinces ruled by the king, who governed the central province Khorasan or Irān in Persia, and Jambudwipa (Central India) in India, while the six surrounding kingdoms were ruled by his six most trusted chiefs. The king was called, as the Emperor of China now is, the ruler of the middle kingdom, and this was the position held by Shahryar, the world-king of the *Arabian Nights,* ruler of India and China. Under this form of government every individual, and each village, city, province, and kingdom looked to trade in the produce of their crops, herds, mines, and manufactures as the best means of increasing their wealth,[2] and among these confederacies, in which trade was, as it still is among the Chinese, chiefly carried on by guilds, wars were but of rare occurrence, for individual ambition was kept in check by public opinion, and strictly enforced rules, while the whole community were united in their determination to put down all systematic robbery as the greatest hindrance to trade. Hence the Fish-sun, who ruled the age when this system of peaceful trade reached its fullest developments was rightly called by the Assyrians the God of the foundations of peace.

It was this people who, starting from the forest-clad coasts of Western India, which yielded the only ship-building timber close to the sea in the Indian Ocean, founded the Kushite Semite maritime commerce of the early Phœnicio-Turanians, at Turos or Dilmun in the Persian Gulf and afterwards in Egypt,[3] and along the coasts of the Mediter-

---

[1] *The Ruling Races of Prehistoric Times,* Essay iii., pp. 145, 146, 253.
[2] *Ibid.* Preface, pp. lvi-lxi ; Essay ii., pp. 99-101.
[3] *Ibid.* Essay iii., pp. 281-284 ; iv., pp. 346, 347.

ranean, whither they carried the worship of the twin gods, the common meals of the Indian Dravidians, and the other Dravidian customs of the Dorian Greeks, which I have described,[1] and whither they also carried the Dravidian reverence for law which was the foundation of the long-enduring power of Rome.

One of the great trading centres of those early merchant princes, whose kings bore in India and Egypt the sign of the Nāga snake on their foreheads, was Southern Arabia, which furnished ample supplies of trading materials in the products of the land of Oman at the mouth of the Persian Gulf, of the Minyo-Sabæan country on the south-east of the Red Sea now called Yemen, and in the mineral wealth of the Sinaitic Peninsula, the Akkadian Magāna sacred to Māga, the goddess-mother of the Magi, or magicians. This peaceful community of tillers of the soil, artisans, merchants, and merchant princes was broken up by the invasion of the fighting races of the North, who called themselves sons of the sun-horse, the Turcoman of the *Arabian Nights*, who brought southward the theology of the Northern Nibelungs, sons of the mist (*Nebel*), and worshipped Sigurd, the victorious son of the gnomon-pillar (*urd-r*), riding on the sun-cloud-horse Grani. This sun-horse appears in the *Arabian Nights* in the flying mare given by Sakhr to Bulukiya, of which I have already spoken, the mare given to Gharib, the sun-god, by Mura'ash the fire-god, and finally in the sea-horse, the fish-sun-god, also a gift from Mura'ash to Gharib.[2]   These invaders introduced the epoch of anarchy, called in the Zendavesta, the reign of Kereshāni, the Vedic Krishānu, the rainbow and thunder-god, and in popular Hindu mythology, the reign of Kansa, the Northern moon-goose (*khans*) son of the antelope-doe Kalānemi, and of Ugrasena, the king of the ogres, the Skoya of the Mexican Indians, of the race of the Bhojas or cattle-herdsmen, who

---

[1] *The Ruling Races of Prehistoric Times*, Essay iii., pp. 297-299.
[2] Burton, *Arabian Nights*, vol. v., ' Story of Gharib and his brother Ajib,' pp. 245, 246, 253.

were sons of the Druhyu, or sorcerers. It was during their reign that the old religion was overthrown, priests and cattle ruthlessly massacred by the Northern invaders, who said, according to the Zendavesta, 'No priest shall walk the land for me as a counsellor to prosper them. He would rob everything of progress, he would crush the growth of all.'[1] The ruler who thus expressed his scorn for the territorial priest-hood of the Southern earth-worshippers was the Northern king who ruled his people as his family, as each of the fathers of the families who were his subjects ruled their own house-holds. He and his people worshipped the sun-god Rā or Rai of the Lithuanian Wends, whose worship passed through the stages in the religious history of India, Assyria, and the Persian Zends, which I have traced in Essay v., and who became finally the Semitic father Abram, who brought the rain and the sun which gave life to the seed of life, the barley-seed, which made all those who lived on it sons of the seed of Ram. It was during the reign of this Northern king, Kaṇsa, that Kṛishṇa, who became the Hindu Vishnu, was born, according to Hindu tradition, as the youngest of the eight children of Vasadeva, the creating (*vasu*) bright god (*deva*) of light, and the light goddess, Devaki, ruling the eight sections of the eight-rayed star, the tortoise earth bearing the divine seed, and his first avatar was a fish, the fish-sun-god.[2] It is this sun-god who appears in the *Arabian Nights* as the hero Gharib, who preaches the faith in one God, and in Abram as his friend, the sun-messenger sent to earth to carry out the behests of the invisible god of light, the god of the revolving pole, and the last proselyte he makes and conquers in his career of victory is Ra'ad Shah, the thunder-king.[3] These conquering sons of Rā were the barley-growing sons of the

---

[1] Mill, *Yasna*, ix. 24 ; S.B.E. vol. xxxi. pp. 237, 238 ; *The Ruling Races of Prehistoric Times*, Essay v., p. 462, 463 ; iii., pp. 240.

[2] *The Ruling Races of Prehistoric Times*, Essay iv., pp. 467-69; ii., p. 126.

[3] Burton, *Arabian Nights*, ' The History of Gharib and his brother Ajib,' vol. v. p. 255 note 1.

sun-horse, who totally revolutionised the ancient ritual, both
in Irān and India, and who, while they maintained in both
countries the old animal sacrifices to the Northern totem-gods
of generation,[1] abolished in that part of the Soma sacrifice
offered to their sun-god Rā, called Indra in the Soma ritual,
the slaughter of animals, and the former orgiastic festivals on
the flesh of the totems, washed down by draughts of mead
and strong drinks. In place of these festal orgies they, when
their government was consolidated, and conquerors and conquered were united, learned that the fruits of the earth offered
by the Southern races were more acceptable offerings to the
gods than living animals. And as they had discarded the
totemistic fiction of animal descent, and had become sons of
the barley and the tree, they ate and drank at their sacrificial
feasts to the creating-god Su, who had been once the bird
Khu, the holy food of barley mixed with running water from
the parent-rivers, the Soma of the Hindus, the Haoma of the
Zends, the sacred cup, and cake eaten in Greece at the
worship of Demēter, the barley-mother and the corn-meal,
pollen, and water of the Mexican Sia. As the spirit of moral
development gradually triumphed over the old ritualistic
spirit, and the idea of the common brotherhood of all the
corn-growing races superseded sectional tribal exclusiveness,
the old tribal feasts, only joined in by those descended from
the same totem, became accessible to all who made blood-
brotherhood with the confederacy of the sons of the antelope
by cleansing themselves from sin in the baptismal bath,
clothed, like the Hindu Soma neophytes, in the Jaraiyu, or
afterbirth of the antelope skin, and by fasting during the
sacrifice from all food except milk, and to this discipline that
of confession of sin was subsequently added.[2] The original
central stem of these confederated human ants was that of
the sons of the bird and of Danu, the judge, whose sons were
the Hus or Hushim and the Shus or Shuham, the worshippers

[1] *The Ruling Races of Prehistoric Times,* Essay iii. pp. 172, 270, 271, 321-323.   [2] *Ibid.* Part I. Preface, pp. xliv-xlviii.

of the god Hu or Shu, whom I have shown to be the male
element in the Hebrew conception of Jehovah, and it was
with the children of the mother-bird, the sons of Danu, the
judge, that the races who derived their descents from the
mother-wild-cow Gauri, and the mother-sheep, sons of
Varuṇa, the ram, were amalgamated. They were, in the
Hebrew genealogy, the sons of Leah, the wild-cow, the
Hittite Le, and of Rachel, the ewe. They became the race
called at Telloh in the Euphratean Delta, and in India, the
Gaurians, and in Phrygia the Satyrs,[1] and it was they who
were conquered by the Northern sons of the sun-horse.
These Gaurians are depicted in the Telloh bas-reliefs, and in
the features of the Turano-Dravidians, who in India worship
Gauri, the wild-cow, as having round heads, low, wide, and
straight foreheads, wide and fleshy faces, with a large but
not aquiline nose and curly hair. Their conquerors, the
ruling race who built the temples of Telloh, appear from
their statues, as described by M. de Sarzac, to have been a
stout, short-necked people, with square large heads, large
open, projecting eyes, with eyebrows meeting over the nose,
well-modelled cheeks, and strongly-marked aquiline noses. On
the very ancient seal of Sargon, depicted by Professor Hommel
in his *History of Babylonia and Assyria,* these people are
depicted with beards, like the Mexican god, Quetzacoatl.
They again appear in Egyptian sculptures, described by Mr.
Petrie as the men of Pun or Punt, the people of Southern
Arabia, 'with aquiline noses and fine expression,' from whom
the Egyptians of the higher class traced their descent, while
the earlier trading races, whom they ruled, have a 'large but
snouty nose,' like the Indian Dravidians.

These united races were the sons of the nut, whose birth I
have described in telling the story of the Poshai-yänne, and
their sign of union was not the spear Shelah, which led
them when they first came southward from Ararat, the magic

[1] *The Ruling Races of Prehistoric Times,* Essay iii., p. 254 ; vi., pp. 544
note 2, 545.

pole of the American-Indian nomads, but the prophetic
gnomon-stone, the pillar (*urdr*) of victory (*sig*), the Northern
sun-god Sigurd, the rider on the sun-horse. He was Jacob,
the supplanter of Esau, or Uz, the goat-god of the twin-
Hittites; the Khati, or joined people, of the Akkadians; and
Hindu Vaishya, the god who watched the sun-disc. The
god Jacob, worshipped in Canaan as Yakob-el,[1] was consecrated
as the national gnomon-stone under the almond tree (*Luz*),
as Beth-el, the house (*beth*) of god (*el*), telling of the daily
and hourly progress of the sun in completing his annual
journey. The almond tree was looked on by the Jews as the
harbinger of spring, and they still carry its branches at their
spring feasts to represent the palm branches of the temple,[2]
which proclaimed them sons of the Indian and Babylonian date
palm. It was this gnomon-stone, the sacred obelisk of the
Egyptians, which became to the unitarian Semites, who had
united the tri-une year-gods, the Iru of the Esthonians and
Iberians, into the one god, El, and the bi-une female and
male elements of the Hittite or twin races into the Supreme
God, Yahveh, the sign of the creating god, and it was this
worship which led to the destruction of the totemistic idols
of Leah and Rachel, the winged bull and ram, which Jacob
is recorded to have destroyed at Luz.[3] The worshippers of
the one god of light, whose sign is the prophetic gnomon-
stone, were the Semites of Harran, the road (*Kharran*)
between the Euphratean countries and the Mediterranean,
who introduced the year of the sun-fish of thirteen lunar
months, and who, under his guidance, became rulers of the
whole civilised world. Their rule degenerated into the Semitic
despotism, which was afterwards, as I show in Essay VI., overthrown by the revolt headed by the Aryan Celts.

[1] The worship of Jacob as Yakob-el is proved by Yakob-el and Iseph-el
being mentioned among the list of cities in Canaan taken by Thothmes III. in
about 1600 B.C. (Sayce, *Hibbert Lectures for* 1887, Lect. i. pp. 51, 52).
[2] Burton, *Arabian Nights*, vol. xii. p. 4 note 5.
[3] Genesis xxxii. 1-7.

This story of the birth of the gnomon sun-god, from the nuts of the garden-fruit he ripened, assumes another form in the Akkadian genealogy of Gud-ia, the Patesī, or priest-king of Telloh, whose name, meaning the bull (*gud*) Ia, proclaims him to be a parent-god. The inscription on his statue describing his parentage says, 'Mother I had none, my mother was the water deep; father I had none, my father was the water deep.'[1] This proves that he was the son of the flying bull of light, the Kerub of the Assyrians and the Gudura of Hindu mythology, who had become the son of the sea, the fish-god Ia. The Semite flying bull of the Assyrian sons of the wild cow, has, as his colleague, the flying ram of the Greek story of Jason, the father of the children of Rachel, the ewe. These latter were the prophet race of the sons of Joseph, from the root *asip*, ' the diviner,' the father of the Babylonian Asipu, the god of the oracle spoken by the prophetic gnomon-stone, and his priests, inspired by the divine Bhang of Zarathustra. This god of the oracle was throughout the ancient world, both in Delphi and Assyria, the fish-god, and his worshippers, the new conquerors, who founded the more spiritual prophecy of the age of moral regeneration, were not only, as their predecessors were, sons of the barley, but also sons of the sheep, the paschal lamb of the Jews. It was under the guardianship of the ram, the constellation Aries, that the young fish-sun left his nurse, the moon, in February, and proceeded on the independent course he pursued during the last ten lunar months of his annual existence,[2] and the ram was the sign of the god Ia in early Akkadian script. But these inspired prophetic sons of the ram had been first the warrior sons of the deer, who became, as the fighting men of Judah, sons of the moon-lion, a genealogy which, as I have already shown, proves them to have been sons of the parent of fire, who was the dog first, and afterwards the lion, the constellation Leo, the Lig of the Akkadians, meaning both

---

[1] Hommel, *Geschichte Babyloniens und Assyriens*, p. 320.
[2] *The Ruling Races of Prehistoric Times*, Essay iv., pp. 374-394.

dog and lion. It was these people who made their king and
father the Patesī, or priest-king, of their confederated armies,
and who, by this new creation, superseded the old professional
rain-makers and territorial priests, who were elected, like the
Hindu Ojhas, the men of knowledge (odh), and the Ooraon
Pahans from the priestly families of each district or parish.[1]
This change caused the popular discontent expressed in the
denunciations of the innovators in the Zendavesta, and in
Hindu folk-mythology. They were the great building race
of antiquity, the bearded rulers of Telloh, sons of the bearded
white Mexican god Quetzacoatl, the 'feathered twin or
serpent,' the flying dragon of the Chinese, who appears in
the American rock-sculptures in Illinois, as the Piasa, or bird
'which devours men,' who had a bearded human head, deer's
horns, wings, and a body covered with scales, and ending in
the caudal tail of a fish.[2] It was they who framed the
Akkadian year, beginning with the months of the two
foundations (te-te), and of the pair of bricks (Masmas),[3]
and ending with the Baraziggar, the altar of the Almighty,
on which Abram offered up the totem ram, in lieu of his
son Isaac, the ear of corn. They changed the stars of Gemini,
which, in the Hindu constellation of the Simshumāra, the
dragon, were the physicians whose hands turned its fourteen
stars round the pole,[4] into the two foundations, the pair of
bricks, and the two pillars or door-posts. These formed the
Bab-el or gate (bab) of God, the entrance of the earthly
temple. This, in its earliest architecture, was, as we see from
the temple of the pole-star worshippers, described in Essay
VIII., placed in the south of the building. It was from the
South, the home of the mother-buffalo of the West, that
the worshippers passed into the temple, guarded by the sun-

---

[1] *The Ruling Races of Prehistoric Times*, Essay ii., pp. 76, 92.

[2] *Publications of the Bureau of Ethnology*, vol. x.: Mallery, 'Picture-Writing
of the American Indians,' Fig. 40, p. 70.

[3] *Mas* also means 'an antelope.'

[4] Sachau's Alberuni's *India*, vol. i. chap. xxii. p. 242.

warders of the East and West. The architectural myth of these primæval temple builders is thus shown to symbolise the birth of the regenerated sons of God, baptized in the sacred pool standing before the south door, from the mother-ocean of the South, ruled by the mother-constellation Argo, the heavenly ship, called Mā by the Akkadians, whose yearly voyage is, as I have shown in Essay VIII., traced in the original legend of Jason. These parent-pillars, the sexless twins, who, as gods of time, ruled the days of the solar revolution from East to West, and from West to East, were the pillars of the temple of Solomon, the fish-god, called in the Bible Jachin, the establishment of Jehovah, and Boaz, or Booz, strength or beauty, the originally female Night and the male Day.

The conquests of South-western Asia and India by the armies led by the fish-sun god, in his various manifestations, beginning with the belief in one invisible ruling god, who turned the pole-pillar, the fire-drill of the world, is told in the *Arabian Nights*, in the story which describes the conquering career of Gharib, the poor (*gharib*) outcast, born in the deserts, who preaches while he fights ' the belief in one god, and in Abram, the friend of God,' and in this story we find a graphic picture of the phases of the great unitarian uprising, which in its final career of conquest, was led by the warrior confederacy of the Semites, the sons of Assar, the sun-fish-god, another name of Ia. This is the same story which, in Zend mythological history, is told of the conquests of Keresāspa, the horned (*keres*) horse (*aspa*), son of Sāma, who married the Pairika Kñathaiti, meaning the wandering star (*pairika*) adored by men (*kñathaiti*), the moon-goddess, and also in Hindu history of Karṇa, the horned (*keren*) son of Ashva, the horse-river,[1] both of whom were the mythical leaders of the age which made the year of thirteen lunar months the year of the Turkomans and the fish-god, the national method of reckoning time in Persia and India, the countries which were, to

---

[1] *The Ruling Races of Prehistoric Times*, Essay iv., p. 395; ii., pp. 306, 307. With the horned horse, cp. the horned bird Piasa of the Illinois Indians, p. 321.

the story-tellers of the *Arabian Nights*, the mother-lands of
civilisation. Their story, which tells how Gharib, the sun-
god, leads the armies of the believers from one victory to
another, enables us to understand how these people, who
believed in the regenerating power of pure running water,
imposed their faith on those whom they subdued, founded
the purified form of the Soma and Haoma sacrifices of the
Zends and Hindus, and of the Greek mysteries of Demēter
and made the belief in expiatory ablutions in water instead
of in the earlier lustral bath of blood,[1] an article of faith
preserved in the Hindu, Zend, Semite, and Greek ritual. These
conquering warriors, who subjected the whole world to the
rule of the Semites, also made abstinence from strong drink
binding on all followers of their creed, a rule which has, since
the revolt of the Celtic Aryans, and the establishment of
the worship of Dionysus, the wine-god, and his Semitic
mother, Semele,[2] been obliterated in other countries, but
which still survives in the prohibition of wine in the Kurān,[3]
and in the customs of the Hindu twice-born castes, the Brahmins, Kshatryas and Vaishyas, who think it disgraceful to
drink any intoxicating drink, and carefully guard all water
they drink from contamination.

But besides the story of 'Gharib and his brother Ajib,'
telling of the conquests of the sun-god, there is the still more
archaic and comprehensive history, called the 'Queen of the
Serpents,'[4] which, as I have already pointed out, tells, under
the guise of the adventures of Hásib Karim-al-Din, the Arab
form of the Persian Jamáspa, the story of the religious
history of South-western Asia, up to the age when the
Zoroastrian reformed worship of Ashura Mazda, the fish-god
of knowledge, was made the established religion. Hásib, or
Jamáspa, is said, in the story in the *Arabian Nights*, to have

[1] *The Ruling Races of Prehistoric Times*, Essay iii., p. 181.
[2] *Ibid.* Essay vi., pp. 539-562.
[3] Palmer, *The Qurān*, v. 90 ; S.B.E. vol. vi. p. 110.
[4] Burton's *Arabian Nights*, 'The Queen of the Serpents,' vol. iv. pp. 245 ff.

been the posthumous son of Daniel, a Grecian sage, that is, of the father of the trading races, called by the Hindus, Akkadians, and Turano-Zends, Danu, the Turanian judge, who was also Dan, the father-judge of the Jews, and of the Greek Danaoi. He was left almost penniless by his father ; his only possessions being five leaves of his father's books saved from the shipwreck preceding his death ; that is to say, the learning enshrined in the national knowledge of the Hindu maritime race, led by the heavenly twins, the people who added the two seasons of the rainy season and the autumn to the original Northern year of three seasons of spring, summer and winter, and thus made the year of five seasons, which preceded that of the Zoroastrian Zends, the year of Ahura Mazda, and of the Southern and Western Hindus, the sons of the palm, in which six seasons, the six threads of the Parsi Kôsti, or sacred girdle, were reckoned.[1]

Hásib, or Jamáspa, gained his living as a woodcutter, and was thus the founder of the race of artisans, the tree and sun worshippers, who, like the Druids and Zends, called their father-god Hu, and who, like the Hindus, made the goddess Ka-drû, the tree (*drû*) of Ka, the queen of the serpents, and the mother-goddess, who received from the Shyena, or frost-bird, the divine Soma, or seed of life, which she gave to India, the eel rain-god, and Agni, the fire-god. She was also, as I have pointed out before in this Essay, the thirteenth wife of Kashyapa, the mother of the year of thirteen lunar months, and is named in the Mahâbhârata as the thirteenth month.

Hásib found in a woodland cave a trap-door leading into a chamber filled with honey, the food of the prophets of the earlier barley-growing races, to whom mead was the magic drink of their prophetesses, called by the Jews Deborah, ' the speaking bee,' and by the Greeks Melissai, ' the bees.' This honey was taken away by his companions, who became wealthy with the proceeds, that is, became mead-inspired

---

[1] *Ruling Races of Prehistoric Times*, Essay ii., pp. 78, 115 ; iv., pp. 405-8.

prophets, and they shut up Hásib in the pit. He was shown by a scorpion the track leading to a door by which he was able to escape, and this scorpion was the sign borne on the banners of the tribe of Dan, denoting the star Antares (or Scorpionis, called by the Assyrians the 'star of the lord of seed in the month Tisri.'[1] This was the first month of the Semitic year, beginning with the autumnal equinox. The door through which he passed under the star of the scorpion led him to a hill where he found the serpent queen, surrounded by her subject snakes, the Nāga or cloud snakes of Hindu and Kushite mythology, the snake borne on the banners of Khorasan, and worshipped as Susi-nāg, the snake of the Shus, the sign of royal dignity both in India and Egypt. The queen was a white serpent borne on a golden tray, showing her to be the sacred sun-snake of the barley-growing race, who from time immemorial have begun their year of three seasons at the autumnal equinox, the time when Hadding, the hairy (*haddr*) sun-god of the year of Orion, wedded Ragnhold, the maiden of the twilight (*ragna*), or of the dying-sun, the sun of the six months of the decreasing day, the season of the Pitriyāna, which began the year of Tvashtar, according to the Rigveda. The queen welcomed Hásib or Jamāsp, as the young sun of a new yearly reckoning, which I have shown in telling the conclusion of the story to be that of the year of four seasons, measured by observing the annual path of the sun through the stars. She insisted upon his staying with her, and learning from her the history of Bulukiya, his predecessor in the search after the true interpretation of nature's secrets. He was an Israelitish king of Egypt, who found among his father's treasures a book telling him of one god and his prophet, and who forthwith determined to seek throughout the world for further knowledge of the true faith. He was taken in a ship to the island of the queen of the serpents, who told the story.

---

[1] R. Brown, jun., F.S.A., Star xxiii. 'Tablet of the Thirty Stars,' *Proceedings of the Society of Biblical Archæology*, February 1890.

Bulukiya, after leaving her, went to Jerusalem, where he found a learned sage called Affan, whose books told him that only the queen of the serpents could show them the magic plant, which would enable them to cross the sea and reach the tomb of Solomon, the Akkadian Sal-manu, the fish-sun-god, on whose finger was the year ring. They went together to the queen of the serpents, whom Bulukiya stupified with wine and milk. He thus succeeded in shutting her in a cage, and taking her with them to show the magic plant, the Soma-tree of the divine sap of life, whose changes, as the sprouting-seed, the flowering, seeding, and withering plant, first taught men how to cross the seas of time by numbering the stages of the year, marked by the changes wrought by the sap given to the Indian queen of the serpents Kadrū, the tree (*drū*) of Ka by the Shyena or frost (*shya*) bird. When they found the plant they let the serpent-queen go, and anointed themselves with its juice, which enabled them to cross the seas between them and Solomon's tomb, or, in other words, to learn from plants the laws ordaining the course of the revolving year.

Affan, who tried to get Solomon's ring from his finger, was burned up by the guardian serpent, but Bulukiya was saved by the angel Gabriel, who set him on the way to cross, by the help of the magic plant, the seven seas or days measuring the periods of the lunar phases, between the mother-mountain, where Solomon lay, and the realms of Sakhr, the great mother-bird, who stole the ring of Solomon, according to the story already told in this Essay. Sakhr gave Bulukiya, the sun-mare, or in other words became instead of the mother-bird, who brings the rain, the sun-horse Pegasus, who, as I have shown in Essay viii., discovered the holy wells containing the water of life. The mare took him to the land of Barakhya, the lightning-god (*bharakh*), who sent water to the holy wells, and who, as Solomon's Wazir Barkhya arrested, that is, superseded the mother-bird Sakhr as the ruler of time. From the heavenly kingdom of

the lightning-god Bulukiya went northward as the sun of
the winter solstice, to Kaf, the Caucasus, the mother-mountain
of the Semites and of the pole-star, crowned with the mother-
tree of the Kushite races. Under this tree Michael showed
him the four angels representing the four quarters—one in
the form of a man, who is the dark warrior of Chinese
cosmogony, representing the North, the second a wild beast,
the Chinese tiger of the West, the third a bird, the
Vermilion-bird of the South in China, and the fourth a bull,
the Chinese dragon of the East.[1] They are the four beasts
of Ezekiel's vision, each looking four ways.[2] From Kaf he
went westward to the land of the setting sun, which begins
the Semite day, till he came to the great gate, the Bab-el or
gate of god, through which the day enters. This was
guarded by the lion of the West, and the bull of the East,
the seraph, the winged fiery serpent or lion of fire, the moon-
sphinx lion, and the Kerub or winged-bull of the rising-sun
of the East. On entering the gate he came to the edge of
the mother-ocean, the sky of night, and over this he passed
to an island where he found Janshah, the king (*shah*) of life
(*jan*), the rising-sun weeping between two tombs.

We see in this analysis how the pioneer prophet of the
worship of one god, the god of light, came, in his search
after truth, by the help of the earliest form of the sun-
serpent, the mother-tree, to the knowledge of the revolving
year indicated by plant life, but this knowledge did not
suffice to empower him to wake the fish-sun-god, and learn
the secrets which revealed the true measurement of the year
by the ring of months. He was, therefore, obliged to leave
the quest, and to go to the land of the mother-bird, who
marked the course of the year by coming in the spring and
migrating in the autumn, and from thence to the land of
the lightning-god, whose winds and storms marked the

<hr>

[1] J. O'Neill, *The Night of the Gods*, vol. i. ' Axis Myths : The four living
creatures,' p. 185.
[2] Ezekiel i. 5 ff.

seasons of the year, and the path of the sun-mare, who fills
the springs with living water, the water-goddess worshipped
as Dharti, the wet (*dhar*) mother, by all Hindu-Dravidian
tribes. He next went to the mountain of the North, consecrated to the pole-star, the dark warrior who lights up the
northern night without ever setting; there he learned the secret
of the upright right-angled cross, marking the four quarters
of the heavens by which the annual course of the sun can be
measured, and times calculated more certainly than by the
flight of birds and the prevalence of winds, indicating the
seasons. It was when he was armed with this knowledge
that he could enter the gate of heaven, guarded by the
western and eastern sun, and study the secrets of solar
astronomy, passing also to the central mountain of the
world, where he found the king of life, and learned from him
the past history of the world.

This king Janshah, who told his story to Bulukiya, was
the son of the king of Kābul, the mother-land of the Kushite
race, where their father Kavād was found in the reeds of the
river, called in the Zendavesta Haētumaṇt, the modern
Helmund, on which Kandahar stands, and where Gaṇdhārī,
the bird-mother of the Indian Kushika, laid the egg whence
the race was born. His mother was the daughter of the
king of Khorasan, the centre of the seven Irānian kingdoms,
so that Janshah was the heir of the seven kingdoms of Persia,
ruled by Susi-nāg, and of the seven Nāga or Kushika kingdoms of India, and the era marked by his birth was that of
the confederacy of the sons of the bull of the East, called
Irā in Persia, and the Hindu Go, the cow, both being united
under the common form Gud, the Akkadian bull, father of
the Lu-gud, the race (*Lu*) of the bull (*gud*), meaning the fair
Northern people, and his history tells the story of their
southward march.

When Janshah grew up he, one day when hunting, chased
a gazelle which led him to the sea, where the gazelle took
refuge in a fishing-bark, in which the whole party went to an

island they saw in the distance. On their attempting to return they were driven by the wind into mid-ocean, passing by two islands till they came to a third, and living, during the voyage, on the flesh of the gazelle. The gazelle was the sun-antelope Dara of the Akkadians, a form of the god Ia, Terah of the Jews, the son of Nahor, the river (*Nahr*) Euphrates, who led the Semites, as the sons of Eber and Joktan, or Jokshan,[1] from the land of Armenia, called in Genesis Arpachsad, that is, Arpa-kasidi, the land (*arpa*) of the conquerors (*kasidi*), to the sea by the Euphrates and Indus. This sun-antelope was, as Hindu mythological astronomy shows us, the star Orion, called Mriga-sirsha, the deer's (*mriga*) head (*sirsha*), father-star of the sun-worshipping Brahmins, and the star which, in the cosmogony of the worshippers of the pole-star, was thought to lead the stars during the year of three seasons, the three stars in his belt, in their daily revolution round the pole.

The island to which Janshah was led by the gazelle was the kingdom of the apes, ruled by the wind ape-god of Hindu mythology, called Maroti, the god of the tree (*marom*), by the Gonds, and Hanuman by the Gonds and Hindus. He was the Egyptian Set, god of the South, called in his earliest form Kapi, the ape, whose thigh was the constellation of the Great Bear, called the Thigh of Set. His head was, as I have shown in Essay vii., in dealing with the myth of Perseus, the constellation Cepheus, in which the pole-star was situated from 21,000 to 19,000 B.C.[2] The apes made Janshah their god and king, and he, as receiving his life from the gazelle who brought him to India, was the antelope father-god of the Brahmins. He led the apes in their war with the Ghuls, a name derived from the Akkadian Gul, ' a demon.' They were the tree-worshippers of Maroti, the tree (*marom*) ape, for, in the story of Gharib and

---

[1] For the history of Jokshan, the Yakshus of Indian history, see *The Ruling Races of Prehistoric Times*, Essay v., pp. 471, 480.

[2] J. O'Neill, *Night of the Gods*, vol. i., ' Polar Myths,' diagram, p. 500.

Ajib, the Ghul king, Sa'adan, son of Hindi, king of India, fights with an uptorn tree for his weapon. This tree was the cognisance of the tribes, just as in the Mahābhārata the plough-god Valarāma is described as he who has the plough for his weapon, and the plough marked his banner in war. After conquering the tree-worshippers and establishing the rule of the wind-apes, who measured time by the circuit of the stars round the pole-star, the head of the ape, Janshah came with them to the land of the ants, the myrmidons of Greek mythology the sons of Peleus, the god made of potter's clay ($\pi\eta\lambda\delta s$), on the revolving-wheel of heaven, and father of Achilles, the sun-god. The valley of the ants, the greatly increased population which had grown up under the rule of the sons of Orion, the antelope-star, led to the river which dries up every seventh day, the river-father of the barley-growing sons of the rivers, which divide the land of the ants from that of the Semite Jews, the stream of working time which ceases it labours on the day of rest.

In the battle which took place between the apes and the ants the latter were victorious, but Janshah escaped from both and settled among the Semite Jews, who measured time by the equinoxes and solstices, marking the course of the sun-god Rā, the Hebrew Abram, instead of the antelope-star, and used as their measure the lunar periods of seven days. This was, as we shall see presently, the land of the extreme East, beyond Turkestan, the home of the equinoctial sun. There Janshah was employed by a Jew merchant to get jewels from the mountain of precious stones which no one could climb, that is to say, he was changed from an antelope star into the sun-god who alone could rise to the summit of the heavens, whence he was to send down at the bidding of the unseen father-god the jewels indicating the the sun's yearly work on earth, the jewels of Poshai-yänne, the annual flower-garland of the Greek goddess Koronis. Janshah was sewed up in the belly of a mule, and was thus carried up by the Rukh, the bird of the breath (*ruakh*) of life, the mother-bird, to the top of the mountain, the home of the pole-star Vega, the vulture-star, where he picked up the jewels scattered on the ground and threw them down to his employer, as Poshai-yänne had given the jewels of his bracelets to the Tiämoni. The merchant refused to help him to come back, and he was left as the sun-god in the heavens. He wandered on till he came to the castle of Sheikh Nasr, meaning the elder victory (*nasr*), the vulture-god worshipped by the Sabæan Arabs as El Nasr, the pole-star, the star Vega, which was the pole-star from 10,000 to 8000 B.C. In his castle Janshah found the room containing the carpet of Solomon, the record of the changing patterns of the year woven by the fish-god, who knows its unvarying course. To this room the three bird-maidens, the three stars in Lyra, called the weaving sisters by the Chinese, the swan-maidens of Northern, the three Harpies or time-devouring vultures, and the three Charites or Graces of Greek mythology, came once every year to bathe in the fountain of life which welled forth in the chamber of the spring and autumn sun, the sun of the vernal and autumnal equinoxes, the two periods when the Semites began their year. Janshah was hidden in the room when they came, saw them and fell in love with the youngest sister the Princess Shamsah, the Babylonian sun-god Samas, and, by the advice of the vulture-king, stole her feather garments, the hawk-garment of Freya, the sister of Frey, the deer-god, while she and her sisters were bathing in the fountain in human form. She married Janshah and flew with him to his father's palace in Kabul, thus becoming the Southern sun of winter who had flown southwards from the equinoctial east; but after Janshah had built her a palace and buried her feather garments under its foundations, she stole them and escaped to Takni, the castle of jewels in the North beyond Kaf, the Caucasus. It was then that the Kabul was attacked by Kafid, the King of Hind (India), whose name appears to be connected with that of Kapi the

Ape, the Phœnician sacred apes of Southern Arabia called Keftenu by the Egyptians, who wished to regain the power taken from him by the deer-antelope sun-god allied with the bird-mother, the hawk of the sun-worshippers. But Janshah took no part in resisting the invaders, and escaped during the siege to look for his lost bride, and to find out where Takni was. He went to the city of the Jews and found it in the extreme East, the dwelling-place of the equinoctial sun beyond Chorasmia (Turkestan) or, in other words, in China, the land of jewels, where jade has always been a divine stone. There he became servant to the same Jew jeweller he served before, the invisible god of the revolving-pole who collects, at the vernal equinox, the jewels which are to adorn the earth during the spring, summer, and autumn. He was sewed up this time in the belly of a mare. We have in the mule and mare used to lead Janshah to the mountain home of the year's jewels, the Shu stone of Adar, the Babylonian fire-god,[1] a fresh instalment of primæval history, for they tell us of the ages when the sons of the ass and of the twin-gods Day and Night, the Hindu Ashvins, whose car was drawn by asses, called the ass which bears on its back the Latin cross of the worshippers of the pole-star, their father, as the barley-growing Ooraons still do,[2] and of the later time when they joined with the sons of the mare, the sun-horse, to form the mule race, sons of the Hindu Ashvatara, the mule king, one of the seven Nāga kings, who was the mythological father of the Gandharvi, the heavenly-cloud warriors, the Arab Jinns, the breath of life (ji), the rain-clouds, who were the warriors of the land (gan) of the pole, the Dhruva or pole star. This was placed by Hindu mythologists in Janshah's land of Kabul, to which the mother rain-bird brings the rains every year at the summer solstice. This mule race, worshippers of the pole-star, the moon, and the sun, the

---

[1] *The Ruling Races of Prehistoric Times*, Essay iii., p. 144.
[2] *Ibid.* Essay iii., pp. 255, 256.

Hittite worshippers of the six-rayed star and the lunar crescent, were succeeded by those who made the sun-maiden Savitar, the sun-mare, given by Sakhr, the mother-bird, to Bulukiya, their mother-goddess, in place of the bird ; and it is this mythological age which is depicted in the story of Janshah, the sun-horse, born from the sun-mare. When he was taken by her to the top of the mountain of jewels, and was born in the sun-horse, he refused to obey the commands of the god of the revolving-pole, ruled by the mother-bird, but he went again to the castle of the vulture-Sheikh Nasr. He failed in getting any news there, and Shah Badri, the king of the beasts, the parent-god of the Northern races, who made the beasts, and not the birds of the Finnish races, their totems, to whom Sheikh Nasr sent him, was equally ignorant.[1] But King Shimakh, whose name in this Persian story is probably an Arab reproduction of the Persian Simurgh or Simmurgh, the moon (*sin*) bird (*murgh*), the bird he gave him as his bearer, sent him to the monk Yaghmus, living in the valley of diamonds. This bird, like the Sin-murgh, had four wings, each thirty cubits long, and represented the year of four seasons, the year of the corn-growers when they had become fruit-producers. The history of civilisation, as shown in the stages passed through, from the days when the grove was the mother, is epitomised in the magic staff or tree of Yaghmus. From the base of this, when planted, flesh and blood issued as in the ages of the totem animal sacrifices of the hunting-tribes ; from its middle pieces, milk, the age of milk-offerings to Indra, the eel-god ; from the top, wheat and barley, the age when the barley-cake was eaten and the barley-cup drunk, as containing the seed of divine life. In the hermitage of Yaghmus, Janshah learned from a great black bird, the Northern raven, the

---

[1] Major-General Forlong has proved, in the *Journal of the Royal Asiatic Society*, January 1895, pp. 203-210, that Badr is still worshipped everywhere throughout Eastern Asia as the animistic-spirit dwelling in the mother-mountain.

prophet-bird of Odin, that he must go beyond Kaf to reach
Takni. This bird, the constellation Corvus, took him to a
hill opposite Takni, the Western land ruled by Corvus,[1]
whence a Jinn, sent by Shamsah to look for him, took him
to her father's palace. Janshah and his wife then flew back
to Kabul as the Northern rider on the sun-horse, the sun of the
summer solstice, when the festival of the Soma and Haoma,
the germ of life fostered by the summer sun and rain, was
held. They took with them an army of flying Jinns, or
cloud-warriors, the rains of the rainy seasons. When they
came to Kabul, Janshah sent a Jinn, called Karatash, the
black (*Kara*) stone (*tash*), the storm-cloud ushering in the
rains, to bring back as a prisoner Kafid, the Indian king, the
ruler of the dry hot rainless summer, who was still besieging
Kabul, and who was assisted by Fakun-al-kalb, who had
joined him after the siege began, and who seems, by the name
Kalb, the dog, to be the star Sirius, the dog-star, who, as
Tishtrya in the Zendavesta, brings up the rains at the
summer solstice. Kafid, when arrested, confessed himself
vanquished and went back to India.

After this victory, Janshah and his wife passed their time
in going, as the sun-gods of the year beginning with the
winter solstice, from Kabul in the South, where the young
sun of the new year was born from the sun mother-bird of
the North, ruling the six months, beginning with the summer
solstice, to Takni, the home of the Northern sun, and returning to Kabul for the six months of the dying sun. This
year was closed with the death of the sun-mother Shamsah,
who died, like the Māyā, the mother of the Buddha, when
her son, the sun-god of the new year, was born. She was
buried in the grave in the mountain island, in the centre of
the mother-ocean, where Bulukiya found Janshah waiting to
be buried in the other tomb, when his career as sun-god of
the year, born of the mother-sun of the winter solstice was
closed.

[1] *The Ruling Races of Prehistoric Times*, Essay iv., pp. 330-336.

The story of Janshah tells us of the epochs when the gods worshipped as the rulers of heaven were the wind-apes, who, under the rule of Barakhiya, the storm-god, the Rāvana or storm-god of the Rāmāyana, who cut off the wings of Jatāya, the mother-vulture, drove the antelope-sun, the star Orion, round the four quarters consecrated to the four parent-beasts. The directors of the course of the sun were the twin-gods, Night and Day, who were the hands of the ape, who made the pole the fire-drill of heaven, visible in the pole-star, to revolve. They became the mule-stars, the sexless Ashvins, the great twin-brethren of Greek and Roman mythology, the twin-stars Gemini, when the constellation of the ape was superseded as the ruler of heaven by the fourteen stars of the earliest form of the constellation Draco, called by the Hindus Siṃshumāra, the alligator. The stars Gemini were the hands of the Northern dragon,[1] who superseded Set or Kapi, the ape, who was god of the Southern sun, wedded to the corn-mother, the Akkadian goddess Sala of the copper hand, and it was they who, as the heavenly physicians, turned the stars round the pole. They made it to grind out, as the heavenly fire-drill, the heat and rain which produced life and thought, and brought to earth the Soma, the sap or seed of reproduction. This filled the constellation Kratēr, the cup, and thence descended in the rains of the constellation Hydra, called by the Akkadians ' the divine foundations of the prince of the black antelope,' the prince Janshah, who was saved from death and brought back to life by the flesh of the gazelle, or antelope. This year of the black antelope is that of four seasons, which I have described in Essay IV., and which succeeded the earlier year of three seasons, denoted in primæval astronomy by the constellations Corvus, Kratēr, and Hydra.[2]

This year of four seasons was the year symbolised by the cross of St. George, called, in the story of Bulukiya, El Khizr,

---

[1] *The Ruling Races of Prehistoric Times*, Essay iii., pp. 258, 259; Sachau's Alberuni's *India*, vol. i. chap. xxii. p. 241.

[2] *The Ruling Races of Prehistoric Times*, Essay iv., pp. 332-335, 370-372.

the Arabic form of his Syrian name, El Khudr, or the water god,[1] who was found by Bulukiya, after he left Janshah, and El Khizr took him to Egypt. That is to say, the water-god, ruling the year of four seasons, brought Bulukiya to Egypt as the god Horus, the supreme (*hor*) god. This confirms M. Clermont Ganneau's identification of St. George, the rain-god of Cappadocia, with Horus, son of Hat-hor, the armed knight, with the head of a sparrow-hawk, who, in the Egyptian statue in the Louvre, is slaying the dragon of drought with his spear. This rain-god Horus, son of the hawk-mother Hat-hor, is the god born, according to Egyptian mythology, in the papyrus marsh of Buto, whether his mother was led by the seven scorpions, just as Hásib, or Jamáspa was shown the way to the queen of the scorpions, by the scorpion of the tribe of Dan.[2] These seven scorpions were the seven sacred days of the week, consecrated to the parent-constellation Scorpio, and reckoned by the Semite races as the units of the lunar-solar year, the year sacred to the rain-god, who vanquished the dragon of drought. The history of the year of the rain-god, inaugurated by El Khizr, and ruled by the Egyptian Horus, the god of the year of

five seasons[3] of the five-rayed star $\times$, is told in the remain-

ing portion of the story of Hásib or Jamáspa, who succeeded Bulukiya, as the favourite of the queen of the serpents, and which I have already related. He was the god of the year of the solstitial sun, as shown by the St. Andrew's cross, in the sign of the five-rayed star, and the fish sun-god sanctified and regenerated in the baptismal waters of the sea of life, the constellation Aquarius, and his transformation from the year-god of the races who measured their year by the seasons and the lunar phases, to be the reborn god of the ring of months, begun by his conveyance from Aquarius, is told in

---

[1] *The Ruling Races of Prehistoric Times*, Essay i., pp. 9-12.
[2] *Ibid*. Essay ii., pp. 66, 67.   [3] *Ibid*. Essay iii., p. 271.

the story of the queen of the serpents, by her making promise
Hásib, when he was leaving her, that he would never bathe
in the Hammam public bath, for, if he did, he would cause
her death.   That is to say, the life of the serpent queen
Kadru, the tree (*dru*) of Ka, the mother of the years of the
seasons of the sprouting, growing, and ripening of plants,
would cease when the year was reckoned by the path traversed
by the moon and sun, through the stars, beginning with the
star Aquarius, the star of the winter rains of Babylon.
Hásib was, after his return home, dragged forcibly into the
bath, stripped of his clothes by the bath servants, and when
he came out he was taken to the Wazir of King Karazdan,
who had, by his magic art, recognised him as the man who
could cure his master, the sun-god, of his sickness.   This, as
I have already shown, led to the killing of the queen of the
serpents, and the cutting her up into three pieces, the three
seasons of the year of the growing plant, which were eaten
by the king, and cured him.

This story of the close of the old year-reckonings by
seasons, and the birth of the new astronomy, tracing the
path of the fish-sun through the stars, has a parallel in that
of Gharib and his brother Ajib,[1] telling of the victorious
career of the sun-god, and his conquest of the idolatrous
worship of visible symbols of the hidden god, who makes the
pole revolve, whose chief messenger and prophet is the sun
of light.   Gharib, the poor (*Gharib*) sun-god, was born in the
wilderness, of a concubine of his father, the king Kundamir,
cast out by his brother Ajib, after his father's death.   He
began by making the Ya'arubah clan of Oman, descended
from a Sil'at, or female demon,[2] the pole-star mother-bird,
the ruling clan among the Arabs of Oman.   He then, after
he had been converted to the belief in one god, and in
Ab-ram, the sun-god, as his friend and prophet, by an old

---

[1] Burton, *Arabian Nights*, 'The History of Gharib and his brother Ajib,'
vol. v. pp. 162 ff.

[2] Robertson Smith, *Religion of the Semites*, Lect. ii. p. 50.

Sheikh-conquered Sa'adan, the ghul-tree worshipper, of whom I have already spoken, and rescued from him Fakhr-'Taj, the daughter of Sabur, the king of the Persians, Turkomans, and Medes, who introduced the Turkoman year of thirteen lunar months, the year of the fish-sun, and who became the mother of Gharib's only son, Marad, the acknowledged sun-god. He then defeated and converted Samsam, and the idolaters who adored the three moons, Allāt, the light moon, the Akkadian goddess, ruling the nether world, Manat, the dark moon, worshipped as a black stone, and Al Uzzá, the full moon,[1] Wadd, the Arab form of Odin, or Bodh, the man-god of knowledge, Su-waa, the mother, and Yaghus, the lion, the gods of the totemists. When he had restored Fakhr-'Taj to her father, he conquered for him, and converted Jamrkan, the worshipper of the god of dough, made of dates, butter, and honey,[2] the sacramental food of the sons of the date-palm, the cow, and the mead-drinking prophet. The incidents of the war with Ajib and Jaland are too complicated to be dealt with here, and I will only add a short account, proving Gharib to be the fish-sun-god, of the next part of his career, beginning with the dealings of Gharib, and his brother Sáhim, the moon-god, with Mura'ash, the king of the Jinn, with four heads, showing him to be a counterpart of the angels, of the four quarters, of the story of Balukiya, and of Ezekiel's four beasts. One head was a lion, the second an elephant, the third a panther, and the fourth a lynx. Gharib and his brother had been carried to Mura'ash by Jinns, while sleeping in the valley of springs, the home of the water of life. They converted Mura'ash, a worshipper of the fire-god, by praying for rain, which put

---

[1] Tiele, *Outlines of the History of Ancient Religions*, 'Primitive Arabian Religion,' §40, p. 64 ; Allāt or Alytta was the fish-goddess, the heavenly ship of light of the year of Orion, the ship on which he embarked on his year's voyage. *The Ruling Races of Prehistoric Times*, Essay i. p. 23.

[2] Burton, *Arabian Nights*, 'The History of Gharib and his brother Ajib,' vol. v. pp. 215, 216.

out his sacred fire. By the help of Mura'ash Gharib obtained Al Mahik, the sword of Japhet, the Egyptian god Ptah, whose names mean the opener, the sword of light, made by the master smith, which, according to Turkoman legends, brought the rain.[1] He thus became the sun-god, Rā. Armed with this sword, the sword of Galahad,[2] and mounted on the sun-horse, given him by Mura'ash, he and the Jinns conquered Barkan, the lightning-god of the South, the Barakhya, of the queen of the serpents, visited by Bulukiya, mounted on the sun-mare, given him by Sakhr, the mother-bird. In the story of Gharib, Barkan had taken refuge with the blue king, the king of the Northern sun dwelling in Kaf. Gharib then visited Kaf as the god of the Northern solstitial sun of the summer solstice, and there he wedded the daughter of the blue king, the morning-star, the planets of the new astronomy of the fish-sun. For it was as the fish-sun that he left Kaf, taking the morning-star with him, mounted on the sea-horse, the bearer of the fish-sun, given to him by Mura'ash, to fight his brother Ajib and his ally, Tarkaman, king of Hind, the counterpart of Kafid, in the story of the queen of the serpents. He defeated him and his son Ra-ad Shah, whose name means the thunder-king. He converted Ra-ad, and was installed as the sun-god of the circling sun, supported by Sáhim his brother, the moon-god, whom he made king of Bab-el, the gate (bab) of God (el), the opening months of the year, during which, as I have shown in Essay IV., the fish-sun-god was nursed by the moon. His wives were Fakhr-Taj, the mother of his heir Murad, Mahdiyah, the heiress of the Ya'aruba, or people of the mother-bird, the pole-star, and the Morning-star. His emergence to power as the full-grown fish-sun, entering the constellation Aries, after he had been kept in the obscurity of the three winter months, by his nurse, the moon, is told in the last part of the story, telling of Fakhr-Taj being taken by her father's order

[1] Burton, *Arabian Nights*, 'The History of Gharib and his brother Ajib,' vol. v. p. 242 note.　　[2] See previous Essay viii., pp. 224, 225.

to the banks of the river Jayhun, the Oxus or creator (*hun*)
of life (*ji*), the Gihon of Genesis, when he found out her
connection with Gharib.   Her son was born there, and
Gharib who, in revenge, had conquered and deposed Sabur.
Fakhr-Taj's father, converted his general Rustam, and made
himself master of his capital, Isbanir al Madain.   He was
there attacked by Khirad Shah, an aspirant for the hand of
Fakhr-Taj.   He defeated Khirad Shah, but was afterwards
taken by a Jinn, sent by a sorcerer employed by Khirad Shah,
who was told to cast him into the river Jayhun.   The Jinn,
after stupefying him with Bhang, which was, as I have
shown, the agent of divine inspiration in the theology of the
worshippers of the fish-sun, took him there, but instead of
drowning him, put him on a raft on the river, whence he was
carried to the Indian Ocean, for in ancient geography, the
Oxus flowed into the Caspian Sea, and was thought to pass
thence into the Tigris.[1]   He was there imprisoned by a heathen
king of Karaj, worshipper of a fire-breathing copper idol,
called Mingkash.   He was released by the Jinn Zal-zal, who
had charge of the idol son of the fire-king, Al Muzalzil, the
earthquake-maker, who worshipped a spotted calf, the
Egyptian spotted-bull-god Apis, the god of the star and
moon worshippers.   Zal-zal took him to his father, who
imprisoned Zal-zal, and told a Jinn to take Gharib, who had
converted his son, to the valley of fire ; Gharib killed the Jinn
on the way, and after landing on an island, was carried away
by a Jinn, who took him up to heaven.   Then a thunderbolt
struck the Jinn, and Gharib fell into the sea, and was thus
baptized as the sun-god, in the constellation Aquarius.
From thence he swam to the island of Janshah, the lord of
life, who was a man in the story of the queen of the serpents,
and is here a woman, the mother-bird.   She tried to induce
him to remain with her, but he broke her neck, just as
Hásib killed the serpent-queen, and became the ruling sun-
god of the winter solstice, born from the southern mother-

---

[1] See diagram, *The Ruling Races of Prehistoric Times*, Essay iii., p. 220.

bird. Gharib was carried away by Zal-zal,[1] who had been released from prison, and had slain his father, the earthquake-god. He took Gharib back to Isbanir al Madain, the Persian capital of Khorasan, and to his wives, with all the treasures, and the creating power of Janshah, the lord of life. He was there besieged by Murad his son, and Fakhr-Taj, who had come, after conquering the East, thinking that his grandfather Sabur, who had tried to kill his mother, was on the throne. When Gharib was recognised by Murad as his father, peace was restored, and Gharib, his three wives, and his son, ruled the Turkoman Empire, the eastern world, as the fish-sun-god, of the Turkoman year of thirteen lunar months.

I have now completed the survey of the history of the worship of the all-wise fish-god, and of the successive beliefs, as to the nature of the god who made time. This shows that the theology of all the ruling races of the ancient world, who from America in the east to Europe in the west, traced the reign of peace and of universal knowledge to the fish-sun-god, was framed in a series of stages, from beliefs generated in all quarters of the globe, and united in successive national creeds in south-western Asia. That it was in the Indian Ocean and the Mediterranean Sea that the age of maritime enterprise introduced by the Hittites, the Khati, or joined races of India, called Tur-vasu, Tursene, and Tursha originated, and that it was these people who, when they called themselves the Semite sons of Ia, or Yah, the fish-god, extended the work begun by their predecessors, and became the Phœnician rulers of the ancient world, who explored all seas, and traversed all countries, accessible by the Indian and Pacific Oceans, the Mediterranean and the Atlantic, as far as Britain and Ireland. But, in working out

---

[1] Zal-zal is apparently an Arab form of the Akkadian Tal-tal, or Dadal, meaning 'very wise,' a name of the Akkadian fish-god Ia. He plays the part of the chosen Wazir of the fish-sun-god in this story, the part assigned to Jamâspa in that of the queen of the serpents. Sayce, *Hibbert Lectures for 1887*, Lect. i. p. 28, note 2 ; Sayce, *Assyrian Grammar Syllabary*, No. 15.

this history and retaining it in the memory, it is necessary to have certain definite marks, denoting the different ages of belief, and marking them off from one another, as separate strata in the object-map of past time. These, as we have seen, can be supplied, by the successive year reckonings of the agricultural races, who were the founders of organised civilised life, and the worshippers of the god who made time and arranged the seasons, so as to bring the crops of each year to maturity, in regular and unvarying order. They may be divided into :—

I. The age of the year of the two seasons, marked by the Pleiades, the year-stars of the agricultural people, of the southern hemisphere, the sons of the buffalo, who worshipped the evening sun of the West, the sun of the equinoxes of the equator.

II. The age of the year of three seasons of Orion, the year-star of the East, originating among the mining and fighting sons of the mother-bird of the North, and the deer-god of the East. This was the age of the magicians, when the worship of the pole-star began, which was first, from about 21,000 to 19,000 B.C., one of the stars in the constellation of the Ape, the Egyptian Kapi, the Greek Kepheus, then one of the stars in Cygnus, and from 10,000 to 8000 B.C., one of those in Lyra, first called the constellation of the Vulture.

III. The age of the year of the sons of the Great Bear, who became the seven antelopes, of the race who worshipped the pole-star, guarded by the seven bears, and who looked on the revolving pole, generating the rain and heat, as the invisible father of life. This was the age of the sons of Shelah, the pole, and of the year of four seasons, of the worship of the upright right-angled cross, the ancient emblem of the cultivated earth, the sign of the sun-god and of God the Judge who ruled the world by unvarying laws.

IV. The age of the twin gods, Night and Day, children of the mother of corn, and of the sun-worshipping races, whose totems were the wolf of the East and the bear of the West.

This was the year of the five seasons of the ancient Zends, Hindus, and Egyptians, the year of the solstitial sun and of the five-rayed star, of the Egyptian supreme god Horus, when the two seasons of the summer rains of the Indian Ocean, and the autumn, were added to the previous three. It was in this age that the mother-tree of life was substituted for the mother-bird slain of the Sia twins.

V. The age in which the addition of the St. Andrew's cross of the solstitial sun, the sun of the rider on the sun-horse, to the upright right angled cross of St. George, formed the eight-rayed star, the sign of god and seed. This was the age when totemistic beliefs were discarded, the totem gods turned into stone, and made the gods of the gnomon, when men worshipped the God of time, the invisible ruler, whose messengers and prophets are the sexless moon and sun, ruling the night and the day. This was the age of the wheel-year of the shadow-casting gnomon, and of the tree-mother of flowers, the Greek Athene, goddess of the oil-press, turning as the stars do, round the central pole, the age when mankind were divided by the Kushite rulers, into classes based on community of function, the age of the Ants, ruled by the weavers and potters, and that of the first distant commercial voyages, undertaken by the Twin races, the Hittites, Kheta, or Khati of India, Assyria, and Syria.

VI. The age of the lunar-solar year of thirteen lunar months and of the worship of the fish-sun-god, born of the gnomon-stone and the pair of nuts, the mother-trees of the gardening races who were sons of the mother-bird and the river-eel. It was consecrated to the nine Akkadian Igigi, or spirits of heaven,[1] whose chief was the Akkadian Ia, the Hindu Vishnu, the fish-god, the god of Solomon's Seal, of the two interlocked triangles, the most sacred sign of the building races, which has now become the sign of the Royal Arch of the Freemasons. This was the age

---

[1] Sayce, *Hibbert Lectures for* 1887, Lect. iii., p. 141, note 1.

which depicted the tortoise-earth, round which the fish-sun-god revolved in his yearly course, in the diagram of Solomon's Seal,[1] called by the Hindu Varāhamihira the Kūrma-chakra, or tortoise index of nine vargas, or divisions, the centre of which is Pānchāla, the Gangetic Doab, the corn-land of the Srinjayas or men of the sickle (*srini*), and round this are grouped the eight regions of the eight-rayed star. This was the land of Poshai-yänne as described in the Sia cosmogony, the land of the original Buddhas or knowing ones, the people who had grasped the secrets of the new astronomy, which traced the annual journey of the fish-sun, or Buddha, the all-wise god, who knows the innermost secrets of the universe, through the heavens, the god called Fo-sho by the Chinese, and Poshai-yänne by the Mexican Sia, the god Hue-hueton-catco-acateo cipactli, the fish-god of our flesh, the divine spirit-man, son of the seed-grain, the breath of life of the Almighty Creator. His rule was the epoch of Semite supremacy, of general peace, and of the greatly increased commercial activity which made the Semite merchant-princes of India, Assyria, Egypt, Asia Minor, and Greece, rulers of the world, the founders of maritime cities, and the great builders of the ancient world.

VII. The age of the revolt against Semite legality, headed by the Aryan Celts, votaries of the wine-god, who believed that life is given to man for enjoyment, within due limits, not trespassing upon the rights of others, and not for repressing natural desires, from fear of transgressing the divine law. This was the age intended by its founders to be that of government by the law of liberty, so administered as not to degenerate into unsocial licence, the age of individual effort and aspiration, of poetry, song, and artistic reproductions of natural beauty, and in which the problems, arising out of the conflict between law and liberty, between altruism, or the claims of social law, and egoism, or those of the individual, were first formulated and discussed in theory and practice, as

[1] Sachau's Alberuni's *India*, chap. xxix., vol. i. pp. 296, 297.

those on which the national life and progress depended. It was the age when moral improvements and the elevation of thoughts and desires to a higher spiritual sphere, became the vision looked for, by the prophetic teachers, such as Siddharta Gautama, the Buddha, Confucius, and the Hebrew prophets, who preceded the Indian and Chinese reformers, and who all looked on the attainment of the highest truth in theory and practice as the goal which all should strive to reach.

Ball, magic ball of Medea and of the Russian fairy tales, the rolling ball of the sun, viii. 204

Baresma, the Zend magic or rain staff, viii. 216. *See* Prastara

Barhis, sacred Kuṣha-grass thatching Hindu altar and covering sacrificial ground, vii. 10, 34 ; viii. 147, 156 : substitute used by American Indians, ix. 292

Barley, the plant of life of the Hindu fathers and of the sons of the eight-rayed star, vii. 7, 17, 41 ; viii. 132, 137, 148, 186, 190, 191, 193, 194, 202 ; mixed with the sacramental Soma cup of the Hindus and Greeks, and the sacramental cakes of the Sabæan Mandanites, viii. 160, 162, 178, 179, 219, 229 ; barley mixed with the water poured on the Udumbara (*Ficus glomerata*), the house-pole of the sons of the fig-tree in the Hindu Sadas or house of God, viii. 166, 167 : wheat used by the Parsis, instead of barley, for their sacramental cakes called Drōna, viii. 162 note 1 ; Indian corn used instead of barley in Mexican sacrament, ix. 286, 287, 289

Basque, sons of the forest (*baso*) and of the boar-god, also called Iberians or Ibai-erri the people (*erri*) of the rivers (*ibai*), viii. 109, 129, 130, 149

Bast, Egyptian cat-goddess, form of Hat-hor, the hawk-goddess, ix. 241

Bathilda or *Bothvildr*, she who joys in strife (*both*), the winter-goddess, mother by Völundr (Wieland) of Widonga, god of woods and meadows, viii. 94, 99, 100, 121.

Bear, parent totem of the Finn magicians and parent-constellation of the barley growers of Western Asia and Greece, marking the seven days of the week by which the worshippers of the pole-star and the Latin cross of St. George measured their year of four seasons, vii. 17, 21, 26, 49, 83 ; viii. 93, 97, 99, 137, 201 ; ix. 247, 253, 260; the seven stars of the Great Bear and the pole-star, the eight sacred mother and father-stars of the Kabiri, vii. 58 ; Besla, the bear-

mother of Odin, viii. 102 ; the bear of the west of the Mexican Sia, successor of the buffalo, after Ūt'set, the corn-mother, had gone to heaven, ix. 258, 261, 263 ; bear medicine animal and inspired prophet (*Manido*) of the Ojibwas, ix. 262 ; bear-god of the west, slain by the Mexican twin-gods, ix. 268 ; white sun-bear of the west of the Mexican Sias, ix. 290. *See* Artemis, Deer, Wolf, and Twin-gods.

Bear, the Little, called by Egyptians the constellation of the Jackal, vii. 83.

Bee, sacred divine-bee of the Persians, viii. 215

Beetle, the Mexican sacred beetle, holder of the star-bag, and the Egyptian Scarabæus, ix. 260

Bellerophon, rider on the sun-horse Pegasus, slayer of the sun-ram with the golden fleece, viii. 178

Belt of Orion, the sacred girdle (*Mekhalā*) of the Hindu Brahmins, the Kûsti, or girdle of the Parsis, vii. 20, 21 ; the Kamberiah of the Sufi dervishes of the East, ix. 257 ; its stars in Mexican cosmogony, the sacred stars of the Kat'suna, or men with masks, the totem-worshippers of the North, ix. 260.

Bhang or *Hashish*, made of hemp (*Cannabis Indica*) use of by Zarathustra and his disciples in eliciting oracles and learning divine knowledge, ix. 299, 301, 302, 340

Bird-mother, myth of, the mother rain-stork of India, vii. 4 ; the mother-bird in whose house the year-calf of Vedic time-measurement was born, vii. 8 ; the two birds Day and Night, vii. 9 ; the holy eagle, child of the waters and rain-plants, vii. 10 ; five sun-eagles of the Rigveda, vii. 59 Gaṇ-dhārī, she who wets (*dhara*) the land (*gan*), bird-mother of the Kaurāvya or Khurāvya (*khu*), vii. 3, 42, 47 ; viii. 123 : the storm-bird mother of rain who brings to earth the Waka of the American Indians, the mysterious soul of life called Takoo Wakan, ix. 235, 236 ; the Norse Hreiðmar, who keeps the nest (*hreid-r*), the Shyēna

of the wheel-year of the oil-press, and of the fifteenth Sujata born (*jata*) of Su, son of the hill-bamboo the parent-totem of the Bhāratas, vii. 70-72 ; the four frogs and four snakes worshipped by the Egyptians as creating gods after the eight apes, vii. 71 ; viii. 208 ; the frog-sun slain by the help of the year-ring of the Esthonian mother-goddess Pörga Neitsi, a counter-part of the Arabian Aminah, partner of Sakkr, the Indian Sukra, and of Solomon, the fish-god, in the *Arabian Nights*, ix. 296

Galahad and the Holy Grail, story of, and its meaning, viii. 223-230

Gandhārī, bird-mother of the Hindu Kaurāvya. *See* Bird-mother myth

Gaurias of Telloh and India, the Phrygian Satyrs, sons of the wild cow (*gauri*) and father-goat amalgamated with the sons of the sheep, their features as depicted on the monuments and in their descendants, ix. 318

Gnomon-stone, originally the house-pole of Finn round houses, the sacrificial stake, the Ashēra or rain-pole of the Jews, and of the Hindu and Basque creating-god Vasu, then the obelisk used to measure the sun's daily and yearly course, the German sun-god Sigurd, the pillar (*urd-r*) of victory (*sig*), the Hindu Dhritarāshtra, husband of Gandhārī, the bird-mother of the Kushites, the Hebrew Jacob, vii. 5, 46, 64 ; viii. 118, 122, 123, 137, 166-168, 174 note 1, 200, 224 ; ix. 319 ; the spear of the sun-god, the weapon that cannot be baffled, viii. 185 ; the mother-tree of the united sons of the ash-tree marking time by its shadows, the ash Yggdrasil growing from the Urdar or fountain of the pillar (*urd-r*) in the Edda, viii. 185 : ix. 251 note 1 ; the gnomon-stone of the temple of the sun-horse at Stonehenge, used to mark the rising and setting of the sun at the solstices, viii. 144, 145 ; the sacred gnomon · stone now the church spire, viii. 224 note 3 ;

originally deified by the Sabæans of Central Asia, viii. 175

Goat-god (*see* Esau), the goats which draw the chariot of Thor, viii. 108, 194

Goths, sons of the bull, subjects of Siggeir, conqueror of the Volsungs or woodland people in the Nibe-lungen Lied, viii. 3 ; their union with the Lithuanian sons of the horse, worshippers of the sun-god Rā, when they became the Massagetæ or greater Getæ, viii. 130-132

Groves, sacred village groves of India still existing in Abyssinia, where the church, like the Hindu temple, stands in the village grove, vii. 44

Hadding, the hairy (*hadd-r*) red-haired sun-god of the North, the counterpart of Frey, the deer-god, son of Njord, the god of the north pole-star, who came from Asia Minor to the north, the successor to the wolf-god, whose blood he drank and whose heart he ate, viii. 96, 97, 108, 109 ; ix. 255, 256 ; the year he ruled began, like that of Asia Minor, with the autumnal equinox, viii. 106 ; ix. 325 ; he was the god of the year of the Great Bear who was his guardian called Wagnhofde the wain-head, the age when Odin was made supreme god, viii. 108, 124 ; proves, like Njord, by his feet that he is the circling sun-god, viii. 106 (*see* Njord) : he slew Swip-dag, the star Orion, and thus put an end to the year-reckoning by Orion, viii. 109

Haran, the road (*Kharran*), the city of the wives of Jacob and of the year of the fish-god, viii. 198 ; worshippers of the prophetic gnomon, ix. 319. *See* Sabæans

Heavens, the four Buddhist heavens, (1) of the hundred sons of the mother-bird, (2) of the thirty-three Nāga gods headed by Sakho the rain-god, (3) the heaven of the Twins, (4) of the Tusita gods of wealth (*tuso*), vii. 73

Hekate, Greek counterpart of Hindu

riage compared with that of the Arabs, ix. 312, 313.

Koronis, the flower garland, marking the flower clock of the year, sister of Ixion, and counterpart of Athene, the flower (ἄνθος) goddess, viii. 93, 182, 206.

Krater, constellation of the Cup of Demêtêr, the barley-mother, viii. 179; ix. 335.

Krishanu, rainbow-god of the Rig-veda, the Zend Kereshani, god of the sorcerers, the invading Magi fire-worshippers, ix. 315, 316.

Krishna, the black antelope-god, first form of Vishnu, vii. 18, 40, 70; viii. 157; ix. 277, 294, 316.

Krittakas, the spinners, the Pleiades. *See* Pleiades.

Kshatriya, Indian warriors of the red race, whose clothes are dyed with madder, and who wear deer skins, as sons of the deer, ix. 233, 238, 309, 310. *See* Deer.

Kurrum, Ooraon festival, compared with that of the Dakota buffalo sun-dance, ix. 293.

Kusha grass, sacred grass of the Hindus, the favourite food of the antelope, on which the Fathers were called to sit at the Pitri-yajña annual sacrifice to the Fathers, vii. 7, 37; viii. 229, 272; an ingredient of the sacramental Soma used by the sons of the antelope, predecessors of the sons of the horse, to make the sacrificial Prastara or magic rain-wand, vii. 7; viii. 178. *See* Prastara.

Kushika, Kushites, Kauravya, Kurus, the Nāga sons of Gandhāri, the mother-bird (*khu*), vii. 3, 27, 46, 47; viii. 91, 123, 124, 150, 227; ix. 236, 312, 314; sons of the Helmend lake Zarah, the home of the mountain lotus, vii. 78; ix. 328. *See* Jāts.

Kutsa, priest, and visible form of Ka, the god of the city-building Purus, ix. 304.

Lake, or water reservoir, the sacred pool of the Hindu sun-fish Rohu and of the lotus, in Europe and Northern Asia, the sacred wells of the sun-horse, vii. 75, 78; viii. 176, 177; first the sacrificial pit of the Kabiri and Hekate, viii. 168, 169; Tiphys, the sacred pool, (τίφος), steersman of the Argo, viii. 201, 202; the holy pool of Hekate, whence Medea got the water of life for Jason, viii. 203.

Lapithæ, people of the storm-wind (λαπ), worshippers of the three vulture Harpies, conquered by the Centaurs, sons of the sun-horse, viii. 198, 199.

Leah, the wild cow, wife of Jacob, the gnomon-stone. *See* Bull myth.

Leo, constellation of, ruling the year of Yudishthira, vii. 43, 60; the constellation of the sons of Judah, sons of the dog and lion, ix. 320.

Leto, wolf-mother of Apollo, vii. 26; viii. 97; ix. 281.

Lingal, Gond god of the threshing-floor of Gonds, ix. 249, 254.

Licchavis, sons of the fire-dog and lion (*lig*), Indian confederates of the Malli or mountaineers, ix. 262.

Lion, the father lion of the tribe of Judah and of the Indian Yadavas who was first the dog-father of the Arab Jinns, flying bird-gods, and of the seven sons and seven daughters of the mother-wolf and father-lion, the phases of the moon, vii. 44; ix. 298, 320; year of the lion in Arabia, vii. 43, 44: lion of the west, guarding the gate of God, ix. 327

Loki, wolf fire-god of the Edda, who ruled during the abdication of Odin, after the matriarchal age, viii. 97, 109 (*see* Hadding); Loki, his son Fenrir, the wolf, and the Midgard serpent-gods of the dwarf race, superseded by Odin and the Æsir gods of Vanirheim (Asia Minor), viii. 104; with Loki was Hœnir, the sun-horse, who went to Vanir-heim in exchange for Njord, the god of the pole-star, viii. 105; Loki with Hœnir kills Thjassi, the vulture frost-giant, and gets from him, before the coming of Njord and the worship of the pole-star, the apples of Iduna, the seeds of life, viii. 105, 119; Loki, Fenrir, the

the Shemol, the star of the left hand, viii. 160, 194. *See* Athene *Orion*, year-star of the year of the revolving pole of the sons of the deer-sun, called by the Hindus Mrigasirsha, the head (*sirsha*) of the deer or antelope (*mriga*), and Prajā-pati, lord (*pati*) of former (*pra*) generations (*ja*) the tree-stem whence they sprang, the Droṇa, or holy jar of the Soma or sap (*su*) of life, called Ka, the supreme god, he in India as the hunter antelope-star chased Rohini his daughter, the doe-antelope (Aldebaran), and became by her the father of the household fire-god, the sacred central sacrificial flame of the house, and the national altar called Nābhānedishtha, nearest (*ne-dishtha*) to the navel (*nābhā*), and Vāstoshpati, lord (*pati*) of the house (*vastu*), in the Zendavesta he led the Pleiades, vii. 18-21 ; the three stars in his belt, called the three-knotted arrow of the three seasons of the year, the sacred girdle of the Brahmins and of the Sufi dervishes of the East, vii. 18 ; ix. 257 ; Ut'set, the Mexican corn-mother placed them in the sky as the stars of the Kat'suna or men with masks, the Northern totem-worshippers, ix. 258, 260 ; they were the stars of the hunting deer-sons of the Mexican Indians, ix. 256, 266, 267 ; Orion was the Akkadian Dumu-zi, the son (*dumu*) of life (*zi*) the Egyptian Smati-Osiris the sun-god, who in November launched his moon-bark, in which he circled the heavens and hunted the moon-hare, the constellation Lepus, at his feet, vii. 21, 40 ; he was the sun-god of the age of silent worship, vii. 33 ; his year, first of three seasons, afterwards of five, began at the winter solstice the last day of Mārgasirsha (November-December) sacred to him, and the first of the month sacred to Rohini, the doe-antelope, which became, Push, the bull when Rohini became the red cow, vii. 34, 35 note 2, 43, 65 ; this is still the sacred year of the Hindus, beginning when the sun turns north after the winter solstice, vii. 36, 37 ; Orion's year of three seasons was the year of the sons of the bird, predecessors of the sons of the ass and goat, the Khāti, who worshipped the pole-star first in Cygnus, and afterwards as Vega the vulture star, vii. 22, 24; ix. 342 ; the earlier year of Orion began in the middle of November, vii. 40 ; Orion's year followed, by that of four seasons, the Great Bear, viii. 109 ; ix. 342 ; in the year of Orion the mother-stars were the Pleiades, the Grecian Penelope, weaver of the web ($\pi\eta\nu\eta$) of time the Hindu Krittakas, the spinners, vii. 28 : the stars of the Great Bear, were mother-stars in the theology of the Kabiri, viii. 58

Osiris, the star Orion, the Egyptian year-god of the barley-growing races, vii. 21, 31 ; the god of one eye of the sun, viii. 213

Oxus or *Jihun*, the creator (*hu*) of life (*ji*) which caused the final resurrection to life of Gharib, the fish sun-god inspired by the Bhang of Zarathustra, ix. 339, 340. *See* Bhang

Padumavatī or *Uppalavannā*, the lotus mother of the Buddhists, the third of the thirteen Thēris, her story and its historical import, vii. 74 - 80 ; goddess of miraculous power, vii. 84

Pajāpati, called in the Buddhist scriptures Mahā Pajāpati Gotami, the great (*mahā*) ruler of former generations (*see* Prajā-pati) daughter of the bull or cow (*gau* or *go*), name of the moon-nurse of the Buddha, the fish sun-god, she was the first of the thirteen Thēris, who gave their names to the thirteen lunar months of the year of the fish sun-god, and was the mother of Nanda, who in Hindu mythology is the bull of Shiva, vii. 69, 72, 73 ; her rebirth as the waning moon in the form of Kisā, the emaciated Gotami, vii. 82

Pāndavas, the five sons of the reputed father, the antelope sexless

of the Pulon tree, the mango stones, parents of Jarasandha and those of Jacob's Almond tree, Luz, compared, ix. 285, 286

Pitarah Somavantah, the rice-growers, earliest race of Hindu fathers, served in the Pitri-yajña with rice on six platters, vii. 8

Pitaro Barishadah, the barley-growing fathers of the Kushite race, seated on the Barhis or altar-seats of Kusha-grass, to whom parched barley was offered, viii. 156

Pitri-yajña, annual Hindu festival to the Fathers, vii. 7, 8

Pleiades, measurers of the year of Tvashtar, the god of two (*tva*) seasons of the year of the Western Hindus, and of the matriarchal races of the Southern Hemisphere, they called them Krittakas, Kirttida, the spinners, vii. 14, 15, 16; it was in the Pleiades year that the Hindu custom began of dividing the year into two periods of six months each, parting in the month of two (*vi*) branches (*shākha*), Visakha (April-May), one sacred to the bright epochs (*deva*) called Dev-ayāna and the other to the Fathers Pitri-yāna, vii. 14, 17: Pleiades called Parwe, the conceiving (*peru*) mothers by the Sabæan Soghdeans of Central Asia, who placed them first in their list of lunar stations, vii. 12, 16, 28; the Pleiades mother Penelope (*which see*), wedded to Odusseus, the star Orion, which ruled the year succeeding that measured by the Pleiades, vii. 22; in the next age to that of the year of Orion, that of the year of four seasons, the Pleiades, still called the spinners, were wedded to the Great Bear, vii. 17; viii. 200, 201; the Pleiades became in the age of the worship of the bisexual fish - sun by the patriarchal successors of the matriarchal races, the stars of the doves (πέλεια), and were called by the Greeks Peleiades, and by the Hebrews Kimah, the binders, the stars consecrating conjugal union, viii. 200; ix. 252, 253; the Pleiades

in Mexican cosmogony called Sussistinnako, the creating spinner, the spider, who created the six regions of the earth, the hunter-sun-god Orion, and was the grandmother who built the rainbow bridge of the day-sun-god for the Mexican twin-gods of time, ix. 248, 250, 256-266 (*see* Bridge of Heaven); the Pleiades stars placed in the heavens by Ut'set, the Mexican corn-mother, with those of Orion and the Great Bear, ix. 260

Plough god, called Nagur the plough, by the Gonds Ge-ourgos, the worker (*ourgos*) of the earth (*ge*), and also Elias by the Greeks, El Khudr, the water (*khudr*) god in Syria, and El Khizr in the *Arabian Nights*, he is St. George of Cappadocia, whence his worship spread over Western Asia and Europe, his sacred sign is the upright-right angled Latin cross (*see* Cross), and his festival is celebrated in India, Syria, and Europe, in April, vii. 29; viii. 135, 136; ix. 335, 336

Pole, the mystic pole carried by the American Indians in front of migrating tribes, and the pole or lance, Shelah of the Jews, ix. 236, 237, 243

Pole-star, worshipped as the only visible sign of the unseen god, the fire-drill of heaven, who in the mythology of the age of the mother-bird succeeding that which measured the year by the movements of the Pleiades made the stars led by Orion revolve, the pole-star god is called in the Rigveda Agohya or 'he who cannot be hidden,' vii. 22, 27, 44, 45; ix. 257: the supreme god of the Sabæans of Byblus, the Semite Phœnicians of Palestine, who worshipped him as Eshmun, the eighth Kabir, the other seven being the seven stars of the Great Bear, vii. 58; Sabæan Mandanite ritual of the worship of the pole-star called by them the world of light (p. 150), 'the primitive light, the ancient light, the divinity self-created' (p. 161), viii. 157-164: worshipped as Shemol, the star of the left-hand, by the Haranite Sabæans, who worshipped the rising sun of the East, the lion of Judah, instead of the sun-god of the West ruled by the star of Dan, Antares in Scorpio, the setting sun, which still begins the Hebrew day, vii. 55 ; viii. 163, 164, 194 (*see* Dan and Judah) ; calendar of the successive pole-stars beginning with those of the Ape, Cepheus pole-star from 21,000 to 19,000 B.C., viii. 214 ; the pole-star Vega, called Ma'āt by the Egyptians, vii, 45 ; the subsequent Egyptian and Buddhist pole-star of the Jackal, vii. 82, 83

Potters, the ruling race, with the weavers, of the age which reckoned time by the year of three seasons led by Orion, the Hebrew sons of Shelah, the pole, the grandson of Japhet, the Greek sons of Mount Pelion, the hill of the potter's clay (πηλός), they worshipped the pole star and the fire-drill of heaven, which makes the stars revolve when turned by the creating potters, the Egyptian Ptah, the Hebrew Japhet, the dwarf-god of the Kabiri, who also bears the hammer, and the architect Egyptian god Chnum, viii. 93, 94, 183, 195, 201 note 2 ; were descended (*see* Minyæ) from the eastern (*kedem*) Kadmus and from Tyro or Turo, the revolving pole (*tur*), the potter's wheel of heaven, and were ruled by Tyro's son Pelias, made of the potter's clay (πηλός), viii. 183 ; they were the Iberian sons of the rivers of the early Bronze Age, viii. 193, 194 ; who took with them the potter's wheel and the oil-press in all their migrations to India, Assyria, Egypt, and Greece, viii. 196 ; they were the Turanian sons of Dan, and confederates of the weavers, also the sons of Shelah, who worshipped the Pleiades as the spinners, viii. 197, 200, 201 note 2 ; their year was one of three seasons, the three daughters of Pelias, the ram sun, slain by his daughters in the form of a ram, viii. 208, 209 ; their age, that of Pelops, Peleus, and Pelias,

preceded that of the worship of the solstitial sun-horse, and the institution of horse races in honour of the dead, viii. 227 ; the Sia artistic potters, ruling in Mexico, ix. 244

Poshai-yänne, the Mexican fish sun-god, ix. 244 : his birth from two Piñon nuts, like that of Jārasandha, the Māghada sun-god, from the mango stone, iv. 272, and the story of Jacob, the gnomon-stone, called Bethel, or the House (*beth*) of God, set up, that is, born under the almond-nut tree Luz, ix. 285, 286 ; he won the rule of the North by gambling like Yudishthira, the sun of the Pāndavas, ix. 277, 278 ; he was, like Sigurd the German, stabbed to death, and rose again from the lake into which he was thrown as the eagle sun of a new year, ix. 275

Prajāpati, lord (*pati*) of former (*pra*) generations (*ja*). See Orion *and* Pajapāti.

Prashastri, teaching priest of the Hindus, priests of the twin-gods Mitra-Varuna, who ruled the year of four seasons, of the twin races who kept the national Shasters, or records, Preface, ix ; viii. 167 ; became the Zend Frashaostra, the eldest of the Zoroastrian triad, the first of the creating-fires, ix. 301, 302

Prastara, the magic rain-wand of the Hindu sacrifices, made by the sons of the horse (*ashva*) of Ashva-vāla, or horse-tail grass, instead of the earlier Kusha-grass, vii. 7 ; viii. 139, 162 note 1

Ptah, the Egyptian and Kabiri dwarf-god of the hammer, the Latin cross, the Hebrew Japhet, turner of the potter's wheel, vii. 52, 53 ; viii. 195. *See* Cross, Kabiri, Potters

Purūrāvas, the eastern (*puru*) roarer (*ravas*), the thunder-god, the fire-drill of the Soma sacrifice, husband of Urvashi, the fire-socket ; they were parents of Ayu, the god of recorded ages, vii. 99, 100 ; ix. 277, 305

Purus, the Eastern city-builders of pre-Vedic India, the race formed by

of god of the Hindu Soma sacrifice, viii. 167 ; the Mexican tribe of the sons of the reed, ix. 245 ; birth of the sons of the reed when Ût'-set, the Mexican corn-mother, was brought to the corn-growing land of the upper-earth by the reeds, ix. 248, 249

Ribhus, the Greek Orpheus, the German Elves, makers of the seasons in the Rigveda, and the gods of the Kabiri, vii. 50-52

Rohinī, the antelope-doe and the red cow, the star Aldebaran. *See* Aldebaran *and* Bull myth

Sabæan Arabs, worshipped the Pleiades as Turayyā, the stars of the revolving-pole (*tur*), vii. 17 ; they and the people of Western Asia were the first worshippers of the gnomon-stone, viii. 175 ; they reckoned their year by the constellation Leo, called the year of the Lion, vii. 43, 44

—— *Mandanites* of Mesopotamia, sons of Manda, the word of God, worship the pole-star, and begin their year, like the Semites, with the autumnal equinox, viii. 150, 151, 154 ; ritual of their New Year's festival begins with baptismal purification and the sacramental eating of the sacred cake made of barley and oil-seed sanctified by the sacrificed pigeon ; ends with the sacrifice and sacramental eating of their totem-wether, the sexless sun-ram, the sun-god of the west, who begins the Hebrew day, viii. 157-164 ; their reed temple and national sacred dress, viii. 158, 159 ; Sabæan temple compared with the Hindu Sadas, the house of the gods of the sons of the fig-tree, viii. 166, 167, 171-173

Sabæans of Byblus, Phœnician Semites, first called the Turvasu, whose creating-god (*vasu*) is the Tur, or revolving-pole, who settled at Turos the island of Dilmun (*Bahrein*) in the Persian Gulf, and went thence to North Palestine ; they worshipped Sakkun, the Pali-Indian god Sakko, the Sanskrit Sukra, the

wet (*suk*) god of the Rigveda and Mahābhārata, and the eight Kabiri, the seven stars of the Great Bear, and their supreme god Eshmun, the eighth, the pole-star, vii. 54, 56, 58 ; viii. 220, 221

Sabæans of Haran sacrifice children at their New Year's feast, and mix the blood in cakes eaten throughout the year, worship Shemol the pole-star of the left-hand, and the Slavonic god Boga, the Sanskrit Bhaga, the tree of edible fruit, they eat at their New Year's feast sacramental bread and wine, vii. 15, 55 ; viii. 153

—— *of Persia*, the Soghdeans, the early fire-worshippers, worshipped their dead in November, and placed the Pleiades, called the Parwe, the conceiving (*peru*) stars at the head of their lunar stations, vii. 16

Sacrifices, first of totem animals, afterwards changed into sacrifice of totem plants, and those possessing magical virtues, vii. 37, 38 ; historical ages denoted by different methods of sacrificing animals, first by breaking the frontal bone, secondly by cutting the jugular vein, and thirdly by strangling them, vii. 52, 79 ; viii. 169, 179

Sal-tree (*Shorea robusta*), parent-tree of the Dravidians, under which the Buddha, the sun fish-god, was conceived and born, vii. 68, 70 ; it and the Champaka parent-tree of the oil-growers were the first parent-trees of the Bhāratas, vii. 72

Sal-fish of Indian mythology, viii. 212

Sal-manu or *Solomon*, the sun fish-god, *which see*

Sar, Shar, cloud-mother of the Hindu autumn season, vii. 29 ; of the Armenians, and mother of corn of the Akkadians and corn-growing sons of the tree, viii. 188 ; mother of the twins Day and Night, the dog-goddess Saramā, ix. 261 note 2 ; the husk-mother of the seed-grain. wife of Abram, the sun-god, vii. 188 note 1

Set, Egyptian god, originally Kapi, the ape, ix. 335

Seven, sacred number of the races

who made their week of seven days instead of five, viii. 138 note 3: and who, like the Kabui and worshippers of Artemis the bear-mother, divided their year into four seasons ruled by the Great Bear, vii. 21, 20, 58; viii. 93; the seven-wheeled car of the year in the Rigveda drawn by seven horses, vii. 7; seven strings fixed by Hermes on the lyre, the constellation of the Tortoise and of the pole-star Vega, viii. 142; seven Mexican Sia societies of the ants and the sacred river of the Semites which dries up every seven days, ix. 274, 330; the seven kingdoms of Irān and India, the Kushika arrangement of the state as the six subordinate kingdoms surrounding the seventh in the centre, ix. 314, 328. *See* Fourteen

Sigmund, in the Nibelungen Lied an equivalent of Wieland the master-smith. He was the tenth son of the king of the Volsungs, or woodland sons of the mother-tree from which he alone was able to draw the sword of light. He became the wolf fire-god by slaying the mother-wolf and drinking her blood, viii. 111, 112; his union with Signy, as the Finn witch-wife, made him father of Sinnfiötli, the wolf sun-god of the united Finn and matriarchal tribes, viii. 113, 114; escape of Sigmund and Sinnfiötli as the shining fire and sun-gods, who had regained the sword of light temporarily taken from Sigmund by Siggeir, king of the Goths, viii. 115; marriage of Sigmund as the sun bearing the sword of light to Borghild the mountain mother, viii. 116; and his second marriage after the death of Sinnfiötli to Hjordis, moon-mother of the herds, and the death of Sigmund, who bequeathed the shards of the sword of light, the lunar phases, to his son Sigurd, to be born from Hjordis as the rider on the sun-horse, viii. 116, 117

Sigurd, the pillar (*urd-r*) of victory (*sig*), the sun-god of the gnomon-stone, son of Hjordis, the cow moon-mother, and Sigmund (*which see*);

he circled the heavens as the rider on the grey sun-horse Grāni, the clouds, given to him by Grip (Sirius), the eastern keeper of the sun-horses, viii. 117, 118; the sword of light made of the shards of that of his father Sigmund, the lunar phases forged for him by Regin, the master-smith, god of rain (*regn*) and evening twilight (*ragna*), the sun of the west, viii. 118; he goes after being blessed by Grip (Sirius) through the realms of darkness, followed by Regin to the Glittering Heath, where he slays Fafnir, the guardian snake of the sun-light of night, brother of Regin and the wise Otr, the otter river-god, sons of Hreidmar, the mother-bird, keeper of the nest (*hreid-r*), gets from Hreidmar's treasure (1) The helm of Awing, the sun-god's cap of invisibility (*see* Perseus); (2) the golden coat of mail; (3) the golden year-ring and the red rings of Andvari's treasure (*see* Andvari), viii. 118, 120; after killing Fafnir, he cooked and ate his heart, thus acquiring his strength, and then slew Regin, the god of the gloaming, and became the god of the light of day and the master-smith, viii. 121; as the rising sun of day he went up Hinda-fjall the hill of the deer (*hinda*) the sun-hill round which Orion, the deer-hunter, led the stars, and then he awoke Brunhilda, the hawk mother-goddess of the springs (*brunnen*), and plighted his troth with her by giving her Andvari's ring, and then he became the sun-god of a new epoch of time, viii. 121, 124; after descending the hill of sunrise he spent the spring and summer with Heimir, the earth (*heim-r*) god, whose wife was Brunhilda's sister, and after the summer solstice he went to the western land of the Niblungs, children of the mist (*nebel*), viii. 124; he there forgot Brunhilda as his wife, and wooed and won her disguised as Gunnar the moon-king, son of Giūki, king of the Niblungs, the offspring of the boar-sun, and got from Brunhilda Andvari's ring,

the Sufi Dervishes, vii. 20; ix. 257; the Akkadian and Egyptian year of Dumu-zi and Osiris-Smati, who are both the star Orion, vii. 21, 40; the Mexican year of the Kat'suna Northern totem races, the men with masks, who worshipped the Wild Hunter of Germany, the first form of the hairy sun-god of the North, the Norse Hadding and Frey, the deer sun-god of the Edda (see Hadding and Twin Gods), vii. 22; ix. 256-258; (iii.) the next year was that of the sons of the ass, the earliest barley-growing races, the nucleus of the Hittites, beginning with the orgiastic dances of the Hindu Moondas and Ooraons in Māgh (January-February), which still survives in our St. Valentine's Day, vii. 41; this, and the year of Orion, were the years of the earliest worship of the pole-star, that of Cepheus, the Ape, and Cygnus, the Swan, ix. 342; the annual circles of the stars led by Orion were measured by the phases of the moon-hare, the constellation Lepus, at his feet, vii. 21; it was the year of the three Harpies, the three seasons of the year of the storm-bird, vii. 51; during this age weeks were reckoned as periods of five days, a reckoning afterwards used in computing the months of the year of 360 days, viii. 138 note 3, 139, 142; (iv.) the fourth year was the year of four seasons beginning at the summer solstice, the year of Sirius, dog of Orion, before he was the sun-horse Tishtrya, and of the four-eyed dogs of the Rigveda and Zendavesta, vii. 23-28; this was the year of the pole-star Vega, the Gridhra, or vulture, of the Rigveda, vii. 27; the year of Hermes, the dog of the gods, who gave seven days to the week, called the seven strings of the Lyre of the tortoise, the constellation of Vega, and of the four dogs of the Babylonian fire-god Bel, the Akkadian Bil, the god of the upright right-angled cross, vii. 50-65; the year of the black antelope of Artemis

as the deer-goddess in which the ship of the sun was steered by Agastya (Canopus) from the east to the west and back again by north to east, the year in which the sun circuits were marked by St. George's cross, vii. 26, 27; ix. 335, 336, 342; this year of four seasons, beginning with the summer solstice, was first the Syrian year of the cross of St. George, beginning with the autumnal equinox, the year of the Semites and Sabæans of Western Asia, and Asia Minor, and of Macedonia on the Peloponnesus in Greece, vii. 28; viii. 106, 122; it was also the year of the Norse Hadding, the hairy (hadd-r) god of the North, who slew Swipdag or Orwandil, Orion, and was brought up under the guardianship of the Great Bear, viii. 106, 108, 109; the year beginning, in Hindu chronology, with the birth of Krishna, the black antelope, and his twin - sister Durgā or Subhadra (which see), at the autumnal equinox, viii. 157 (see Kansa); it was the Mexican year of the rise of Ut'set, the mother of corn, to the upper world through the seed, and of the rule of the wolf of the East, the elk of the North, the buffalo or bear of the West, and the badger of the South, the four quarters of the world marked by St. George's Cross, it was then Ut'set placed in the upper heavens the Pleiades, the three stars of Orion's belt, and the Great Bear, the constellation ruling this year, ix. 260; (v.) the fifth year-reckoning was the year of five seasons of the Hindus, called by the Zends the year of the Varenya Devas; it began, like the last phase of the year of four seasons, with the summer solstice, vii. 28-30; it was the year of the Gonds of the thirty-three gods of time of the Hindus, Buddhists, and Egyptians, and of the Egyptian five-rayed star of Horus, vii. 28, 29; the year of Bhishma and of the five sons of Yayāti, which began with the Indian rains of the summer solstice

shaostra the Hindu Prashastri, the teaching priest Jamāspa, the twin-horses, the sun-horse, and the sun-fish, and Vistāspa, the king of the Naotaras, the people of the new (*nao*) star (*tara*), the planets made rulers of time instead of the fixed stars by the astronomers who traced their year by the path of the sun and moon through the stars, ix. 300, 302 ; these three formed the sacred Bāhrām fire, the sacred fire of the immortal and invisible god formed of two essences, ix. 302, 304 ; the middle fire which united the two creating essences was Jamāspa the twin-horse rider on the sun-horse and the sun-fish, ix. 304-305 ; Zarathustra and his missionary priests used Bhang, that is, Hashish, as the elixir of wisdom, which gave them the power of divine insight and inspired the priests of Assur, the fish-god of Nineveh, ix. 209, 301, 302

THE END

Edinburgh : T. and A. CONSTABLE, Printers to Her Majesty,
at the Edinburgh University Press

www.ingramcontent.com/pod-product-compliance
Lightning Source LLC
Chambersburg PA
CBHW032311280326
41932CB00009B/776